INTRODUCTION TO
COMPUTER SIMULATION:
THE SYSTEM DYNAMICS
APPROACH

NANCY ROBERTS
Lesley College

DAVID F. ANDERSEN
*State University of
New York at Albany*

RALPH M. DEAL
Kalamazoo College

MICHAEL S. GARET
Stanford University

WILLIAM A. SHAFFER
Data Resources, Inc.

Addison-Wesley Publishing Company
Reading, Massachusetts • Menlo Park, California
London • Amsterdam • Don Mills, Ontario • Sydney

To our teachers and friends,
JAY W. FORRESTER
EDWARD B. ROBERTS

Library of Congress Cataloging in Publication Data
Main entry under title:

Introduction to computer simulation.

Includes index.
1. Computer simulation.   I. Roberts, Nancy,
1938–
QA76.9.C65158   001.4'34   82–6800
ISBN 0-201-06414-6   AACR2

This work was developed under a grant from the Department of Education. However, the content does not necessarily reflect the position or policy of that Agency, and no official endorsement of these materials should be inferred.

ISBN 0-201-06414-6
    HIJK-MA-8987

CONTENTS

FOREWORD

We live in a changing world. People grow, business conditions fluctuate, personal relationships develop, major industries come and go, population increases, water shortages worsen, and scientific discoveries increase. Our principal concerns arise from change—growth, decay, and fluctuation.

Change is a central concern in every human activity. The doctor tries to improve a patient's health; a parent tries to enhance a child's character and ability; a student tries to increase his or her competence; an engineer tries to raise the efficiency of a manufacturing process. Everywhere, people are taking actions to accomplish change.

But the processes of change have not been presented in an orderly way in our educational institutions. The dynamics of change have seldom been taught as a basic foundation that underlies all fields. The processes of change have not been organized so that they can be taught at all educational levels, even though a child, from his or her earliest awareness, begins to cope with change and to build an intuitive awareness of change. There are several persuasive reasons for teaching the processes of change as a formal academic subject.

First, the dynamics of changing conditions can become a universal foundation underlying all fields of endeavor. With a solid understanding of the structures that cause change, a person acquires a degree of mobility between fields. If the behavior of a particular structure is understood in one field, it can be understood in all fields. The same structures, each with its own characteristic behavior, are found in medicine, engineering, economics, psychiatry, sociology, management, and the everyday experiences of living.

Second, although life equips us with an intuitive feel for the dynamics of change, our intuition is reliable only in very straightforward situations. In the more complex dynamic structures, which increasingly dominate our lives, the intuition carried over from simple systems is misleading. As an example, in simple systems we learn that cause and effect are closely related in both time and space; in touching a hot stove, the hand is burned now and it is burned here. We repeatedly learn to expect a close association between action and the result. In more complex systems, however, the cause of a symptom may lie far back in time and in a remote part of the system. Only through study of structure and behavior can we develop intuition that is reliable when confronted by complexity.

Third, studying the dynamics of change is fun. It is challenging. It is meaningful because it couples with each person's life and problems. To understand the processes of change is to understand the surrounding world. If we understand our world, we can hope to improve it.

This book is a door leading to understanding of change. It should provide a glimpse of how things are interrelated, how humans and nature evolve, and how we can influence our future if we understand how that future is being shaped. With this book may come a deeper awareness of political processes, an improved grasp of rising and falling economic activity, and a clearer perception of physical and social behavior. A better understanding of change in turn raises the hope for an improved society, a more favorable relationship between humans and nature, and prospects for greater international understanding.

JAY W. FORRESTER

PREFACE

Development of an introductory curriculum in computer simulation has been a project spanning almost ten years and passing through three different stages. The initial work, essentially a dissertation project of Nancy Roberts, was the development of materials to teach the basic concepts of system dynamics.

The second stage was initiated when several people joined together in an informal group to consider the possibilities of teaching both system dynamics concepts and computer simulation to a broader audience. The broader audience had two important dimensions: people who were younger than the graduate students currently being introduced to system dynamics, and people not necessarily having a strong mathematics background. High school students and college undergraduates were chosen as appropriate target groups. In its final year, this second stage had funding to write and pilot-test these materials from the U.S. Department of Education, Program for Environmental Education (Grant #G007903439).

During the last stage of the project the materials were carefully revised, based on additional use at several educational levels and reviews from many readers. The authors' major revisions have therefore attempted to make this text appropriate for anyone, from a variety of backgrounds, wanting an introduction to system dynamics concepts and computer simulation.

Pilot testing of the materials has demonstrated that a mathematics background of algebra is all that is necessary for developing models in the DYNAMO computer language. No prior computer experience is needed.

An Instructor's Manual for this text is available which includes: suggestions for different ways to structure a course based on this material; an explanation of how to write system dynamics models in the BASIC language if DYNAMO is not available, as well as the BASIC program needed for the modeling; and suggestions for more advanced research topics outside the scope of this introductory text. The answers to the exercises not included in Appendix B are also available on request.

ACKNOWLEDGMENTS

The pilot-testing of these materials during the grant year was made possible by the generous donation of DYNAMO simulation software to the cooperating institutions by Pugh-Roberts Associates, Inc., Cambridge, Massachusetts. We are especially appreciative of the assistance of Jack Pugh in helping us with many of the installations.

In addition to this financial support, the project had the intellectual and moral support of many people during the years of conceptualization and development. In particular, three system dynamics faculty members have supported and encouraged the authors over a long period of time.

Edward Roberts, professor at the Massachusetts Institute of Technology (MIT), has been teaching, tutoring, and giving confidence to the first author for twenty-six years. It was through his advice that the original elementary level materials were developed. His continuing advice, guidance in conceptualization, and meticulous editing have enabled this book to be completed coherently. Ed wrote Chapter 22, as well as several smaller sections throughout the text to fill voids he perceived. In addition, Ed served as dissertation advisor for three of the other four authors during their doctoral years at MIT.

Professor Jay W. Forrester, as founder and head of the MIT System Dynamics Group, has often stated that some of the most insightful questions and comments about system dynamics works have come from high school and undergraduate students. Jay has long been supportive of the possibility of writing an introductory curriculum. He worked with the group for three years while Nancy Roberts was a Research Associate with the MIT System Dynamics Group. During this formative stage of the project, Jay guided the group in its initial notions of how system dynamics might be introduced to people of various backgrounds and ages.

Donella Meadows, professor at Dartmouth College, former member of the MIT System Dynamics Group and a founder of the Dartmouth System Dynamics Group, has long been concerned with pedagogy. Some of the computer models used in the last two parts of this book are variations of models developed by Dana over the years for her own classes.

Several additional people have been involved in this undertaking at various times during the past years. George Richardson, who taught system dynamics for many years at Simon's Rock Early College, worked with the group in conceptualizing the project, helped in writing the grant proposal, and developed several of the examples throughout the text. In addition, George wrote much of Chapter 24, as well as the SYSDYN software package (included in the Instructor's Manual) that permits writing system dynamics models in BASIC if DYNAMO is not available.

Joining the staff for the grant year were George Hein, professor at Lesley College, and Nancy Dyer, Lesley College graduate student. They were responsible for developing the grant evaluation instruments, and collecting and analyzing the evaluation data.

Also joining the staff during 1980–1981, the funded year, were Tanette Nguyen McCarthy and Marian Steinberg, graduate students at the State University of New York at Albany. Tanette and Marian helped David Andersen as co-authors of Part Three, contributed many ideas at meetings, and did writing for some of the other parts of the text as well.

Helping in formulating and writing the Urban Growth modeling project, Chapter 20, was Khalid Saeed, currently on the faculty of the Asian Institute of Technology.

The dedication of the secretarial staff enabled the curriculum to be developed and the book written. Headed by Barbara Woodring who, on numerous occasions, took initiative that turned chaos into a semblance of order, the staff successfully met a variety of deadlines. Aiding Barbara in times of overload were Mary O'Reilly and Ann Davis, Lesley College department secretaries who most competently and willingly gave helping hands.

The authors convey a deep sense of appreciation to the several faculty members who criticized, edited, discussed, and finally pilot-tested these materials with their students during the spring of 1980. Some of the most meaningful and practical comments came from these cooperating teachers and their students. One of these teachers, Jonathan Choate from Groton School, Groton, Massachusetts, accepted the responsibility for organizing and writing the initial draft of the Instructor's Manual. Particularly helpful were the suggestions of Michael Goodman who used the materials in a graduate course he taught at Lesley College during the summer of 1981.

Finally, the authors thank two people who greatly added to the final readability and appearance of this material. Diane Senge's exceptional quality graphics work has generated all the text figures. The delightful sense of humor embodied in Carole Roberts' illustrations will keep the students' feet on the ground as the power of creating computer worlds goes to their heads.

In spite of some tense moments, the curriculum-development and book-writing project has been a truly rewarding experience for the authors. We hope these materials will be of benefit and pleasure to others.

August 1982

Nancy Roberts
David F. Andersen
Ralph M. Deal
Michael S. Garet
William A. Shaffer

PART I

BASIC CONCEPTS OF SYSTEM SIMULATION

OBJECTIVES

Part I introduces the concepts key to this text, composed of:

1. Computer simulation, a method for understanding, representing, and solving complex interdependent problems;

2. The system perspective, including three critical aspects: cause-and-effect thinking with causal-loop diagrams, feedback relationships, and system boundary determination.

CHAPTER 1

SIMULATION, MODELS, AND SYSTEMS

SIMULATION AND MODELS

Originally, the word *simulate* meant to imitate or feign. This meaning suggests one important characteristic of simulation: to simulate is to imitate something. For example, children playing house are simulating family life; fighter pilots flying a training mission are simulating actual combat.

Simulation generally involves some kind of model or simplified representation. During the course of a simulation, the model mimics important elements of what is being simulated. A simulation model may be a physical model, a mental conception, a mathematical model, a computer model, or some combination of all of these. For children playing house, their "model" is the toys they are using, along with imaginary characters and settings. For an air force pilot in training, the model might be a mock-up fighter plane.

Many simulations involve physical models. For example, the United States Army Corps of Engineers has constructed a small-scale physical model of the Mississippi River, which is used to study ways of reducing the impact of flooding. The behavior of a major river is quite complex and cannot be studied through direct experimentation on the actual river. Therefore engineers have to rely on experiments using the model. Wind tunnels and wave tanks are other forms of simulation in which a physical model is used to imitate a larger system. For instance, a scaled-down model of a plane or ship can be constructed out of wood or other material, and then placed in a wind tunnel or wave tank. Using a wind tunnel, air is blown past a scale model of a plane to examine the plane's aerodynamic properties. Similarly, by using a wave tank a ship model can be subjected to waves to see how it performs.

Since physical models are often relatively expensive to build and unwieldly to move, mathematical models are often preferred. In a mathematical model, mathematical symbols or equations are used (instead of physical objects) to represent the relationships in the system. To perform a simulation us-

3

ing a mathematical model, the calculations indicated by the model's equations are performed over and over to represent the passage of time (in one-second intervals, for example). If these calculations have to be performed by hand, simulation can be extremely tedious and costly. In the sixteenth century, men spent years performing numerical simulations for creating navigation tables, but great importance of navigation to trade and naval power made the extensive effort worthwhile.

In the last forty years, computer simulation has replaced simulation using hand calculations. By World War II, adding machines had reduced the cost of simulation so that it could be used to design radar, gun turrets, and other military equipment, but the cost of simulation was still so high that usually the only simulations performed were those related to important military engineering projects. Not surprisingly, several engineers who pioneered in the development of electronic equipment during the war became involved in the development of computers, because computers could rapidly perform the calculations needed for simulation. Since the development of computers, the cost of arithmetic computations has halved approximately every two years and is likely to decline at this rate for at least another decade. This decline means that simulation, once a rare and expensive way of solving problems, is now very inexpensive.

Computer simulation is currently used in a wide range of applications in the physical sciences, as well as in the social sciences and economics. For example, much of what is known about the likely behavior of nuclear reactors during accidents is derived from computer simulation models. Needless to say, testing actual reactors or even scaled-down reactor models under emergency conditions would involve excessive risks. Thus computer simulation is of critical importance.

Computer simulation is also used in far more mundane circumstances. Meteorologists, for instance, use computer simulation to forecast the weather. In recent years, meteorologists have constructed models that represent flows of heat, pollutants, moisture, and air—models that are at times extremely large and detailed. In fact, some weather models used for research purposes are so large that the time required to run them is longer than the time period over which they will predict. A detailed model to predict the weather for the next day might take twenty-eight hours to run on the computer. For more practical purposes, smaller, less detailed models must be used.

Computer simulation has recently been applied in the analysis of a number of interesting and important social and economic problems. Perhaps one of the most widely publicized simulation models is the WORLD3 model discussed in the book by Meadows et al. (1973), *The Limits to Growth,* which analyzed possible relationships between population, pollution, natural resources, and economic growth. Jay Forrester, in his book *Urban Dynamics* (1969), employed a computer model to analyze some of the causes of urban growth and decay, and to examine the effects of a number of urban renewal programs, including low-income housing, job training, and new enterprise construction. Levin, Roberts, and Hirsch, in their book *The Persistent Poppy* (1975), formulated a simulation model of factors influencing heroin addiction in a large city. Models have been developed to study a variety of problems, such as school finance, project management, United Nations peacekeeping, natural gas discovery, business growth, and national economic planning. The chapters that follow draw on a number of these models in examples and exercises. The purpose of this book is to enable the reader to understand these computer simulation models and to develop their own original models.

THE SYSTEMS APPROACH

Many different approaches can be taken by model builders in choosing the issues they model and the contents of their models. This book presents one perspective for computer simulation modeling—a dynamic feedback systems perspective.

A system may be defined as a collection of interacting elements that function together for some purpose. For example, the human heart, lungs, and bloodstream are a physiological system whose purpose is to provide oxygen for the body. Of course, the circulatory system, like any system, may at times fail to achieve its purpose. One motive for studying issues from a systems perspective is to gain an understanding of some of the reasons for poor system performance.

Similarly, a city can be viewed as a system whose purpose is to provide employment, housing, and other social benefits for its inhabitants. A firm may be viewed as a system that produces and sells products, maintains inventories, hires employees, and performs other functions to survive and grow economically. An airline may be thought of as a system of personnel, as well as electrical, mechanical, and hydraulic components designed to transport passengers comfortably through the air.

The systems approach to studying systems such as these emphasizes the connections among the various parts that constitute a whole. Systems thinking is concerned with connectedness and wholeness. By its nature, a systems view of a problem cuts across disciplinary boundaries as defined in many traditional sciences, in a search to understand a problem from an integrated vantage point. For example, viewing a firm as a system might involve integrating the economics of the marketplace with the sociology of the employees' work environment and the technology of manufacturing. A systems view of a city might involve aspects of political science, geography, economics, and sociology. Exactly how these various disciplinary perspectives can be integrated is one of the major intellectual challenges of the systems approach.

System science emerged as a serious field of study after the second world war. Originally, the field was rooted in the biological and engineering sciences, only more recently branching out to become involved with social and economic problems. Ludwig von Bertalanffy, one of the early 1930s pioneers in systems thinking, was trained initially as a biologist and posited principles of system structure and behavior drawn from his observations of biological organisms functioning as a system. At about the same time, engineers working at the Massachusetts Institute of Technology and elsewhere began to think in terms of systems engineering, often for military applications, designing flight control, fire control, and other engineering systems.

The early analyses of biological and engineering systems shared a common emphasis on the ways in which system components work together to perform some well-defined function. For example, the mechanism that the human body uses to maintain a constant body temperature of 98.6 degrees Fahrenheit resembles the control machinery designed to stabilize an airplane's altitude during flight. Both of these systems function to keep some quantity (blood temperature in one case, and altitude in the other) within narrowly defined limits under a wide range of possible disturbances (temperature variation in one instance, and atmospheric turbulance in the other).

The pioneers in biological and engineering systems theory also considered the implications of their work for problems outside biology and engineering. In his early work, *Cybernetics,* Norbert Wiener (1948) both named and sketched the outlines of a new field of inquiry. Cybernetics became the study of how biological, engineering, social, and economic systems are controlled and regulated. Wiener proposed that the same general principles that control body temperature and the altitude of airplanes may be at work in the market mechanisms of economic systems, the decision-making mechanisms of political systems, and the cognitive mechanisms of psychological systems. Wiener and other systems theorists argued that all aspects of human behavior, ranging from the economic to the political to social and psychological, may be governed by a single set of governing principles. If this claim proved to be true, great advances in human knowledge could be made through the systematic study of those general organizing principles. This was the premise of the early system scientists.

Although the grand vision of the early cyberneticists has yet to be realized, in the decades since these bold claims were first articulated researchers have made impressive progress in applying a systems perspective in many diverse fields. For example, Karl Deutsch laid out a cybernetic view of political processes in his now classic work, *The Nerves of Government* (1963). Herbert Simon (1965) proposed a cybernetic view of human intelligence. And Jay Forrester first applied the broad principles of cybernetics to industrial systems in his path-breaking work, *Industrial Dynamics* (1961).

Forrester's initial work in industrial systems has been subsequently broadened to include other social and economic systems and is now known as the field of system dynamics. The field of system dynamics, one of several possible variants of the systems approach, forms the basis for the simulation modeling techniques presented in this book. Relying heavily on the computer, system dynamics provides a framework in which to apply the idea of systems theory to social and economic problems.

RATIONALE FOR THE CONTENT AND ORGANIZATION OF THIS BOOK

The same forces that make the systems approach attractive also make such systems studies somewhat slippery and elusive. For example, consider for a moment what might be involved in completing a holistic, interconnected and cross-disciplinary analysis of the causes and effects of rising prices of petroleum products within the United States. In both the short run and the long run, the prices of petroleum products in the United States appear to be linked to the availability of crude oil supplies thousands of miles away in the Middle East. In turn, availability of such supplies depends upon a complex melange of economic, political, and geological factors in the Middle East. Economic stability in the Middle East is at least partially linked to the stability of the U.S. dollar on world markets. This line of reasoning could go on and on. To completely take into account all possible factors in thinking about rising petroleum prices is an impossible intellectual task. Human minds simply overload when trying to take into account so many factors at once. Some specific techniques are needed for helping to simplify and impose order upon the thousands of possible relationships that could be included in a study of a complex problem. In fact, three types of intellectual capabilities, two of which are developed in the chapters that follow, are needed to analyze complex social and economic problems.

First, a complex problem cannot be solved without a deep substantive knowledge of the problem at hand. No matter how skilled analysts may be in systems thinking and techniques, they will not be able to tackle problems involving economic issues, for example, unless they understand economics. A manager cannot solve a company's sales problems unless he or she understands what is going on within the firm. And an analyst cannot understand a city's housing problem without knowledge of housing construction, urban migration, and so on. In the chapters that follow, no special attempt is made to teach the needed substantive knowledge of specific problem areas in any depth. Instead, the examples and exercises used are familiar to most people. (Selected answers to some of the exercises appear in Appendix B.)

A second requirement for analyzing a complex problem is a method of structuring and organizing knowledge about the problem. What factors are important and must be included in the analysis? What factors are less important and may be omitted? How can specific knowledge be written down and communicated effectively to someone else? These are basic questions about how to conceptualize and represent a system, whether or not one intends eventually to utilize formal computer analysis techniques. These critical questions are treated in the first four parts of the book; specific techniques are presented that are helpful in differentiating important from unimportant facts and relationships, and in representing on paper a large body of connected information.

The third intellectual tool required in analyzing a complex problem is more technical in nature. It involves keeping track simultaneously of all of the important relationships once they have been sorted out. This is less an issue of human judgment and intuition than it is of simple capacity to remember and reason through the implications of many relationships all at once. At this point, the computer plays a critical role. In the system dynamics approach to problem solving, a large number of structured relationships are "fed into" a computer. As already pointed out, deciding which relationships to include in an analysis is a difficult issue, requiring judgment and intuition. But once the relationships have been established, the computer can be used to determine and keep track of their implications. The code written into the computer is referred to as a *simulation model* because the computer program simulates or mimics the system under study. The last three parts of the book present system dynamics computer simulation tools, as well as numerous examples of and opportunities for computer simulation model-building and analyses.

Figure 1.1 shows, in more detail, the processes of using the conceptual and technical tools presented in this book. It represents six types of activities that are usually involved in the process of constructing a computer simulation model using the system dynamics approach. The organization of the book mirrors, to an extent, this process. Below, each of these major phases in model building is discussed in more detail with reference to where these materials are covered in this book.

Problem definition. The first phase in the model-building process involves recognizing and defining a problem to study that is amenable to analysis in systems terms. Important properties of dynamic problems are that they contain quantities that vary over time, that the forces producing this variability can be

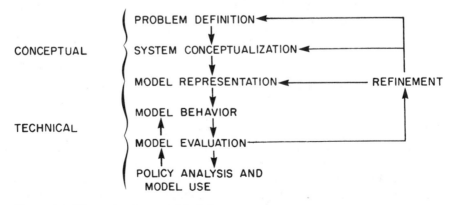

Figure 1.1 Phases in the model-building process

described causally, and that important causal influences can be contained within a closed system of feedback loops. These three defining characteristics of dynamic problems are discussed in more detail in the following chapter. Defining problems that meet these characteristics is not a trivial matter. The two most important skills in recognizing such problems is knowing how to infer causal relationships and knowing how to interpret graphs of variables plotted against time. These are the key skills discussed in Parts II and III. Part IV provides an opportunity to apply some of these ideas in open-ended case studies.

System conceptualization.   The second phase in the model-building process involves committing to paper the important influences believed to be operating within a system. Systems may be represented on paper in several fashions, the three most common being causal-loop diagrams, plots of variables against time, and computer flow diagrams. These three ways of thinking about a system are taught in Parts II through V.

Model representation.   In the third phase of the model-building process, models are represented in the form of computer code that can be "fed into" the computer. Part V presents the basic skills needed to represent systems in equation form and to run computer simulation models using the DYNAMO simulation language. More advanced treatment of these skills appears in Parts VI and VII.

Model behavior.   In the fourth phase of the model-building process, computer simulation is used to determine how all of the variables within the system behave over time. The process of using a computer simulation model to generate the behavior of major system variables over time is introduced in Part V and developed further through the detailed examples of Part VI.

Model evaluation.   In the fifth phase, numerous tests must be performed on the model to evaluate its quality and validity. These tests range from checking for logical consistency, to matching model output against observed data collected over time, to more formal statistical tests of the parameters used within the simulation. Although a complete discussion of model evaluation is beyond the scope of this book, some of the important issues involved are presented in the case examples in Parts VI and VII. For a more complete discussion of the many issues involved in model evaluation, readers are urged to investigate the supplemental readings and exercises mentioned in Part VII.

Policy analysis and model use.   In the sixth phase of the modeling process, the model is used to test alternative policies that might be implemented in the system under study. For example, in an energy supply model for the United States, it might be possible to simulate the impact of a sudden embargo on oil imports from overseas, or the impact of an unexpected discovery of new oil reserves within the continental United States. Furthermore, analysts might be able to investigate the possible impacts of government policies, such as taxes or import quotas. Designing and testing policies using a computer simulation model required the full skill of experienced model builders, and implementing the results of computer-based analyses is often even more challenging. This final phase of the model-building process is perhaps the most difficult. Policy design issues are introduced for causal-loop analyses in Part IV. The exercises developed in Part VI introduce the reader to questions involved in designing policies for a sample of several computer models. Persons interested further in the complexities involved in the process of getting a model used are referred to the supplementary exercises and readings in Part VII.

Figure 1.1 might seem to imply that the model-building process is an orderly one, progressing from the problem recognition phase through system conceptualization, and finally ending in policy analysis and model use. In fact, any real model-building and analysis endeavor is filled with false starts and frequent recycling back to initial stages of problem recognition and system conceptualization. That is, analysts often get a "first cut" computer simula-tion model up and running on the computer, only to discover that one or more major influences have been omitted and that the model-building process must return to an earlier and more conceptual phase. These continuing attempts to improve a model's formulation are indicated by the "refinement" cycles sketched in Figure 1.1. Even though this book introduces the conceptual phases of model-building before the technical ones, in any real modeling effort these two types of activities are co-mingled in a complex way. The exercises presented in Parts VI and VII are designed to help integrate both the concep-tual and technical skills taught in the earlier units.

CHAPTER 2

CAUSATION, FEEDBACK, AND SYSTEM BOUNDARY

This chapter introduces three critical aspects of the system dynamics approach to developing computer simulation models: thinking in terms of cause-and-effect relationships, focusing on the feedback linkages among components of a system, and determining the appropriate boundaries for defining what is to be included within a system.

UNDERSTANDING CAUSE AND EFFECT

Causal thinking is the key to organizing ideas in a system dynamics study. Typically an analyst isolates key causal factors and diagrams the system of causal relationships before proceeding to build a computer simulation model. However, the notion of causation can be subtle, and using the concept requires careful attention. Consider the following examples.

EXAMPLE I: NEWTON'S LAWS

The clearest examples of causation are those involving physical laws. In positing laws that govern the motion of objects, Newton asserted that a "pushing" force will cause an object initially at rest to begin moving in the direction of the force, and a continuing force applied to an object will cause continung acceleration. In diagrammatic form, this causal relationship can be shown as:

PUSHING ————————————→ ACCELERATING
FORCE   MOTION

where the arrow between the two phrases can be read as "causes." Hence, a pushing force causes an accelerating motion. Examples of this type of causal relationship abound. An automobile engine exerts a force that causes an automobile initially at rest to be set in motion; and the continuing force of the engine causes continuing acceleration (up to a point where wind resistance and friction prevent further acceleration). A pitcher's arm exerts a force that causes a baseball to be set in motion, and so on.

However, even these straightforward examples drawn from the physical sciences have subtleties. For example, a human being applying force against the side of a building will, in all likelihood, not set the building in motion (at least not noticeably). That is, applied forces are not always accompanied by accelerating objects. Why does the cause fail to produce its anticipated effect in this case? The reason for this seeming puzzle is not difficult to discover. The causal law involves the provision: "other things being equal." When a human being exerts a force against the side of a building, all other things are not in fact "held equal." Instead, the wood, steel, or concrete from which the walls are made exert counterforces that resist minor forces (such as someone leaning against the building) or even somewhat larger forces (like the wind).

In social and economic systems, causal statements usually include an "other things being equal" provision. To correctly diagnose a causal influence, one must perform the mental experiment of asking what would happen if the particular causal influence under consideration were the *only* influence to act upon the affected object. For example, consider the causal relation, "Births cause the population to grow." This causal statement is certainly always true, in spite of the fact that countries with a positive birth rate may nevertheless show a decline in population. Obviously, in this case, births are not the only causal influence on population. Deaths are another, and thus "all other things are not equal." A positive birth rate is not always accompanied by a growing population, but, all else being equal, births cause the population to grow.

Exercise 1: Logic of Causal Links*

Included here are some hypothesized causal relationships that might explain observed behaviors. Sometimes these arrow diagrams make more sense if you use the word *affect* or *influence* rather than cause. For example, "Food intake influences weight" makes more sense than "Food intake causes weight." The words *cause, affect,* and *influence* are used here to mean approximately the same thing.

Write a sentence or two stating why you agree or do not agree with each suggested causal relationship:

---

*Selected answers to some of the exercises appear in Appendix B.

a. FOOD INTAKE ⟶ WEIGHT

b. MONEY ⟶ HAPPINESS

c. BRAINS ⟶ GRADES

d. LEAVES ⟶ WIND

e. FIRE ⟶ SMOKE

EXAMPLE II: SEATBELTS

A second example shows how the concept of causality can be difficult to observe and measure in some circumstances. A common claim is that the use of seatbelts causes a decrease in highway fatalities. Put diagrammatically:

USE OF ⟶ REDUCED HIGHWAY
SEATBELTS   FATALITIES

Obviously, this causal relationship does not stem from any physical or logical law that must always be true. Doubtless cases exist where highway fatalities have been reduced without increased use of seatbelts (such as when the national speed limit was reduced to 55 miles per hour to conserve gasoline), and cases exist where the use of seatbelts has increased the chances of a highway fatality (such as freak accidents in which it might have been better had an occupant been thrown clear of the vehicle). The relationship between the use of seatbelts and fatalities is an aggregate causal relationship that summarizes a fairly complex situation. Seatbelts can be shown to restrain occupants during a crash, and available evidence suggests that being restrained increases one's chances of surviving a potentially fatal accident. However, notice that this type of relationship cannot be ascertained entirely by logical reasoning. Some evidence has to be collected and analyzed to determine whether or not a strong statistical relationship exists between seatbelt usage and fatalities, "on the average." Even with properly designed empirical research, it may be difficult to disentangle the effects of seatbelts and other influences on traffic fatalities.

EXAMPLE III: SUICIDE

A third example illustrates how questions of causality, when applied to social and economic problems, often touch upon issues that run deeper than the empirical ones discussed previously. Consider, for example, a sociological study of the causal determinants of suicide within a given population. A researcher proposes that shortened daylight hours occurring during the winter months is one of several causes for suicide in some climates and at certain times of the year. Put in diagram form:

```
SHORTENED                    INCREASED
DAYLIGHT HOURS               SUICIDE RATES
```

Postulating and then verifying causal relations such as this one are difficult tasks for several reasons, very different indeed from causal statements explaining why apples fall from a tree. Even the most skilled clinical or experimental psychologists would not claim to be able to predict individual suicides.
So when the level of analysis is *individual* behavior, causal statements are difficult, if not impossible, to make. However, statistically-minded researchers
have been able to predict aggregate suicide rates using causal models. Thus
causal modeling may be possible when the level of analysis is aggregated to a
population as a whole.

Even if a statistical relationship is found to exist between the number of
daylight hours and suicide rates, interpreting this relationship is still difficult.
Several competing or overlapping interpretations are possible, involving differing intervening effects. For example, one researcher may argue that the on-
set of winter restricts outside activity, leading to seasonal underemployment of
farmers, construction workers, and other persons who traditionally work outside. According to this view, underemployment leads to financial stress, which
ultimately is the cause of increased suicide. Which mechanism is at work
within the overall aggregate causal relationship is important to determine,
both for proper detailed model representation and for effective policy.

EXAMPLE IV: INFLATION

A final example involving economic analysis illustrates even another level of
difficulty that exists in identifying causal relationships. Throughout the decade
of the 1960s and the early 1970s, economic policy on the part of the United
States government was guided in part by an inverse relationship hypothesized
to exist between inflation and unemployment. Known as the Phillips curve
(named after the economist who proclaimed this relationship), this relation
said that increases in the unemployment rate were accompanied by decreases in
the inflation rate. One implication of the Phillips curve is that economic decision makers are faced with a difficult paradox: decreasing unemployment
seems to mean increasing inflation and vice versa. One might be tempted to
draw the causal diagram

```
DECREASING                   INCREASING
UNEMPLOYMENT                 INFLATION
```

However, the relationship between unemployment and inflation now appears much more complicated than that indicated by the above diagram; in
fact, the diagram may be such a dramatic simplification as to be seriously misleading. For one thing, the direction of causality is unclear. Have changes in

unemployment caused changes in inflation, or have changes in inflation caused changes in unemployment? Or, as is more likely, have both unemployment and inflation responded in similar but opposite directions to other forces acting within the economy (such as aggregate consumer demand, business investments, or the state of the supply of money)?

In fact, during the latter portion of the 1970s the previously observed inverse relationship between changes in unemployment and changes in inflation dissolved as the economy exhibited both increasing inflation and increasing unemployment. Policies designed to trade-off inflation for unemployment, driven by an apparent causal relation between the two, no longer worked. For the case of inflation and unemployment, the level of causal analysis has to go much deeper than mere analysis of an observed association between two variables.

Exercise 2: Diagramming Causal Links

Complete each of the following sentences to produce a plausible causal relationship. Then draw an arrow diagram to represent causation.

a. Difficult school work causes me to _____.

b. Stress causes me to _____.

c. High gasoline prices cause _____.

d. A shortage of engineers causes _____.

e. Inflation causes _____.

f. Unemployment causes _____.

g. An increase in the number of rabbits on a field causes _____.

Throughout the chapters that follow, the examples and exercises involve specifying and diagraming causal relationships similar to those just illustrated. As these illustrations indicate, drawing causal inferences is sometimes tricky, and different types of causal relationships are possible. Some causal links involve direct physical causation (as in the force and acceleration example). Other links are aggregate representations of individual human behavior (as in the seatbelt illustration). Some causal links involve the interpretation of meaningful human action (as in the suicide example), and still other links are simply based on observed associations (as in the unemployment and inflation example).

When causal links are specified in the examples and exercises to follow, it is worth giving them careful thought. For instance, what type of causality is implied? Are the assumed relationships plausible? What kinds of evidence might be gathered to support or refute a hypothesized causal link? What alternative assumptions might be possible? Although these questions are seldom raised explicitly in the exercises and examples, they are of central importance in assessing the causal assumptions underlying a model.

FEEDBACK

While thinking in terms of causal relationships is necessary to cast a problem in a form that can be analyzed using system dynamics, it is not sufficient. Causal chains can often be linked together nearly endlessly to create an undisciplined morass of causal relationships.

Consider, for example, the earth's water system. One key element in the water system is, of course, the ocean. The sun's rays cause water to evaporate from the oceans, and evaporation from oceanic surfaces causes moisture to build up in large air masses. When moved by air currents (also caused by the sun's heat), moist air masses can collide with cooler air masses, causing rain. Rain causes plants to grow, abundant vegetation is an important causal factor in the growth of animal species, and so on, and so on. Unending causal chains such as these can be found in social and economic systems, and, if left unchecked, they produce an unintelligable web of detail with no clear purpose or point of view.

One way to clarify the representation of a system is to focus on circular chains or causal loops. Within a causal loop, an initial cause ripples through the entire chain of causes and effects until the initial cause eventually becomes an indirect effect of itself. This process whereby an initial cause ripples through a chain of causation ultimately to reaffect itself is called *feedback*.

As a simple example, consider an initial disturbance in the temperature of a room caused by a sudden cold spell. This drop in temperature might in turn cause various types of activities. For example, persons in the room might put on sweaters or move to a warmer spot in the house. Also, the thermostat might turn on the furnace. The activity of the furnace might in turn cause a number of things to happen. Furnace activity might cause the level of fuel oil in the storage tank to go down, which in turn might cause the future purchase of more fuel oil. Furnace activity might also cause wear and tear on the burner unit, which might cause future repairs to be made. However, none of these causal chains feeds back to influence room temperature. The important effect of furnace activity, for our purposes of analyzing the control of room temperature, is the heating up of the radiators in the room, which eventually causes the room temperature to rise.

A diagrammatic representation of all of the causal influences previously discussed is shown in Figure 2.1. In this figure, the only chain of causal influences that contains feedback is the causal loop that runs through room temperature, thermostat activity, furnace activity, radiator activity, and ultimately back to room temperature. This is the causal loop that acts to control temperature within the room, and is the only set of causal relationships that is directly relevant to the immediate problem of temperature control. (Notice, however, that if the fuel oil were to run out, then the partial chain, starting with furnace activity and going through fuel oil level in the tank, would eventually feed back to influence furnace activity and hence room temperature; i.e., when the oil runs out the room might get very cold.)

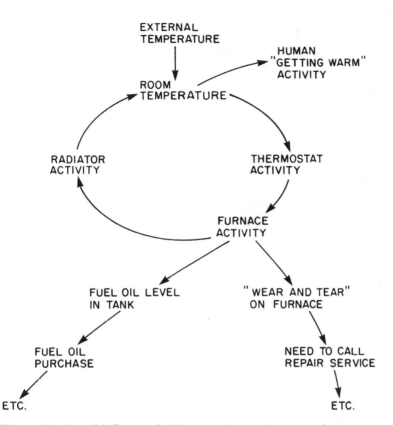

Figure 2.1 Causal influences in the room temperature example

One of the key elements of the system dynamics method is a search to identify closed, causal feedback loops. The emphasis on causal loops can be a powerful tool to help define a system's boundary, to sort out what should and what should not be included within the study of a social or economic system.

The reasons for looking for closed-loop feedback effects go much deeper than just simplicity in including or excluding factors from an analysis. If one is interested in problems of control (controlling room temperature, controlling inflation, controlling insects, controlling worker productivity), *the most important causal influences will be exactly those that are enclosed within feedback loops.* Two dramatic things occur when limiting attention to closed-loop feedback in the analysis of social and economic problems. First, the number of factors or variables to be included within a system's definition can be drastically reduced to a manageable level. Second, and more important, attention can be focused on those variables that are most important in generating and controlling social and economic problems.

To the extent that these two assertions are true, then thinking in terms of causal relationships that feed back through causal loops should be an extremely important and powerful way to organize analyses of social and economic systems.

EXAMPLE V: DESSERT AND WEIGHT-GAIN

Dessert-eating and weight-gain can be viewed as a simple feedback system involving three causal links. First, the number of rich desserts I eat *affects* the amount I am overweight.

RICH DESSERTS ⟶ AMOUNT I'M
EATEN   OVERWEIGHT

Second, the amount I am overweight *influences* my concern about my weight.

AMOUNT I'M ⟶ CONCERN ABOUT
OVERWEIGHT   MY WEIGHT

Finally, my concern about my weight *influences* the number of rich desserts I eat.

CONCERN ABOUT ⟶ RICH DESSERTS
MY WEIGHT   EATEN

These three links can be combined to form the following *causal loop.*

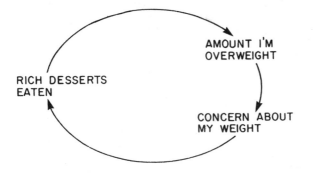

The circular form of the diagram indicates that a *closed loop* or *closed system* has been identified, which might account for changes in my dessert consumption over time.

EXAMPLE VI: PIANO PRACTICE

Piano-practicing provides another illustration of a simple feedback system. First of all, practicing piano *affects* how well I play.

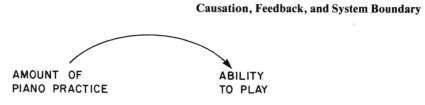

AMOUNT OF
PIANO PRACTICE   ABILITY
                                          TO PLAY

How well I play *affects* how much I enjoy playing the piano

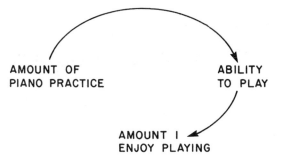

AMOUNT OF   ABILITY
PIANO PRACTICE   TO PLAY

                    AMOUNT I
                    ENJOY PLAYING

My enjoyment of the piano *influences* how much I am willing to practice.

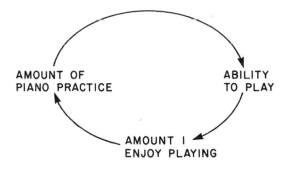

AMOUNT OF   ABILITY
PIANO PRACTICE   TO PLAY

                    AMOUNT I
                    ENJOY PLAYING

Exercise 3: Closed Systems

Complete the following diagrams (drawn from Exercise 2) so you produce a logical closed causal loop or closed system for each. Write a short explanation of the expected behavior of the loop you have identified.

a. Difficult school work causes me to _____.

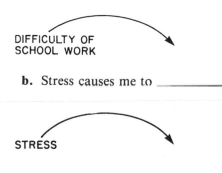

DIFFICULTY OF
SCHOOL WORK

b. Stress causes me to _____.

STRESS

c. High gasoline prices cause _____.

HIGH GASOLINE
PRICES

d. A shortage of engineers causes _____.

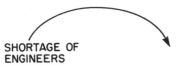

SHORTAGE OF
ENGINEERS

e. Inflation causes _____.

INFLATION

f. Unemployment causes _____.

UNEMPLOYMENT

g. An increase in the number of rabbits on a field causes _____.

NUMBER OF
RABBITS

CLARIFYING THE IMPLICATIONS OF CAUSAL-LOOP DIAGRAMS

Consider once again the relationship between piano practice, ability, and enjoyment (Example VI). According to the example, the amount I practice influences my ability to play.

AMOUNT OF
PRACTICE

ABILITY TO
PLAY

The nature of this relationship can be clarified by describing the relationship in the following terms: the *more* I practice, the *better* my ability to play. And, of course, the reverse is also true: the *less* I practice, the *worse* I am able to play. Another way of saying the same thing is: an *increase* in the amount I practice causes an *increase* in my ability to play; and a *decrease* in the amount I practice causes a *decrease* in my ability to play.

ABILITY TO
PLAY ————————————▶ AMOUNT I
ENJOY PLAYING

The relationship between my piano-playing ability and my enjoyment can be analyzed in similar terms: the *better* my ability, the *more* I enjoy playing; and the *poorer* my ability, the *less* I enjoy playing. Equivalently, an *increase* in my ability causes an *increase* in my enjoyment; and a *decrease* in my ability causes a *decrease* in my enjoyment.

AMOUNT I   ————————————▶ AMOUNT OF
ENJOY PLAYING   PRACTICE

Finally, the relationship between my enjoyment and the amount I practice can be stated: the *more* I enjoy playing, the *more* I practice; and the *less* I enjoy playing, the *less* I practice. Or, an *increase* in my enjoyment causes an *increase* in the amount I practice; and a *decrease* in my enjoyment causes a *decrease* in the amount I practice.

The main advantage of describing causal links in this fashion is that it permits tracing through the implications of the chain of causal relations around the entire closed loop. For example, the more I practice, the better my ability; the better my ability, the more I enjoy playing; and the more I enjoy playing, the more I practice.

According to this analysis of the causal relations, the loop is self-reinforcing. If something leads me to increase the amount I practice, my ability will improve, my enjoyment will increase, and I will eventually practice even more. Of course, things can also work in reverse.

This same approach can be used to analyze the causal loop relating dessert consumption and weight gain discussed in Example V. According to the loop, the more rich desserts I eat, the more overweight I become; the more overweight I become, the more concerned I am about my weight; and the more concerned I am about my weight, the less rich desserts I eat.

The causal relationships in this loop differ substantially from the relations in the piano practice loop, in that this loop is *compensating* rather than reinforcing. If something leads me to eat unusually large amounts of rich dessert, the loop will compensate by increasing my weight, which in turn will increase my concern about my weight, thus eventually reducing the amount of rich dessert I eat.

It is worth noting that the dessert loop can also, unfortunately, work in reverse. If something leads me to eat an unusually *small* amount of rich dessert, the loop will reduce my concern about my weight, thus eventually reducing my resistance to rich desserts!

As a final illustration, consider the following loop, involving depression.

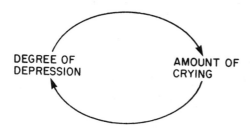

According to the loop, the more depressed I am, the more I cry; the more I cry, the more my eyes turn red; the more red my eyes become, the more time I spend wearing dark glasses; the more time I spend wearing dark glasses, the more embarrassing questions my friends ask me; and the more embarrassing questions my friends ask, the more depressed I become.

While this might be a perfectly adequate account of my pattern of depression, the loop surely contains more elements than are needed to provide a concise analysis. The following diagram might do just as well:

According to this loop, the more depressed I am, the more I cry; and the more I cry, the more depressed I become. Both this loop and the original, more complex loop are self-reinforcing. If for some reason I become depressed, my depression will steadily worsen. But the second loop is certainly easier to read. By omitting some of the intervening variables, it is possible to clarify the underlying causal structure.

One final point needs to be made concerning the tracing of causal connections around a loop. To carry out this sort of analysis, it is necessary to state each variable in the loop as a *quantity that can sensibly increase or decrease.*

Thus, for example, in the depression loop, one element is "redness of my eyes." The formulation indicates that: the more I cry, the more red my eyes become; or, an increase in the amount I cry causes an increase in the redness of my eyes. Thus, according to this formulation, the redness of my eyes is a matter of degree. It can increase (if my crying increases), or it can decrease (if my crying decreases).

"WALKING THROUGH" A CLOSED LOOP

Another approach to better understanding the implications of a closed-loop diagram is "walking through" the links. Take, for example, the "Tired-Sleep" loop shown here.

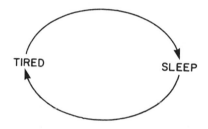

It can be read: "The *more* tired I am, the *more* I sleep. The *more* I sleep, the *less* tired I am. The *less* tired I am, the *less* I sleep," and so forth around the loop.

If this loop were walked through as it is talked through, stepping in the same direction each time saying the same word (*more* or *less*), or reversing direction as the words changed, a distinct two-directional pattern would develop. The tired-sleep loop would produce the pattern of footprints shown below.

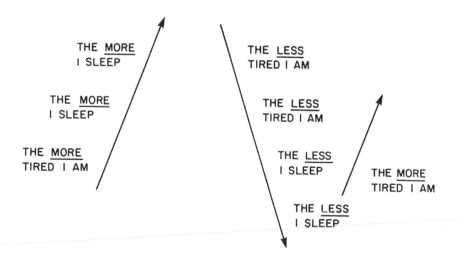

The up-and-down pattern is symptomatic of *compensating* causal loops, such as this loop and the dessert loop.

The "cry-depressed" loop provides another and different type of example for "walking through."

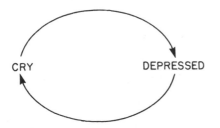

The footsteps produce the distinct one-directional pattern shown below.

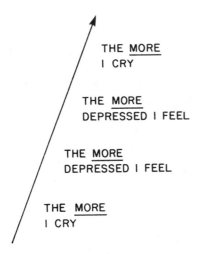

The one-directional pattern is symptomatic of *reinforcing* causal loops, such as this one and the piano-playing example.

Exercise 4: Loop Patterns

Go back to the loops you developed in Exercise 3, and trace (by reading and/or "walking") through the causal relations in each, following the examples just given. If necessary, revise your loops to insure that all of the elements in each loop are quantities that can sensibly increase or decrease. Then attempt to reduce each loop to the most concise form possible.

SYSTEM BOUNDARY

Feedback-oriented thinking helps to indicate how relationships are to be represented in a system model. But further attention is needed to specify the boundary for a system model. Simply put, a system's boundary is a line of demarcation that determines what is included in the system and what is not. Identifying a system's boundary is the complex process of defining the size, scope, and character of the problem being studied. Different analysts may disagree on what a system's boundary actually is, and a slight change in the way the study is defined may lead to a dramatic shift in the system boundary. For example, suppose a business firm that delivers perishable products to a retail market (such as dairy products or baked goods) is interested in the factors influencing the timely delivery of these products to its retail outlets. The system boundary for this study would probably include the firm's manufacturing and distribution facilities. However, if the problem were redefined slightly to focus on the impact of the product freshness on sales, then the boundary of the analysis would have to be expanded to include an analysis of factors related to consumer behavior and the relative freshness of competitors' products.

Perhaps the best way to clarify some of the problems involved in defining a system's boundary is to work through an example.

EXAMPLE VII: ELEMENTS OF A SYSTEM

A superintendent of a school district is interested in taking a "systems view" of the elementary schools under her jurisdiction. She begins her task by listing the major components that she believes make up an elementary school system. Her list is:

students	principal	paper
books	games	classrooms
teachers	sports	friends
work	library	music
desks	tests	blackboards

Implicit in this list is a statement of what the superintendent believes to be the important components of an elementary school. However, we have no information concerning how this list was selected, or for what purpose. The list might be useful in communicating with a parents' group concerning life in an elementary school, since the list tends to emphasize the activities that are relevant to children within the school. However, if the superintendent were interested in the problem of setting next year's budget for the elementary schools in the district, she would probably have elected to include important variables such as last year's budget level, the tax rate, taxpayer attitudes, or enrollment trends in the system under study. If the problem of interest were curriculum reform, the list would again look quite different, perhaps emphasizing the content of textbooks, and the existence of remedial reading or mathematics "tracks."

It is impossible to identify the components of any system without a clear idea of what problem is to be addressed and who is interested in the problem. This last point—that *who* is interested in the problem is a determinant of the system boundary—is worth more consideration. Imagine what the superintendent might stress in a financial analysis designed to be read by taxpayers, and how such a study might differ from one designed for the teacher's union. Taxpayers and teachers might not be interested in the same financial issues (in fact, their interests might run directly counter to each other over some issues such as teacher salaries), and the wise superintendent would recognize this difference in perspective in selecting what elements to include in her study.

The system dynamics approach to modeling uses two important schemes to clarify what to include about a problem. The first scheme is thinking about the problem in terms of how one or more quantities vary through time (the very word *dynamics* implies an emphasis on change through time). For example, a system dynamics study of a school budget would involve examining how and why the school budget changes over time. Such a study would choose to include within its boundary all of the components necessary to explain why a school's budget changes over time (such as declining or expanding enrollments, changes in teacher salaries, or increases in the costs of books and materials).

The second perspective is thinking about whether a substantial feedback relationship is involved. A model's boundary must surround items that are in important cause-and-effect feedback relationships to each other. For example, a simulation model of General Motors' sales, employment, and profits would no doubt need to include within its system boundary many key aspects of the U.S. economy. Not only does the economy affect G.M., but G.M's size makes it a significant influence on the economy. That feedback would require a broad boundary definition for the system. In contrast, a model of a small local manufacturer would need an input regarding economic conditions (technically defined as an "exogenous input"; that is, outside of the system), but surely the economy itself would not need to be included within this system's boundary. Causal-loop diagrams include the major feedback relationships that seem to cause a problem, as well as the few exogenous or external influences that also affect the problem. The next unit, Structure of Feedback Systems, continues to give the reader information and experience in developing these causal-loop diagrams.

Exercise 5: Identifying Systems

    a. List five systems and an interesting problem or question associated with each one.

b. Pick one of the systems from your list above. Write down words or phrases that might define the system's boundary for the question of interest.

c. Can you think of a way of refocusing the question of interest that might shift the boundary of the system that you have defined? State the new question and restate the system's boundary by making up a modified list of the components that you believe to be within the newly defined system.

d. Try developing a causal-loop diagram for one of the system viewpoints stated in (c).

PART II

STRUCTURE OF FEEDBACK SYSTEMS

OBJECTIVES

Part II is designed to extend the reader's ability to understand and diagram dynamic feedback systems by:

1. Introducing a set of symbols used to clarify causal-loop diagrams;

2. Providing practice in reading signed causal-loop diagrams and inferring the likely system behavior;

3. Teaching how to take a verbal description and identify the underlying feedback structure by developing a causal-loop diagram;

4. Applying this tool of causal-loop diagramming to unstructured problems.

CHAPTER 3

READING CAUSAL-LOOP DIAGRAMS

STRUCTURE AND BEHAVIOR IN FEEDBACK SYSTEMS

Part II develops one of the most fundamental ideas in the system dynamics approach to computer simulation and problem solving—namely, system structure. In brief, the structure of a system is the network of causal feedback loops necessary to explain why certain key elements within a system behave over time as they do. Representing system structure in causal diagram form must precede equation writing for computer simulation.

Consider the simplest of all possible investment systems. One hundred dollars is put into a savings account at a 10-percent interest rate compounded annually. Each year, interest is reinvested and added to the original principal. The causal structure for this system is shown in Figure 3.1.

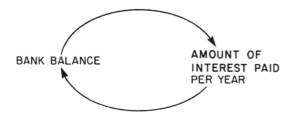

Figure 3.1 Causal structure for the bank balance problem

The interest paid and current bank balance can be computed over time, as shown in Table 3.1. When graphed, the behavior of the bank balance against time looks like Figure 3.2.

Table 3.1 Computing the bank balance for ten years

Deposit	Year	Interest paid	Bank balance
$100.00	0	—	100.00
	1	10.00	110.00
	2	11.00	121.00
	3	12.10	133.10
	4	13.31	146.61
	5	14.66	161.27
	6	16.12	177.39
	7	17.73	195.12
	8	19.51	214.64
	9	21.46	236.10
	10	23.61	259.71

In simple terms, the causal structure of the banking problem as summarized in Figure 3.1 gives rise to the behavior plotted in Figure 3.2. This chapter discusses the structure of feedback systems represented in causal-loop form. Single closed loops, such as the bank balance problem, and then their behavior over time, are treated first. Then combinations of several closed loops as they would appear in more complex systems are considered. A summary sheet explaining all the causal-loop diagram symbols appears at the end of this chapter. Chapter 4 focuses on developing causal-loop representation of less structured examples.

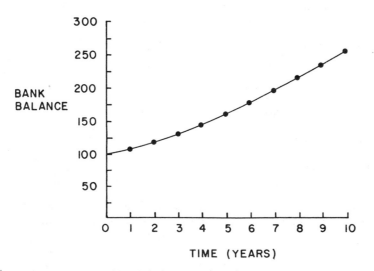

Figure 3.2 Plot of bank balance against time

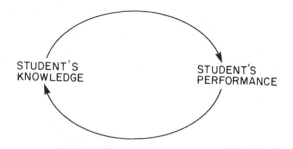

Figure 3.3 The mutual effects of knowledge and performance

SINGLE-LOOP POSITIVE FEEDBACK SYSTEMS

EXAMPLE 1: STUDENT PERFORMANCE

Figure 3.3 shows a causal-loop diagram of one aspect of the dynamics of student performance. This diagram suggests that the amount of knowledge a student has will affect the student's performance, and the student's performance will affect his or her acquisition of knowledge.

More information about student performance can be conveyed by adding either a + or a – *sign* at each arrowhead, as shown in the diagrams below. This is called *signing* the diagram.

This causal-loop diagram suggests that the more knowledge a student has, the *better* her or his performance will be. An increase in knowledge will cause an *increase* in performance. A + is used at the arrowhead to indicate that a change in the item at the tail of the arrow (Student's Knowledge) will cause a change in the *same* direction for the item at the head of the arrow (Student's Performance). Therefore the diagram can also be read, "If knowledge *decreases*, then performance will also *decrease*.

Similarly, a + is used at this arrowhead:

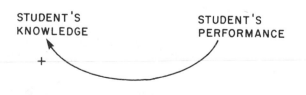

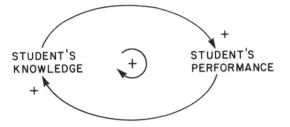

Figure 3.4 Dynamics of student performance

It indicates that an increase in a student's performance will tend to cause an increase in the student's knowledge. The link also indicates that a *decrease* in performance will lead to an eventual *decrease* in the amount of knowledge accumulated by the student. Putting the two linkages together forms Figure 3.4.

A change introduced to either element of the loop in Figure 3.4 will continue to generate other changes in the *same* direction, as the loop is traced around and around. Either the student's performance keeps increasing or the student fails miserably. A ⊕ symbol is placed in the center of the loop to suggest that it is a positive causal loop, one in which behavioral changes are reinforced.

EXAMPLE II: POPULATION DYNAMICS

Figure 3.5 is another signed causal-loop diagram, covering one aspect of the dynamics of population growth, followed by a written description of how to read through the loop. Based on the discussion in Example I, this diagram can be read: "The *more* people there are, the *more* births there will be; the *more* births there are, the *more* people there will be." The + at each arrowhead indicates that a change in the item at the tail of the arrow (Total Population) will cause a change in the *same* direction for the item at the head of the arrow

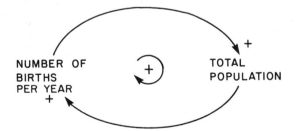

Figure 3.5 Population dynamics—Signed

(Number of Births per Year). Therefore the diagram could also be read: "The *fewer* people there are, the *fewer* births there will be; the *fewer* births there are, the *fewer* people there will be." But is this strictly true? If births go down, does population actually decline? Not unless the number of births per year falls below the number of deaths. Otherwise, if births go down, population will still increase, but at a slower rate. To be precisely true, "The *fewer* births there are, the *fewer* people are added to the total population."

Occasionally, the type of situation encountered in this example arises in interpreting a + sign. "Increase" changes read correctly, but "decrease" changes need slight rewording for accurate reading.

Exercise 1: U.S. Population

Figure 3.6 is a graph of the United States population from 1800 to 1960.

a. Indicate why this graph of U.S. population might be explained by the causal loop shown in Figure 3.5.

b. Bring the population graph up to date by finding the United States population of 1970 and 1980. Is Figure 3.5 still a possible explanation of your updated United States population graph? Why?

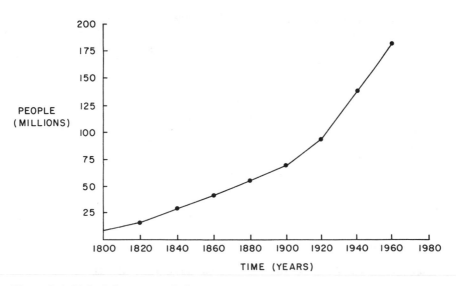

Figure 3.6 United States population

Exercise 2: Plant Dynamics

Figure 3.7 is another positive feedback loop. Examine the loop, and then answer the following questions:

a. Why is this a positive feedback loop?

b. If this is a positive feedback loop, why has the earth not been taken over by plants?

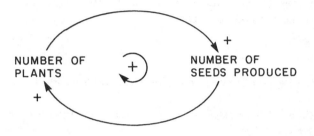

Figure 3.7 Plant Dynamics

Exercise 3: Affecting Sales

Figure 3.8 describes the relationship between the number of salespeople in a business firm and yearly sales.

a. Describe the dynamics of the loop in Figure 3.8.

b. What are some of the other possible important elements of the system that have been omitted?

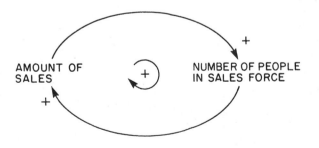

Figure 3.8 Sales loop

SINGLE LOOP NEGATIVE FEEDBACK SYSTEMS

EXAMPLE III: EFFECT OF JOBS ON MIGRATION

Figure 3.9 is an example of a negative feedback loop. It has been "signed" by putting a plus or a minus at the head of each arrow and by indicating the be-

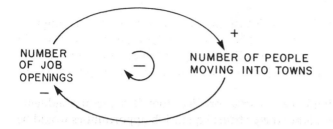

Figure 3.9 Effect of jobs on migration

havior of the loop over time with a ⊖ symbol, identifying this loop as a negative feedback loop. Negative loops tend to keep systems under control by negating or counteracting change.

Figure 3.9 suggests that job openings cause people to move to town. However, as *more* people move into a town, the number of jobs available is affected: The *larger* the number of people moving into an area, the *smaller* the number of job openings available. As the jobs are taken, *fewer* people are attracted to the town.

Taking the diagram one link at a time:

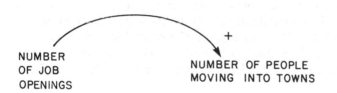

A + is placed at the arrowhead because an *increase* in the number of jobs will cause an *increase* in the number of people moving into town, *or* a *decrease* in the number of jobs will cause a *decrease* in the number of people moving into town. The second link has a − at the arrowhead:

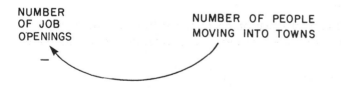

As people move into town, they fill the available job openings. The *more* people that move into town, the *fewer* job openings left; the *fewer* the people that move into town, the *more* the job openings available.

The loop in Figure 3.9 is signed with a ⊖ indicating it is a negative feedback loop. A negative feedback loop is one in which behavior of the loop tends toward reversing its direction as the loop is traversed. In this manner, a negative feedback loop counteracts, over time, a disturbance by some outside force (outside the feedback loop). A new factory moving into town would increase the number of job openings. In turn, increased job opportunities would cause more people to move into town. Eventually, the number of job openings would decrease because the new jobs would all be filled by people who have moved into town.

Notice that this simple causal loop predicts that if a factory suddenly moved out of town, then, other things being equal, job opportunities would be reduced, and people would start to move out of town to seek employment elsewhere. This negative feedback works to keep the number of people seeking work roughly in balance with job openings, whether the number of jobs in the area increases or decreases.[1]

EXAMPLE IV: NEGATIVE FEEDBACK IN THE POPULATION SYSTEM

The earlier Example II involving the growth of a population focused on the positive feedback loop of Total Population and Number of Births. A negative feedback loop involving Total Population and Number of Deaths is shown in Figure 3.10.

This causal loop suggests that the *more* people there are, the *more* deaths there will be. This is indicated by placing a + at the arrowhead connecting Total Population to Number of Deaths per year. The *more* deaths there are, the *fewer* people there will be. This change of direction, *more* then *fewer,* is indicated by placing a − at the arrow head connecting Number of Deaths to Total Population.

Alternatively, the loop could be read: "The *fewer* people there are, the *fewer* deaths there will be; the *fewer* deaths there are, the *more* people there will be." But is this strictly true? If deaths go down, does population actually rise? Not unless deaths fall below the number of births. Otherwise, if deaths go down, population still decreases due to deaths, but at a slower rate. To be precisely true, "The *fewer* deaths there are, the *more* people are left remaining in the total population."

Occasionally, the type of situation encountered in this example arises in interpreting a − sign. "Increase" changes read correctly, but "decrease"

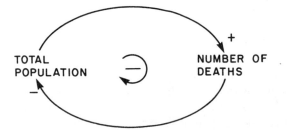

Figure 3.10 Population dynamics—Part II

changes need slight rewording for accurate reading. This is the same qualification as had been made in Example II.

Once the appropriate sign has been placed at each arrowhead and in the center of a closed loop, the signs should not need to be changed. The signs should be correct wherever reading around the loop may be started and whether increases or decreases in loop elements are considered. Proper regard should be given to the slight wording issues identified in this example and in Example II.

Exercise 4: Automatic Heating System

Figure 3.11 is a causal-loop diagram showing how a heating system works to keep a house at a comfortable temperature.

a. Label the arrowheads and loop with the appropriate symbols. (A summary sheet of causal-loop symbols and their meanings appears at the end of this chapter.)

b. This is called a *closed-loop* feedback system. Why?

c. Explain the operation of this loop.

d. Do you think an automatic air-conditioner system works on a similar principle? Why or why not?

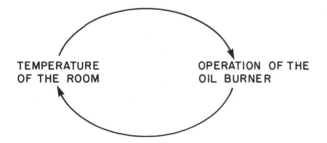

TEMPERATURE OF THE ROOM

OPERATION OF THE OIL BURNER

Figure 3.11 Automatic heating system

Exercise 5: Effect of Attention Paid on School Work

Figure 3.12 is another example of a signed negative causal-loop diagram. Read through the diagram, and then answer the following questions:

a. Explain the behavior suggested by this negative feedback loop.

b. Indicate the changes needed in Figure 3.12 if the following statement were true: "The *more* difficult school work becomes, the less attention is paid by the student." Read around the loop under these conditions. Is the loop ⊕ or ⊖ ? How would loop behavior change if this is viewed as a positive feedback system?

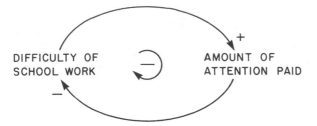

Figure 3.12 Effect of attention paid on school work

MORE COMPLICATED SINGLE LOOPS

EXAMPLE V: MULTIPLE ITEMS IN ONE LOOP

Loops need not necessarily have only two items in them. In fact, a loop may have any number of items in it. What is important is that there are enough items to communicate what is going on, but not so many as to be confusing. Figure 3.13 illustrates a loop with three items. This loop suggests that as population increases, people's concern about pollution also increases. As people's concern about pollution increases they institute more pollution controls. An increase in the number of pollution controls leads to a lessening of the amount of pollution, leading to less concern and fewer controls. This is a negative feedback loop because its implied goal is keeping the amount of pollution under control.

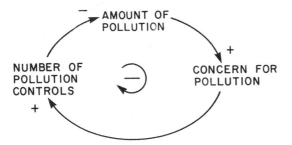

Figure 3.13 Pollution control and public opinion

Another way to identify a negative feedback loop quickly is to see how many − signs there are around the loop. An odd number of − signs (in this case there is 1, which is an odd number) indicates that the loop is negative. An even number of − signs around the loop (or zero − signs) indicates the loop is positive. The effect of an even number of − signs is to cancel each other out.

Exercise 6: Two Loops with Different Behaviors

a. Sign the loop in Figure 3.14 and explain its behavior.

b. Sign the loop in Figure 3.15. Explain how the behavior of this system is different from the system diagrammed in Figure 3.14.

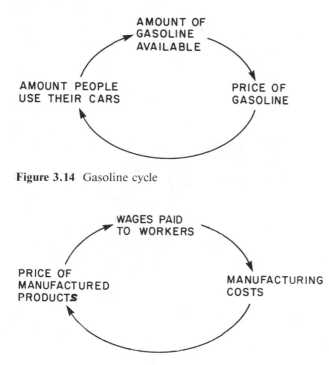

Figure 3.14 Gasoline cycle

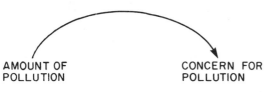

Figure 3.15 Partial explanation of inflation

VERTICAL ARROWS: AN AID TO SIGNING CAUSAL-LOOP DIAGRAMS

Vertical arrows are another aid in determining the appropriate sign to place within a feedback loop. The following example illustrates the method. In Figure 3.13, begin with the link between Amount of Pollution and Concern for Pollution.

To help determine the loop's sign, begin reading the diagram by saying, "An increase in amount of pollution," and use an up vertical arrow to indicate "increase."

AMOUNT OF
POLLUTION

Finish reading this first link by saying, "causes an *increase* in concern for pollution," and place an up vertical arrow at the head of the link to indicate "increase."

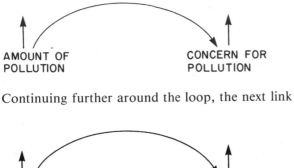

Continuing further around the loop, the next link

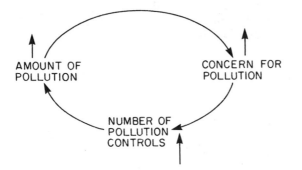

is started with the up arrow on Concern for Pollution. Now, "an increase in concern for pollution causes an increase in the number of pollution controls," indicating that an up arrow should be placed at the head of that link.

The final link begins with the up arrow, or increase, in the number of pollution controls, and leads to an assumed *decrease* in the amount of pollution, or a down arrow at the head of this last link (Figure 3.16). The loop reading was initiated with an up arrow at Amount of Pollution and was closed with a down arrow there. A reversal of arrow directions when the loop has been closed indicates that a negative feedback loop has been formed.

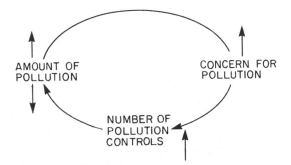

Figure 3.16 Vertical arrows on pollution system—Completed

Exercise 7: Vertical Arrows on Loops

a. Determine the sign of the loop for Figure 3.13 by starting with a decrease in Concern for Pollution. Is the loop sign the same as was found above?

b. Apply the vertical arrow method to determine the sign of the loop for Figure 3.15. Note that if a loop is closed with the finishing arrow pointing in the *same* direction as the initiating arrow, the loop is positive.

INFERRING THE BEHAVIOR OF POSITIVE FEEDBACK LOOPS

Based on the examples and exercises presented, several generalizations may be made about how systems characterized by a single positive feedback loop behave over time. The general pattern is that if some quantity within a positive feedback loop begins to increase, then a "snowball" effect takes over, and that quantity continues to increase. Often the quantity in question increases at a faster and faster rate. One good example of this is the accelerating growth in population in the United States between 1800 and 1960.

One special class of behavior involving unbounded growth at a faster and faster rate is known as positive exponential growth. Exponential growth is characterized by the quantity under study doubling repeatedly, at a time interval called the "doubling time." If a country has an initial population of 1 million persons and a doubling time of fifteen years, then after fifteen years the population would be 2 million; after thirty years, 4 million; after forty-five years, 8 million; and so on. Obviously, no system can continue to exhibit exponential growth indefinitely, but many systems do demonstrate exponential growth at least for a period of time.

The "snowball" effect of a positive loop can also work in reverse. If a quantity in a positive loop begins to decline, that can lead to a continuing decline at an accelerating rate. In Figure 3.4, for example, if student performance starts to decline, it will get progressively worse.

When will a particular positive loop lead to accelerating growth and when will it lead to accelerating decline? Will the student represented by Figure 3.4 exhibit continuously improving performance, or will the student fail miserably? The answer depends on the student's level of knowledge and performance when the analysis begins. If the student's performance is above a particular point, called the *equilibrium point,* performance will show accelerating growth. If, on the other hand, the student begins below the equilibrium point, performance will show accelerating decline. If the student begins exactly on the equilibrium point, performance will remain constant. Unfortunately, the value of the equilibrium point usually cannot be determined from a causal-loop diagram alone. It requires analyzing the model equations, which is discussed in Part Five.

Occasionally, the equilibrium point is not difficult to determine. In the bank balance problem in Figure 3.1, for example, it is easy to see that the equilibrium point is zero. If *any* amount of money *above* zero is deposited in a bank account earning 10 percent interest compounded annually, it will grow without bound. But if *exactly* zero dollars are placed in the account, the balance will remain constant at zero. Suppose a negative sum of money were placed in the account. (This might be interpreted as a bank loan at a 10-percent annual interest rate.) Over time, the account would become more and more negative as the total unpaid interest due on the loan mounts up.

Inferring the behavior of a total system from a diagram alone, even a simple one involving just a single causal loop, can be difficult. One commonly occurring situation is that positive feedback loops are often brought into check by other stabilizing influences that may have been left out of the initial analysis (frequently intentionally left out). The plant dynamics positive feedback loop (Figure 3.7), if left to operate on its own, would produce an extremely large number of plants in relatively short time (assuming that all of the seeds produced finally germinated). However, other forces not shown in that single feedback loop, such as fixed amounts of available ground, would ultimately arrest the continuing growth of plants.

INFERRING THE BEHAVIOR OF NEGATIVE FEEDBACK LOOPS

Previous examples have indicated that negative feedback loops tend to produce behavior over time that is "stable" and "goal-seeking." For example, a heating system in a house (Figure 3.11) always tends to maintain the house at the same temperature under wide variations in the temperature outside. The migration and job opportunities loop described in Figure 3.9 tends (other things being equal) toward maintaining a balance between the number of job openings in a region and the number of persons seeking employment. The situation may, of course, fluctuate over some range of job shortage or job surplus while that balance is being sought.

The behavior of a negative feedback loop can be described more easily by referring to the system's *equilibrium point*. Recall that if a quantity in a positive loop, like student performance in the example in Figure 3.4, is above the equilibrium point, it generally will grow at an accelerating rate; and if it is below the equilibrium point, it will decline at an accelerating rate. Thus the equilibrium point in a positive loop is *unstable*. A small disturbance that pushes the system out of equilibrium will generally take the system further and further away from its equilibrium point.

In a negative loop, the situation is just the reverse. Consider once again the job-migration system in Figure 3.9. If the number of job openings is above the equilibrium point, people will migrate into town, causing the number of job openings to fall toward the equilibrium point. Similarly, if the number of job openings is below the equilibrium point, people will migrate out of town, causing the number of job openings to rise toward the equilibrium level. Hence, the equilibrium point in a negative loop is *stable*. In a negative loop, the system tends to return to equilibrium following a disturbance, although sometimes with fluctuations occurring around that equilibrium.

Generalizations about the behavior of negative loops must be viewed with the same caution as generalizations about the behavior of single positive feedback loops. Given only a causal-loop representation of system structure, the analyst can deduce approximately (but seldom exactly) how that system will behave over time. In the examples and exercises thus far, the causal loops have given no specific information concerning the exact equilibrium point of the system. In addition, the causal-loop diagram does not tell how fast the system will seek its equilibrium or along what sort of path.

Consider a spring with a heavy weight suspended from it. This physical system is a goal-seeking one similar to a single-loop negative feedback system; the goal or the equilibrium point of the system is the position of the weight at rest. If the weight is pulled and then let go, the weight will bounce up and down until eventually it settles into equilibrium. However, depending upon the mass of the weight and the strength of the springs, the weight could bob up and down for some time or it could come to rest rather quickly. If the spring had a shock absorber attached to it (as do the springs in the suspension of most automobiles), then the weight might not bob up and down at all—it might move quickly and directly into equilibrium (just as most suspension systems are designed to do). As with positive feedback loops, the best way to determine the precise behavior of a negative feedback loop is to simulate the system's behavior using a computer.

SYSTEMS INVOLVING MORE THAN ONE FEEDBACK LOOP

Although simple positive and negative feedback loops form the basic building blocks of complex systems, they seldom occur in isolation. Systems associated with complex social and economic problems typically contain multiple interconnected loops. In a complete system dynamics study, analysts begin by tentatively proposing a set of feedback loops that seem to be the most important in explaining the system's behavior. Then computer simulation is used to analyze in depth the behavior of the loops initially identified for analysis. Often the first structure is found to be incomplete or erroneous, and the analyst must then return to the conceptual phase of model building and reexamine the causal structure. The complete model-building process is a reiterative one, beginning with a preliminary hypothesis concerning the important causal loops, moving toward computer simulation, and often returning several times to the process of hypothesizing important causal loops.

The sections that follow present systems involving more than one feedback loop, illustrating that causal-loop diagrams of social and economic systems can easily and rapidly become very complicated. Here, even the best analyst's intuition about how such a system will behave often breaks down, and the computer becomes an invaluable tool for unraveling the relationships between a system's structure and its behavior. The next section is more concerned with drawing out general principles involved in the construction and analysis of causal-loop diagrams, and less concerned with predicting the exact behavior of a set of causal loops.

Exercise 8: Combining Positive and Negative Loops

The positive birth loop and negative death loop associated with population dynamics were previously analyzed. Figure 3.17 shows these two influences combined.

a. Label the arrowheads and feedback loops with the appropriate symbols. Do each loop separately.

b. Based on the information conveyed by the diagram, explain each loop separately.

c. This system is made up of both a positive and negative feedback loop. What kind of behavior would you expect from such a system over time? Look back at the graph in Figure 3.6. Which of these two loops dominates the U.S. population system?

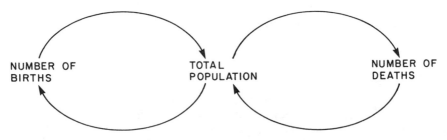

Figure 3.17 Population dynamics—Combined

LOOP DOMINANCE

The word *dominates* suggests that a particular loop in a system of more than one loop is most responsible for the current overall behavior of that system. Which particular loop dominates might shift over time. For example, many people are concerned that the earth's population cannot grow forever. Terms such as "spaceship earth" or our "finite world" express the idea that there are limits to the earth's ability to support human life. People want to know what will cause the behavior of the human population to shift from growth (in this case, dominance by the positive loop) to decline or equilibrium (in this case, dominance by the negative loop). Possible explanations, such as insufficient food or energy, or excessive population, or an orderly policy of zero population growth, have been proposed. The prospect of more people dying than being born is grim. Therefore, understanding the underlying structure of systems, and being able to suggest when or why a shift in loop dominance might occur, is crucial to problem-solving from a dynamic perspective. Obviously, simple causal-loop diagrams, such as those shown in Figure 3.17, will need to be elaborated significantly before they can help solve such important global problems.

Exercise 9: Tired-Sleep-Active System

In any complex system, more than one loop would most likely be needed to explain the observed behavior. Figure 3.17 showed two loops linked together with a common element. Figure 3.18 is another example of two linked loops. Sign the loops, then explain the behavior of the total system. Do so by explaining *each loop separately* and then the general overall behavior of the system.

a. Explain the Tired-Sleep loop.

b. Explain the Tired-Active loop.

c. Explain the overall system behavior.

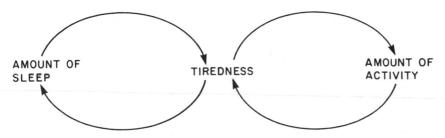

Figure 3.18 Tired-sleep-active system

Exercise 10: Other Possible Causal-Loop Diagram Structures

Symmetrically linked loops, as shown in Figures 3.17 and 3.18, are not always the results of causal-loop development efforts. Elements of systems interact in an endless variety of ways, therefore an endless variety of causal-loop diagram structures are found. Figure 3.19 shows a diagram of another system represented by two loops different in structure from those in Figure 3.18. Sign each loop separately. Explain each loop separately, then comment on the overall system.

a. Explain the smaller causal loop, on the left-hand side of the diagram.

b. Explain the larger causal loop, around the outside of the diagram.

c. Explain the characteristics of the total system. Discusss the characteristics of the system when the positive loop dominates, then when the negative loop dominates. (An explanation of the word *dominates* may be found following Exercise 8.)

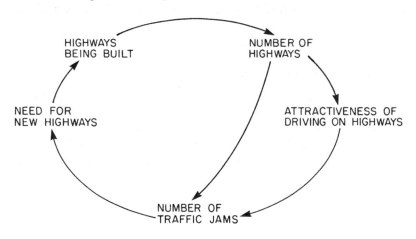

Figure 3.19 Traffic dynamics

EXOGENOUS ITEMS

Most causal-loop diagrams include some items that affect other items in the system, but are not themselves affected by anything in the system. In other words, arrows are drawn *from* these items to other parts of the diagram, but there are no arrows drawn *to* these items. These are called *exogenous* items. "Favorable Location" in Figure 3.20 is an example of an exogenous item. A favorable location of some sort, perhaps a good harbor or a mountain pass, attracts people to settle in a particular geographic area. Figure 3.20 shows Favorable Location as the element that starts off the growth of a city. This diagram represents a *partial possible* explanation of why a city grows and declines.

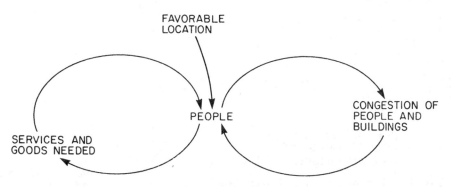

Figure 3.20 Partial explanation of urban growth and decline

Exercise 11: City Dynamics

a. Label the arrowheads and loops in Figure 3.20 with the appropriate symbols.

b. Write a paragraph explaining what Figure 3.20 tells you. Notice that there are *two closed loops*. Explain each one separately.

c. Using the graph in Figure 3.21, tell which feedback loop dominates at which time during the history of Boston.

Exercise 12: Beef Price Cycles—Part I

Figure 3.22 shows one aspect of the dynamics of the cattle industry.

a. Label the arrowheads and loop with the appropriate symbols.

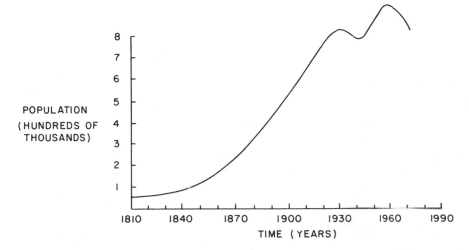

Figure 3.21 Population of Boston, 1810–1975

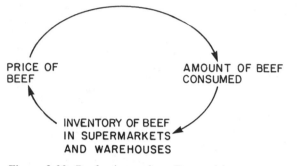

Figure 3.22 Beef price cycles—Demand loop

  b. Write a paragraph explaining this feedback loop. What happens to the price of meat when the amount of meat available in inventories increases?

  c. Suppose the government tried to hold beef prices steady by buying all surplus meat for school lunch programs. How might this affect the dynamics shown in the diagram?

Exercise 13: Beef Price Cycles—Part II

Figure 3.23 explains another aspect of the cattle industry. When the price of cattle feed is low, ranchers keep their herds longer to fatten them up. However, when feed becomes expensive, perhaps due to unfavorable weather, the ranchers cannot afford to keep their herds very long; they send more cattle to the market.

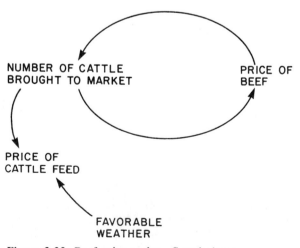

Figure 3.23 Beef price cycles—Supply loop

a. Label the arrowheads and loops with the appropriate symbols.

b. Write a paragraph explaining Figure 3.23. There are two closed loops. Explain each one separately.

c. What is the exogenous item in Figure 3.23?

Exercise 14: Supply and Demand in the Beef Cycle

Figure 3.24 combines the demand cycle from Figure 3.22 with the supply cycle from Figure 3.23.

a. Label the arrowheads and loops with the appropriate symbols.

b. Explain in your own words what affects the price of meat. How might Figure 3.24 explain why prices of such products as meat rise and fall?

c. Name some other things whose price might be affected in a similar way to the price of meat.

d. Pick one of these items and draw a causal-loop diagram illustrating this as another example of a supply and demand system.

A more comprehensive model of commodity cycles is developed in Chapter 21.

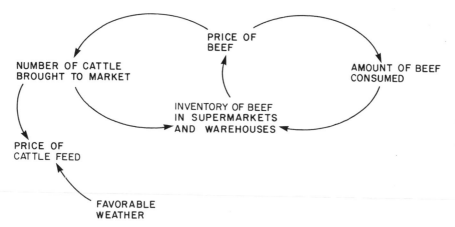

Figure 3.24 Beef price cycles—Supply and demand combined

INFERRING THE BEHAVIOR OF MULTIPLE LOOP SYSTEMS

As the previous examples have shown, inferring the behavior of a system that contains more than one loop is a difficult task. For example, in the simple two-loop system involving Total Population, Number of Births, and Number of Deaths (Figure 3.17), the overall behavior is dramatically different, depending on whether births or deaths dominate. In the first instance, the system grows without bound; in the second, the system decays toward zero, as all of the population eventually dies off. Given only the causal loop representation, there is no sure way to know which of these two dramatically different behaviors will take place.

In the beef supply and demand loops just developed, both the supply and demand loops are negative and will tend to stabilize prices. But at what level will prices stabilize? Which loop, the supply loop or the demand loop, is more important in this stabilization process? What would happen if consumers react much more quickly to the price of beef (by choosing to purchase more or less beef) than ranchers respond (by increasing or decreasing their herd sizes—a process that can take years)? How might this asymmetrical speed of reaction in the supply and demand loops ultimately influence beef prices? These and a host of other questions cannot be answered precisely when working with causal loops alone.

However, a number of interesting inferences can be made . . For example, the traffic dynamics problem (Figure 3.19) shows that the policy of building new highways to relieve traffic congestion touches off two reactions of an opposite nature, one that relieves congestion pressures and one that acts to exacerbate those pressures. Hence, the causal-loop analysis highlights an intrinsic contradiction in any highway building program. Causal loops are useful because of their ability to generate tentative insights and hypotheses that can be explored more fully within the context of a simulation model.

The exercises that follow further develop abilities to think in terms of causal loops.

Exercise 15: Capital Equipment Cycle

Figure 3.25 shows in simplified terms several loops operating in the aggregate acquisition of capital equipment.*

a. Label the arrowheads and loops with the appropriate symbols.

b. Is the capital equipment cycle a system of growth or a system that maintains equilibrium? Explain your answer.

---

*Modified from D. Meadows, D. Meadows, J. Randers, and W. Behrens, *The Limits to Growth* (Washington, D.C.: Universe Books, A Potomac Associates Book, 1972).

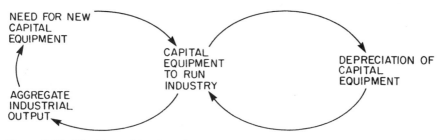

Figure 3.25 World machinery cycle

Exercise 16: Forest Development

Figure 3.26 shows four feedback loops that influence forest development. Three of them are obvious; the fourth one may take some hunting to identify.

a. Label the arrows and loops with the appropriate symbols by reading around each loop individually.

b. Write a few sentences explaining each loop.

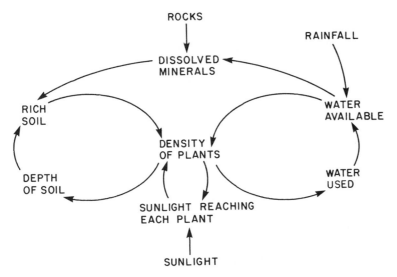

Figure 3.26 Forest development

Exercise 17: Carbon Cycle

a. After determining whether the loops in Figure 3.27 are positive or negative loops, write a paragraph explaining the carbon cycle based on the information shown.

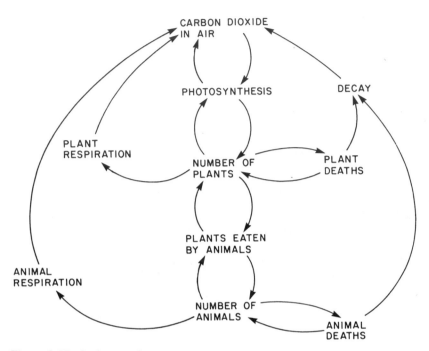

Figure 3.27 Carbon cycle

  b. In terms of Figure 3.27, could available carbon dioxide in the atmosphere ever be exhausted?

  c. What might people do that could change the carbon cycle?

Exercise 18: Dyslexia

Figure 3.28 describes one principal's understanding of a reading problem called "dyslexia," which means a disturbance of one's ability to read. His basic argument is that if children were allowed to begin reading when they are ready to read, there would be no dyslexia. These children would not encounter unnecessary frustrations, which lead to troublemaking, lack of attention, and increased general pressures.

  a. What is the probable dynamic behavior of this problem as shown in Figure 3.28? Read and label each loop separately.

  b. Do you agree with the advice suggested by this principal? Why or why not?

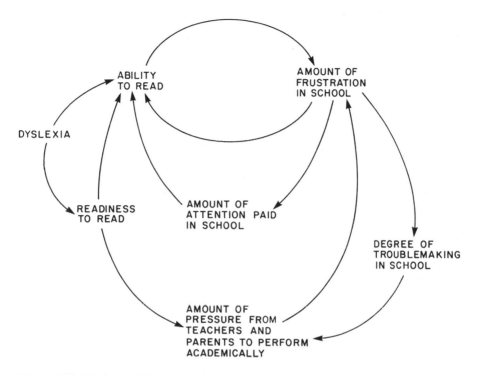

Figure 3.28 Understanding the problem of dyslexia

ENDNOTE

1.   The empirical observation that some regions have unemployment percentages that
      are higher than other regions indicates that Figure 3.9 is clearly a simple and in-
      complete formulation. A more comprehensive representation of unemployment
      issues in a system dynamics model appears in H. R. Hamilton et al., *Systems Sim-
      ulation for Regional Analysis* (Cambridge, Mass.: MIT Press, 1969).

Summary sheet
Causal-loop diagram symbols

Symbol	*Meaning*
(ARROW) (TAIL) (HEAD)	The arrow is used to show causation. The item at the tail of the arrow *causes* a change in the item at the head of the arrow.
+	The + sign near the arrowhead indicates that the item at the tail of the arrow and the item at the head of the arrow change in the *same* direction.
	If the tail *increases*, the head *increases*; if the tail *decreases*, the head *decreases*.
−	The − sign near the arrow head indicates that the item at the tail of the arrow and the item at the head of the arrow change in the *opposite* direction.
	If the tail *increases*, the head *decreases*; if the tail *decreases*, the head *increases*.
(+) OR (+)	This symbol, found in the middle of a closed loop, indicates that the loop continues going in the same direction, often causing either systematic *growth* or *decline,* behavior that unstably moves away from an equilibrium point. This is called a *positive feedback loop.*
(−) OR (−)	This symbol, found in the middle of a closed loop, indicates that the loop *changes* direction, causing the system to *fluctuate* or to *move toward equilibrium*. This is called a *negative feedback loop.*

CHAPTER 4

DEVELOPING CAUSAL-LOOP DIAGRAMS

This chapter presents a series of exercises designed to increase the ability to analyze situations by exposing underlying feedback structures. There is *no one correct diagram* for any of the exercises that follow. Each reading might lead to the development of a somewhat different causal-loop diagram. As long as the diagram helps to clarify the behavior described, and is consistent with the written material, the diagram is "good" or "correct." Part IV provides more guidelines for exercising selectivity in causal diagraming.

The following exercises range from simple descriptions explained by a single feedback loop to situations requiring multiple loop diagrams to explain the behavior reported. Read each case and first develop the causal-loop diagram that explains the behavior *as reported by the author*. To show disagreement with the presentation, draw a second diagram explaining another viewpoint.

EXAMPLE I: THE PIPELINE DELAY

Do you remember the first time you took a shower? The most difficult task for the beginning shower-taker is regulating the water. It is difficult to decide which knob to turn on first: Should you risk burning yourself or freezing yourself? If you start by turning on the cold water, as soon as you feel the first shock of icy water your immediate reaction is to turn down the cold faucet and turn up the hot faucet. However, the water temperature does not improve immediately because of the delay in the water coming through the pipes. You continue to adjust the knobs, probably causing the water to become too hot.

A causal-loop diagram to explain the problem of the beginning shower-taker is shown in Figure 4.1. The terms in that figure are kept neutral. For example, it does not say "Hot Water," but "Temperature of Water," which can then increase or decrease. The figure suggests that the *higher* the water temperature, the *more* cold water you turned on; the *more* cold turned on, the *lower* the temperature of the water.

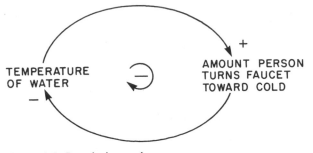

Figure 4.1 Regulating a shower

This negative feedback system might oscillate somewhat as it seeks its goal of a comfortable water temperature. If you were to graph the water temperature over the few seconds in which you attempt to adjust it, the graph might look like Figure 4.2.

Why, then, does the beginning shower-taker usually experience too hot and then too cold water? What one usually forgets is the time required for the water to change temperature. A delay occurs between adjusting the shower controls and the change in water temperature. In causal-loop diagrams, each arrow represents time passing, or delays. Sometimes the delay is a matter of seconds; in other situations, delays could be centuries. Perceiving the length of delays is sometimes quite important in understanding the behavior of a system. In this example, understanding the delay in the water system advances one from a beginning shower-taker to an advanced shower-taker. Chapter 7 provides a variety of experiences in graphing the effects of delays on system behavior; Chapter 17 treats delays through the use of computer simulation.

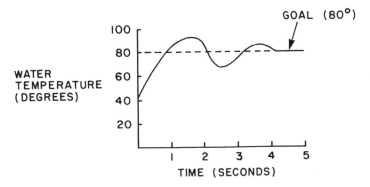

Figure 4.2 Shower adjustment

Exercise 1: Practice Makes Perfect

Often when you get very involved in doing something, and really focus on it in terms of time and effort, you can readily improve your ability. A good example of this might be the ability to do pushups. If you faithfully practice doing pushups, you strengthen your arm muscles, and the number of pushups you are able to do increases, further strengthening your muscles. As soon as you let up on practicing, the cycle begins reversing. Pushups become increasingly hard to do and your arms become increasingly weak. Further, you become discouraged and find it more difficult to force yourself to practice regularly.

    a. Using the items as shown below, draw one causal-loop diagram that might explain the two situations described here.

STRENGTH OF   WILLINGNESS
MUSCLES   TO PRACTICE

NUMBER OF
PUSHUPS

    b. Give an example from your own experience that is similar to this pushup story.

Exercise 2: Diary—Part I

The following is quoted from the diary of a third grader: "The first time I tried to hammer a nail into a piece of wood, it was a sad experience. I carefully placed the nail in the wood and slowly started hammering. Nothing seemed to be happening, so I started banging with more force. The harder I banged, the more difficult it was to keep the nail on the right spot. The more the nail slipped, the more upset I became. Finally, I decided to ask my mother to help me."

    a. Why do you think this third grader had so much trouble hammering the first time?

    b. Draw a causal-loop diagram that explains the problem as told by this child.

Exercise 3: Diary—Part II

"When my mother came in to help me hammer in the nail, she explained, 'You must have patience.' She told me I have to hammer slowly and carefully at first, until the nail is firmly in the wood. Then I can hammer a little harder. She said the important thing was to aim the hammer carefully. The more carefully I aimed the hammer, the more likely it was that I would hit the nail on the head, and the more the nail would go into the wood. As the nail went further into the wood, I could hammer harder and harder. After my mother demonstrated her way of hammering a nail, she told me to try again and I succeeded."

a. What was the difference between what the child did and what the mother did?

b. Draw a casual-loop diagram showing what causes the nail to go into the wood.

c. Is this diagram the same as the one you drew for Exercise 2? Explain how the two diagrams are the same or different.

Exercise 4: Influenza

Some form of the flu seems to go around every year. Of all the common diseases, this is one that most people are likely to experience at least once during their lives.

The flu virus is spread by the emission of droplets of infected saliva into the air. The virus is released as the droplets evaporate, freeing the virus to lodge on another individual.

Influenza epidemics occur periodically because the virus never quite dies out. Once people have the flu, they are immune from another attack for a year or more. However, they eventually become susceptible again. The more susceptible people there are in a town, the more likely it is for a virus epidemic to break out. As people get sick and recover, they become immune again, and the epidemic slowly dies out.

a. Draw a causal-loop diagram that explains the cyclical occurrence of influenza epidemics. The key items in this system are numbers of:
  - susceptible people
  - sick people
  - immune people

  The exogenous item that starts the system going is:
  - virus in the air

b. What is the dominant system characteristic that causes these cycles?

c. Given that the spread of infection is rapid and a person is usually sick for about a week and then immune for a year or so, fill in a graph similar to Figure 4.3, predicting what a town's epidemic history might look like over four years.

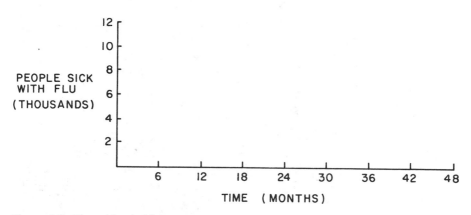

Figure 4.3 Flu epidemic history

Exercise 5: Insulin and the Balance of Body Sugar

Individual life processes such as digestion and heart action have built-in self-regulating mechanisms. The mechanisms either speed up the system or slow down the system to maintain nearly constant body conditions. This state of automatic, dynamic balance in living things is known as *homeostasis.*

Homeostasis is a general term for the action of many self-controlling mechanisms in the body. Some are chemical and some are activated by the nervous system.

An example of chemical control is insulin regulation of the amount of sugar in the blood. When the level of sugar in the blood is higher than normal, cells within the pancreas release insulin into the bloodstream. Insulin allows the sugar to pass through the cell walls for use by the cells, thus lowering blood sugar levels. When the level of sugar in the bloodstream drops enough, the pancreas stops producing insulin, so that the cells stop using sugar. When the level of sugar rises again, the process automatically begins again.

   a. Draw a diagram to show how your body controls its blood sugar level. Sign the arrowheads.

   b. Is this an example of a negative or positive feedback system, one which maintains equilibrium or one which tends to explode?

Exercise 6: The Oil Crisis

The dramatic and sudden rise in the prices of petroleum products in the United States and other industrialized nations during the decade of the 1970s has been referred to as the "oil crisis." According to one theory, the oil crisis involves a "vicious circle." This vicious circle was begun with decisions by the oil-producing countries (OPEC) in 1971 to raise the price of oil. The rise in oil prices

meant that these countries earned so much money that they had large surpluses of foreign currency in their balance of payments, so much that some had difficulty investing or spending all their foreign currency. Realizing this, some of these countries decided to reduce oil production. They knew that eventually their oil supplies would run out, so they might as well make them last as long as possible.

Because less oil was being produced in the world, while more oil was still being needed every day, a scarcity of oil developed. This scarcity of oil forced prices to go up even higher, continuing the "vicious circle."

a. Draw a causal-loop diagram showing the "vicious circle" this economist describes. A vicious circle is an example of a positive feedback loop.

b. Could the mechanism posited above be an adequate explanation of rising petroleum prices in the United States? Why or why not?

c. Extend the diagram of this vicious circle by adding other loops that might eventually impact this system. Things to consider are the limits imposed by the amount of oil left in the ground, the effects of conservation by consumers, and the use of alternate energy sources.

The expansion of the causal-loop diagram to include considerations such as the amount of oil in the ground and the effects of conservation represents an expansion of the boundary of the problem. The discussion describing the vicious circle had focused solely on the behavioral phenomenon affecting the oil producers and resulting in the positive feedback loop. However, others attempting to understand the problem might feel this is too narrow a view. They might therefore extend the boundary of the problem by including other feedback loops, as necessary, for a clearer explanation of the situation. (Boundary issues will be further discussed in Part IV.)

The following exercises require more than one feedback loop to explain adequately the behavior described. As you read through the next example, make a list of all the things you think are important to include in a causal-loop diagram to explain adequately the system discussed.

EXAMPLE II: THE SELF-REGULATING BIOSPHERE[1]

The biosphere is the narrow space around the earth where all life exists, containing the air, water, and soil necessary for living things. The biosphere is made up of thousands of self-regulating systems that keep the earth going.

Consider one of these systems—the system that keeps in balance the amount of fresh water on earth. The earth's water is constantly moving, turning from vapor to liquid to snow to ice, joining a glacier for thousands of years, or flowing up through plant roots during photosynthesis.

Water evaporating into the air begins a long series of steps that contribute to regulating the amount of water on earth. When the sun shines on a water surface, it causes evaporation. The vapor rises and forms clouds, which the wind blows toward land. Eventually the clouds break up, making rain. As soon as the clouds are gone, the sun begins forming more clouds. This system not only regulates the amount of fresh water being carried from the oceans to the land, but also controls the temperature of the earth. As the sun heats the earth, evaporation and cloud formations increase. With more clouds in the sky, less of the sun's heat gets to the earth. As the temperature drops, the clouds release their water and disappear. The sun shines again on the earth, starting the cycle over. This self-regulating system guarantees that living organisms will have enough water and a constant temperature range. This is only one in a long series of mutually dependent self-regulating processes.

A list of key items might be as follows:

    water evaporation

    amount of water on the earth

    amount of sunshine reaching the earth

    clouds

    rain

    temperature of the earth

Reread the story, indicating the causal links as stated by the authors. The first statement they make is that sunshine causes evaporation:

SUNSHINE

$\downarrow$

EVAPORATION

forming clouds:

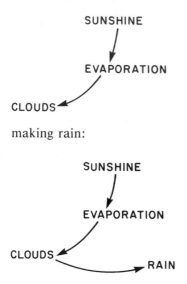

making rain:

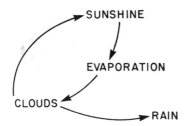

The authors then suggest that when the clouds are gone, the sun shines on the earth again, starting the process over and closing the loop:

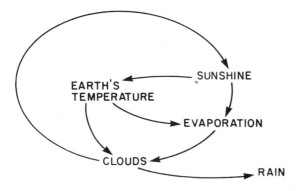

They then add that the temperature of the earth will affect the rate of evaporation of water and cloud formation:

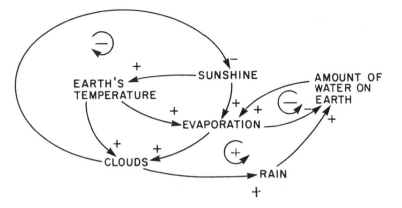

Figure 4.4 The self-regulating biosphere

The only item from the initial list not yet in the diagram is "amount of water on earth." Looking at the diagram, it is obvious that rain will affect the amount of water on the earth. The amount of water on the earth will affect the amount of evaporation that can occur. And the amount of evaporation will affect the remaining amount of water on earth. After adding this item to the diagram, the completed, signed diagram is shown in Figure 4.4.

Exercise 7: The Grape Growers' Quandary—Part I

"For a number of years, an insect known as the grape-leaf hopper had caused considerable damage in the vineyards. Periodically, much to the delight of the growers, the grape-leaf hopper populations in the vineyard were drastically reduced by the activities of anagrus epos, a natural parasite of the grape-leaf hopper. By drastically reducing the size of the grape-leaf hopper population, however, the parasites reduced the amount of food (grape-leaf hoppers) available to themselves. As a direct result of the food shortage, the parasite population also declined in size.

With the parasite population on the decline, the grape-leaf hopper population was able to grow again. Ultimately, the growth of the grape-leaf hopper population was followed by growth of the parasite population and so on" (Melillo, 1972, pp. 103–104).

a. Draw the feedback diagram showing the cause and effect relationships between the grapes, the grape-leaf hopper, and the parasite.

b. Sign the diagram.

Exercise 8: The Grape Growers' Quandary Solved—Part II

"Fortunately for the grape growers, the anagrus epos also feeds on a non-pest leaf hopper which in turn feeds on wild blackberries. By planting small patches of wild blackberries in the vineyards, the growers were able to maintain a stable parasite population that was large enough to control population explosions of both leaf hoppers, the non-pest and the grape-leaf hopper'' (Melillo, 1972, pp. 103–104).

   a. Redraw the causal-loop diagram from Part I of the Grape Growers' Quandary, then add this new information to it.

   b. What kind of feedback system (positive or negative) does the diagram show?

   c. What kind of behavior will this system demonstrate?

   d. Explain how Part II of this story changed the boundary of the problem.

Exercise 9: I Hate Piano—Part I

This was Valerie's third year of piano lessons. She had taken lessons with two teachers during this time. The first year Valerie did not mind playing the piano. The pieces were fairly easy and the teacher was rather pleasant. But by the end of the year, Valerie decided she really did not like the piano, the teacher, or practicing.

   After a serious discussion between Valerie and her parents, they decided that maybe the teacher was not adequate. Perhaps a new teacher would provide Valerie with the enthusiasm she was lacking. After inquiring about teachers from many different sources, Valerie's parents finally took the recommendation of a good friend and neighbor. The new piano teacher was a woman from the next town who came to Valerie's house each week.

   From the very first lesson, Valerie was not sure about this new teacher. She tried to be friendly and joke with the teacher, but she always seemed to say the wrong thing. More often than not, the piano teacher would end up giving Valerie a lecture on manners or politeness as part of the lesson.

   In addition to this clash of personalities between teacher and student, the new teacher was a perfectionist. As a result, Valerie was not given a new piece until she could play the current piece perfectly. After two more years of lessons, Valerie had studied only six pieces, hating each one thoroughly. She also hated the piano, playing the piano, and just about everything to do with music.

   a. Draw a causal loop that might explain what caused Valerie to hate the piano *more and more* each year. The following key phrases might help you develop your diagram:
      • Valerie's piano playing progress
      • Valerie's enthusiasm
      • Teacher's enthusiasm

   b. If this situation can be described as Valerie hating the piano more and more each year, then what type of feedback loop is probably dominating this system's behavior?

c. Give an example of something similar that happened to you. Write a paragraph describing that situation. Then draw a diagram showing what you have described.

Exercise 10: Valerie and the Piano—Part II

After three years of piano lessons, Valerie did not care if she ever touched a piano again. She could not play anything right, she had improved very little, and each half hour of practicing was torture. Her parents, however, still had faith in Valerie and were convinced the teacher was the problem. Once again they searched for a piano teacher.

This second search produced two recommendations: one by a school friend of Valerie's and one by another neighbor. This time Valerie's parents decided Valerie should meet both teachers and choose the one she preferred.

Valerie met both teachers. The meeting with the first teacher was casual and short. The second teacher auditioned potential pupils before accepting them, and would accept only those whom she felt had talent. The meeting with her, which included the audition, was longer and more formal. Valerie liked both teachers, but was more impressed with the second one. The second teacher was impressed with Valerie as well and agreed to take her as a pupil. This delighted Valerie.

This initial mutual admiration carried through Valerie's fourth year of lessons. She was given a new piece almost every week. She memorized pieces in two or three weeks. Moreover, she did not mind practicing. She looked forward to each lesson more and more as the year went on.

a. Explain Valerie's attitude during her fourth year in terms of a feedback diagram.

b. Is this diagram the same or different from the one you drew for Part I? Explain in a few sentences how the diagrams are the same or different.

Exercise 11: Eager Beaver

Beavers are fascinating animals about which there are still many unanswered questions. They are admired by some people and considered pests by others. Trappers, fishermen, conservationists, and hunters seek out beaver ponds because of the great variety of wildlife attracted to these sites. Timberland owners and farmers usually try to get rid of beaver ponds because of the damage done from flooding caused by their dams. However, in spite of the flood damage, most people are on the beavers' side.

Contrary to what many people believe, trapping is necessary to keep a flowage (water formed by overflowing or damming) active for a long period. An area that is never trapped is soon lacking food. Beavers can and do literally eat themselves out of house and home, for they will cut the trees faster than they can grow. Once this food source is gone, the colony must move. Taking a

D. LEONARD-SENGE

certain percentage of the colony out by trapping will insure the colony a longer period of living. Over-trapping can, of course, reduce the population. This has happened in many areas where, under ideal trapping conditions, trappers have caught all the animals in the colony.

a. Draw a feedback diagram showing how the beaver population is controlled naturally.

b. Add to your first diagram to show how people can affect the size of the beaver population.

c. What kind of system have you drawn? How do people affect the system?

EXAMPLE III: THE TRAGEDY OF THE SAHEL[2]

A great tragedy took place in an area of the world known as the Sahel. This narrow area of land, running almost the whole width of Africa, just under the Sahara Desert (Figure 4.5), has been the home of nomads for centuries. It has never been an easy place to live, but the nomads have survived amazingly well. Over the years they have mastered the skill of moving their animals from grazing spot to grazing spot with each season.

Because of the limited amount of food and water available to the animals, the herds never got very large. Due to the severe climate, disease, and poor diet, the same was true of the nomad population. Moreover, every twenty or thirty years a severe drought would kill many animals and people, keeping their numbers from growing too large.

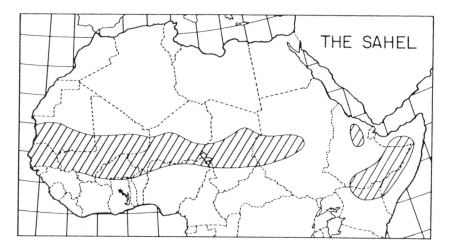

Figure 4.5 Map of the Sahel

   In recent years, people from other parts of the world, acting for organizations like the United Nations, decided to try to improve the life of these nomads. These organizations did two major things. First, they introduced modern medicine. They vaccinated the nomads against smallpox and measles; they brought malaria and sleeping sickness under control. Modern medicine greatly increased the life-span of the nomads. Animal diseases were also controlled.
   Second, more water was made available. There are great supplies of underground water in the Sahel which the nomads' hand-dug wells never reached. Using modern machinery, deep wells were drilled. This large new water supply increased the number of animals possible for the nomads to own. Although the animals had water, they soon ate or trampled the little grass available. A six-year drought further decimated the grass. The animals began to die of starvation. Because of the drought and the loss of animals, many nomads starved. The United Nations was faced with a more severe problem than the one it originally went to the Sahel to solve.
   A list of important items might include the following:

   nomads
   more animals
   limited food
   limited water
   limited herds
   severe climate
   disease
   poor diets

small population
severe droughts
modern medicine
   people
   animals
deeper wells
   increased number of animals
limited grasslands
   eaten
   trampled
animals starved
people starved

Two key items on the list are: the number of nomads and the number of cattle. These elements interact with almost all the items on the list.

NUMBER OF
NOMADS

NUMBER OF
CATTLE

The next step is to add to the diagram all the relationships that affect directly the size of the nomad population.

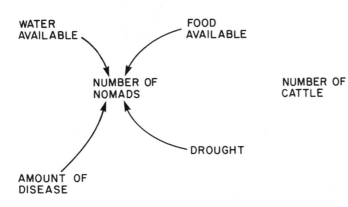

Additional arrows are added to show which items the nomads also affect:

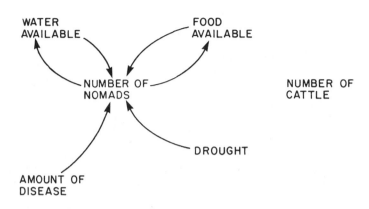

Now everything that affects and is affected by the number of cattle is added:

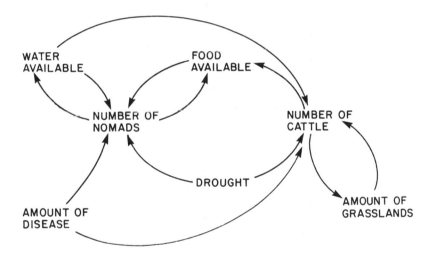

Finally, add the outside intervention of modern medicine and deeper wells, sign the diagram, and you have a diagrammatic explanation of the Tragedy of the Sahel (Figure 4.6).

Before the intervention by outsiders, the one positive feedback loop that could potentially cause the nomad and cattle populations to grow too large in years of favorable climate had never become dominant. Increasing the water supply and introducing modern medicine at about the same time allowed both the nomad population and the cattle population to grow far larger than their ecosystem could support. Before this system was understood, however, the damage was done.

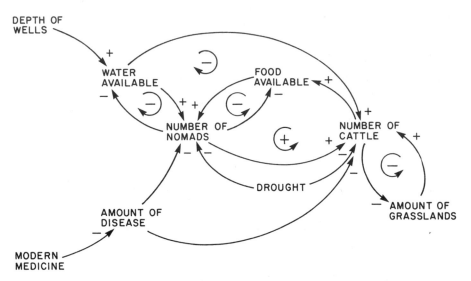

Figure 4.6 The tragedy of the Sahel

Exercise 12: The Body Temperature System[3]

Most healthy people take it completely for granted that their body temperature will stay around 98.6 degrees Fahrenheit. Yet the body is constantly producing heat by being active and by eating and using food. It is exposed to countless weather changes and many extremes of temperature, from very hot to below-zero cold. The wonder is not that the body occasionally "runs a fever," but rather that in such constantly changing conditions its normal temperature varies by only about one degree.

The automatic temperature control mechanism of the body is located in a tiny area just under the brain. When chilled blood from other parts of the body passes through this region, chemical and nervous responses start. The heartbeat increases and shivering begins. These actions cause heat-producing muscle contractions. Nervous impulses cause the sweat pores to close, and other heat-producing mechanisms throughout the body are triggered.

Conversely, when the temperature of the blood is too high, these automatic controls act in reverse. Heat-producing actions stop, and nervous stimuli now cause muscles to relax and sweat pores to open. As much as a gallon of sweat per hour helps to cool the skin's surface.

Draw a diagram showing what happens inside the body when the body temperature changes. (The article explains that when the blood passing under the brain is either too warm or too cold, the heartbeat, shiver rate, and pore size will be affected. Each of these in turn will affect body temperature.)

Exercise 13: A Global View of Natural Resource Usage

One of the main factors that enabled the Industrial Revolution to occur was the availability of seemingly unlimited *amounts of natural resources.* The recent dramatic rise of crude oil prices has caused a general realization that natural resources can be "used up," for all practical purposes, long before their natural geologic replacement time can replenish them.

Many recommendations have been made and strategies undertaken for coping with this situation. Industries spend increasing amounts of money on *developing the technologies* needed to use other more available resources, extract less accessible resources, and use lower-grade resources. Meanwhile, *world demand* for industrial products grows. As more countries progress toward an *industrial base,* more peoples' lifestyles depend on industry.

  a. Draw a causal-loop diagram that shows the dynamics suggested by this description. (The key items are italicized in the story.)

  b. Based on your diagram, what might happen to the price of natural resources in the next ten years? twenty years?

  c. Based on your causal analysis, how might your lifestyle be affected over the next twenty years?

  d. The supply of natural resources, like most other products, is determined by price mechanisms. Add an explicit price effect to the causal-loop diagram that you have developed.

Exercise 14: The Workings of an Ecosystem

"A park, a field, a forest, an ocean, a lake, a pond, or even a classroom aquarium is an ecosystem. Whatever the size, an ecosystem operates as a whole unit. Both its nonliving and living parts are constantly interacting. For example, the constant interaction of nonliving and living components of an ecosystem can be observed in an abandoned field in the northeastern part of the United States. There we can find the soil providing young grass plants with moisture and nutrients needed by plants for growth. Then, in the same field, we find the crickets that feed on the grass, the leopard frogs that feed on the crickets and other insects, and the garter snakes that eat frogs, toads, mice, and other small animals that live in the field.

"When the plants and animals of the field die, the small animals, bacteria, and fungi living in the soil decompose the dead plant and animal parts. The small animals, bacteria, and fungi obtain the energy and nutrients they need for life from the decomposition process. They also perform a great service to the ecosystem by returning the nutrients stored in the dead plant and animal tissue to the soil. With this replenishment, the soil then has an ample supply of nutrients to support new plant growth. Nature, then, has been in the 'recycling' business for a long time" (Melillo, 1972, p. 10).

a. Draw a feedback diagram that shows how nature "recycles" itself. The term "recycle" suggests that what type of feedback loop dominates this system?

b. What kind of feedback system have you drawn?

c. What kind of behavior over a period of time should you expect from this system?

d. What would happen to the system if a farmer came along, decided that the field was perfect for planting corn, so plowed the field?

Exercise 15: Population Growth

One major difference between developed countries like the United States and developing countries like those of South America seems to be their rate of population growth. In the United States, Europe, and Japan (the major industrialized areas of the world), young couples usually have two children. This means that when these couples die, they have left just two people to replace themselves. They have not added people to the total population. This is called *zero population growth*. The population is being maintained at a certain number; it is not increasing.

In the developing countries, like those of South America, Africa, and most of Asia, young couples have four or five children. This means there will be two or three additional people per family in the next generation. When these children grow up, marry, and have four or five children, there will be eight to fifteen people added to the earth's population. In two more generations, there will be added to the earth from this *original* couple who had four or five children, *128–375* more people. This is a huge growth rate, especially when it is happening in poor countries where people are already starving.

People in industrialized countries sense that cities are getting crowded and that it is very expensive to live pleasantly in today's world. Therefore, to provide the kind of life found in industrialized countries, young couples have decided that two children are all they want or can afford. On the other hand, people in developing countries see that their only hope for a reasonable life is to have several children. First, some of these children will die young. Second, the more children who live long enough to work, the more money earned for the family. The hope for these people from the poor countries is to raise several children who can contribute to the family income. If there are enough children working, life for the whole family may be less painful.

a. Draw a feedback diagram that explains what influences the size of the total population in industrialized countries. Since the observed behavior is zero population growth, or maintaining equilibrium, your diagram should explain this.

b. Draw a feedback diagram that explains what influences the size of the total population in the developing countries. Since the observed behavior in the developing countries is population growth, your diagram should explain this.

c. Are these the same diagrams or different diagrams? Explain.

Exercise 16: Growth and Decline of a Yeast Population

Yeast is a yellowish sediment that develops in sugar solutions such as fruit juices. It consists largely of simple cells of a minute fungus and is useful particularly as an agent for fermentation in the making of bread, alcoholic beverages such as wine and beer, and other foods. During fermentation, yeast lives by breaking sugar molecules into alcohol and carbon dioxide. In the baking process, the released carbon dioxide causes bread to rise, making it light and soft. The alcohol, on the other hand, is an effective preservative. In fact, alcohol is one of the oldest methods known for preserving juice and food. Yeast cells reproduce by budding, as shown in Figure 4.7.

During the process of budding, a small bud forms on the membrane of a mature cell. As the bud grows, it breaks away from its "mother" and forms a new plant. When put in a favorable sugar environment, yeast cells keep budding and tend to continue to develop until the sugar on which they feed reaches a critical point. At this point, available sugar is low and the yeast's growth medium has been filled with alcohol and carbon dioxide. Since yeast cannot survive in a medium of alcohol and carbon dioxide, individual yeast cells eventually die.

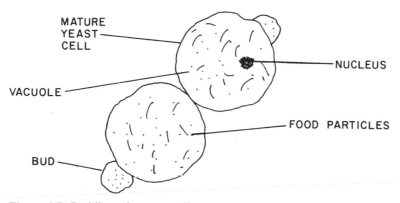

Figure 4.7 Budding of a yeast cell

Draw a causal-loop diagram explaining the dynamics of a yeast culture in a sugar solution, taking into account the growth and decay of yeast cells in the sugar solution. Begin with the positive and negative birth and death loops, then add the factors that affect each of them.

Exercise 17: The Secret of Netsilik Eskimo Survival

The secret of survival of the Netsilik Eskimos is their method of controlling the size and the nature of their population. Netsilik Eskimos value male babies more than female babies because men are expected to hunt for food, clothes, and other necessities. If a female baby is born and the family already has a girl under two years old, the baby is usually given away or left to die. The Eskimos do this because the mother can nurse only one child at a time. A baby is nursed for two years among the Netsilik Eskimos. The family hopes the next child will be male. They cannot risk a three-year delay.

If food is scarce, the Netsilik Eskimo might, out of necessity, let old people die. In these ways, the Netsiliks balance their population. A critical factor that determines the size of their population is the amount of food available.

   a. Draw a feedback diagram showing how the Netsilik Eskimos control their population.

   b. What delay is very critical in the Netsilik life cycle?

   c. Use an axis similar to that drawn in Figure 4.8 to show how you believe the population of the Netsilik will behave over the years.

   d. How might the population change if our modern American values and culture had been brought to the Netsiliks in 1950? Add a second line, in

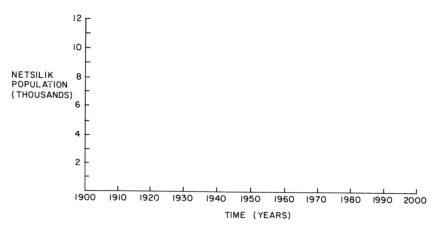

Figure 4.8 The Netsilik Eskimo population from 1900 to present

another color, to your graph to show how the population curve might change under these circumstances.

e. Do you think it would be advantageous for our present values and culture to be brought to the Netsilik Eskimos? Explain.

Exercise 18: Rabbits and Lynx in Northern Canada

The daily life of the lynx is closely tied to the snowshoe rabbit. The lynx has a huge appetite for these rabbits, and its body is particularly well suited to hunting them. In summer the lynx often feeds on mice or ground squirrels, but its mind seems generally filled with visions of Peter Lepus, and its imagination travels beyond a rabbit dinner only with difficulty. This would be all very well, except that through the ages things have come to such a pass that the lynx cannot do very well without rabbits.

Since rabbits do not have planned parenthood, their numbers depend wholly on natural laws. For a period of years the rabbit population grows rapidly. For the lynx this makes a very rosy world. Its food supply increases by leaps and bounds; the lynx flourish and baby lynx abound. A day comes when there are hundreds of rabbits per square mile and a large lynx population. The woods are filled with life and activity. The size and concentration of the rabbit population, however, jeopardizes every rabbit, for now any disease can spread rapidly. And that is what generally happens in Northern Canada unless food scarcity strikes first. Great numbers of rabbits die of disease. In a year or two the lynx find the woods empty. Starvation is now a problem. The lynx roam the woods, capturing what they can. Old hunting grounds are deserted; where there were dozens of rabbits there may now be none. New habitats are explored and hunting of new prey takes place. Lynx become thin and fail to reproduce, and in a year or two are scarce or absent over wide areas.

Little by little, first slowly, then more rapidly, the rabbits come back and grow in numbers again. Lynx recover, too, and become more plentiful. The pattern of the rabbit and lynx cycles continues.

a. Draw a feedback diagram showing the rabbit-lynx cycle. Draw the diagram for the lynx first because it is simpler. (*Clue:* The total number of lynx depends on how many lynx are born and how many lynx die. The number of lynx that die depends on how much food they have.)

b. Next draw the diagram of what happens to the rabbits. (*Clue:* The total number of rabbits depends on the number of rabbits that are born and the number of rabbits that die. The number of rabbits that die depends on the number of lynx, as well as on the amount of food available and the spread of disease. The amount of food available and the spread of disease depend on the density of the rabbit population.)

c. Fit the two diagrams together.

UNSTRUCTURED MATERIAL

The last set of exercises in this chapter illustrates how causal-loop diagraming might be applied to a variety of written materials. These materials are presented in their original form. Develop a causal-loop diagram that extracts the essence from each of the following materials. Keep the diagram as simple as possible, without leaving out any key items.

Exercise 19: The Vicious Cycle

Draw a causal-loop diagram of the newspaper advertisement entitled "The Vicious Cycle" by Friends of Animals, Inc.

Exercise 20: Welfare Reform and the Stock Market

Not everything you read has all the causal links and feedback loops explicitly stated. Sometimes you have to "read between the lines." Exercise 20 is an example of this.

Read the newspaper article "Welfare Reform, the Stock Market, and Politics" by Joseph C. Harsch, and draw a possible corresponding causal-loop diagram.

Opinion and Commentary
Welfare reform, the stock market, and politics*
Joseph C. Harsch   The Christian Science Monitor, August 11, 1977

President Carter's welfare reform program stands an excellent chance of getting through the Congress relatively unchanged for the simple reason that in its final version as unwrapped for the public last Sunday it was more than just welfare reform. It was welfare reform plus job buying.

Had Mr. Carter stuck to welfare reform he would have run into political trouble. It is doubtful that a bill providing specifically and only for welfare reform, no matter how worthy, would have survived the ordeal by Congress. The essential political fact is that more people benefit from welfare programs whether reformed or unreformed than care seriously about getting those programs reformed. Therefore in order to get welfare reform, which is proper and desirable, Mr. Carter had to pay the price of more job buying which is politically popular but not encouraging to Wall Street investors.

Originally, Mr. Carter intended that his reformed welfare program would cost no more than the existing programs. As sent to the Congress it adds another $6.2 billion to a budget deficit already estimated at $48 billion for the current fiscal year. Any addition to the current budget deficit is likely to increase doubts in the investment community about Mr. Carter's dedication to the idea of a balanced

The Vicious Cycle.

Your cat gives birth. It's a joyous and thrilling experience, right? Hopefully. But too often, the miracle of birth is a tragedy in the making. Here's how it happens.

The kittens are beautiful. You'd love to keep them. But you simply can't. Too many mouths to feed. Not enough room. The usual problems. What do you do?

Naturally, you try to find good homes for them. Friends. Relatives. Neighbors. Friends of friends. After all, who wouldn't want a cute, cuddly kitten?

Well, good homes don't always work out. Many kittens face the same situation all over again: too much effort, too much expense. And too many people who don't know how to care for a pet.

So you hear all kinds of stories. This kitten ran away from home. This one was sold. Whatever. The point is, too many kittens wind up homeless, unwanted and alone.

The life of unwanted pets is heartbreaking. They sleep in dirty alleys and vacant lots. They become filthy and miserable. They suffer hunger and abuse. You name it.

Worse yet, strays continue to breed. At an astonishing rate. In one year's time, more than 40 million stray cats and dogs roam our cities and towns.

Every day of their lives, stray cats are not only starved for food and nutrition. They're also starved for love and affection.

Let's reverse the fate which awaits homeless animals. If they're not picked up and destroyed, they linger on the street, starving. How many are yours?

This vicious cycle, unfortunately, is not make-believe. The fact is that every hour nearly 2,500 cats and dogs are born across the U.S. Our pet population is 100 million. As a responsible pet owner, you should have your female cats and dogs spayed. Your males neutered. Not only will you have a happier, more loving pet, you'll prevent uncontrolled breeding and senseless suffering.
At Friends of Animals, we're working on a national spaying program. It provides low-cost spaying and neutering in many areas of the U.S. But we still need help. Join us. Give whatever you can. Let's stop this vicious cycle. Now.

Dear Friends of Animals,
I, too, am a friend of animals and therefore pleased to enclose my contribution (tax deductible) of $_____ to help stop their suffering. Please add my name to your mailing list.

NAME_____ ADDRESS_____ CITY, STATE, ZIP_____

Please send this form with your check or money order to: Friends of Animals, Inc. 11 West 60th Street New York, NY 10023

FRIENDS OF ANIMALS, INC.
11 West 60th Street, New York, NY 10023

This advertisement paid for by Regina Frankenberg, Director of Friends of Animals.   Illustrations by Wally Neibarth

Reprinted with permission, Friends of Animals, Inc.

budget. Hesitation in the investment community is the weakest item in the American economic picture.

The American economy itself seems to be remarkably sturdy. The latest recession reached its low point in March of 1975. Since then the American economy has generated 6½ million new jobs. This is a net gain of nearly a million. The work force went up by 5.6 million at the same time.

Any economy which generates jobs faster than people who want jobs is essentially healthy.

Yet investors continue to show doubt about the American economic future. Wall Street continues to sag. The Dow Jones is below 1966 levels. Investors are not willing to invest in the future because the future is clouded with too many doubts. The biggest doubt is over the future effect of inflation on prices and profits.

Mr. Carter took office promising to balance the budget by the last of the four years of his present term, 1980. There is no reason to doubt the sincerity of his desire to do precisely that. But what he, like many an eager prospective president, failed to appreciate before he reached Washington is the pressure which bears on any president to provide easy money.

Inflation is easy money. American history is marked by a repeated struggle between the bankers who want a "sound dollar" and politicians who want to help out those in economic distress. In Andrew Jackson's day the Bank of the United States was highly unpopular out on the western frontier. President Jackson broke that bank.

Ironically, the effect was the opposite of what was intended. To quote Samuel Eliot Morison in his history of the American people:

"Poor farmers, mechanics, and frontiersmen gained nothing by this bank war; the net result was to move the financial capital of the United States from Philadelphia to New York."

But unemployment, sagging land values and decline in agricultural prices always lead to pressure for easy money in some form or another. William Jennings Bryan brought the Democratic convention of 1894 to a frenzy of excitement and won its presidential nomination with his famous "Cross of Gold" speech. "You shall not press down upon the brow of labor this crown of thorns, you shall not crucify mankind on a cross of gold." Bryan and his followers wanted free coinage of silver, the equivalent in its day of today's easy money inflation.

The fact is that throughout American history there has been a persistent assumption among the less privileged that unemployment and economic decline can be reversed by easy money. Mr. Carter does not face a new problem, but one as old as his country, when he tried to draft a welfare reform bill without any easy money attached to it. He had already disappointed organized labor and the "liberal" establishment by pulling back on earlier job-buying programs and on the $50 tax refund.

Perhaps Mr. Carter has held out better than could have been expected. Perhaps this retreat is minor and will make little difference. But the Dow Jones industrial average sagged by another 9.27 points on the Monday after.

Exercise 21: Tidal Waves and Lobster Beds

Read the newspaper article entitled "Tidal Wave of Fishermen Endangers U.S. Lobster Beds" by Stewart Dill McBride,* and draw the corresponding causal-loop diagram.

---

*Stewart Dill McBride, "Tidal Wave of Fishermen Endangers U.S. Lobster Beds." Reprinted by permission from *The Christian Science Monitor*. © 1975 The Christian Science Publishing Society. All rights reserved.

Tidal Wave of Fishermen Endangers U.S. Lobster Beds
From coast to coast, Americans savor lobsters from the famed lobster beds of
Maine. But now the lobstermen, romanticized figures of history and literature,
face competition from teachers and factory workers who have lost their jobs.
Newcomers find licenses easier to get, costs are rising, and the overall catch is
growing smaller and smaller.

By Stewart Dill McBride
Staff writer of *The Christian Science Monitor*
August 15, 1975

Long Island, Maine   "A fisherman from the mainland pulled a rifle on me because he thought I was cutting his traps." twanged Kenny Davis over the drone of
his 22-foot lobsterboat as it bobbed toward choppy Atlantic fishing grounds.

"It's getting tougher to make a living these days and the fishermen are starting to
crowd each other." He squinted into the brine spray that spilled down his yellow
rubber apron as he continued to grumble about Maine's "lobster wars."

The lobsters off Maine's coast are becoming scarce, and the state's fishermen—who catch the majority of the American lobster (Homarus Americanus)
harvest—are growing more desperate. Toting firearms and cutting traps now are
commonplace.

Normally lobstermen like Mr. Davis spend the foggy July "slack season" on the
wharves repairing and drying their wooden ribbed traps. But this year he can't
afford to stay in port; it hurts his pride and his pocketbook.

Tough Job Grows Harder   Maine lobstermen have never had an easy haul. For
more than a century they have risked their lives on rolling seas to harvest a deli-
cacy that is served in restaurants from Portland, Maine, to Portland, Oregon.

But these days the politics and economics of this much romanticized business look
as ominous as a "smoky sou'wester" storm, say the fishermen.

High unemployment along the Maine coast has crowded the traditional fishing
grounds with schoolteachers, factory workers, and welfare mothers looking for
extra cash. Consequently full-time fishermen who want to keep a steady income
are trapped in the vicious circle of having to work harder to harvest an ever diminishing supply of lobsters.

Gone are the days when lobster was considered a "junk" fish, so plentiful that
housewives could gather them at low tide. And you seldom hear a Maine lobster-
man utter the old saying: "A crate [100 pounds] by 8 [o'clock], two by 10."

In 1957 Maine's fishermen harvested 24.7 million pounds of lobsters using some
320,000 traps. Last year using 1.8 million traps, they hauled in only 16.5 million
pounds. Maine marine scientists say the fishermen are harvesting more than 90
percent of the legal catch and are in serious danger of wiping out the lobster supply entirely.

"It used to be you could wait until December to go out to the winter fishing
grounds," says Ed Blackmore, president of the Maine Lobstermen's Association.
"Now the fleet goes out in October, and those who wait till December get what's
left."

Regulations are needed to protect the fisheries, but the independent-minded Yankee lobstermen can't agree on restrictions. The result is a highly competitive battle for the fishing grounds, one that threatens to empty the lobster beds.

More and more traps   "Two hundred traps used to be a lot to haul. Nowdays you can't make a living on that," says Kenny Davis, leaning over the side of his boat to snare a white buoy marking one of his traps.

He wraps the slimy nylon line around a winch pulley which in moments hoists to the surface a lobster pot filled with crabs, sea urchins, and a few lobsters. He throws back most of his catch.

The crabs are not worth enough to keep. The prickly green sea urchins do nothing more than steal the herring bait and nibble at the trap's nylon netting. He also throws back three of every four lobsters he catches because they are either "short" or fertile females.

"Sure I could find an easier job," says the curly-haired 25-year-old who held car repair and construction jobs before he married into an island fishing family. "But I would have to look a long time before I found something to trade with this job."

Expenses eat up income   Mr. Davis has been fishing for four years. Between lobstering and scallop fishing in the off-season, he grosses about $15,000 a year. After payments on his boat, house, traps, bait, gasoline, and property taxes, he says he has about $4,000 left to feed and clothe his family.

Added to the rising cost of equipment has been the pressure of a recent Internal Revenue Service investigation into the lobster industry. Now many fishermen who thought they had kept honest books are further burdened with payments of back taxes and fines—in some cases as high as $8,000.

Nevertheless, the hardy, hard-working Maine lobstermen are proudly gritting their teeth and clinging to their "I'm my own boss" independence and dreams of some day "making big money."

"You can work whenever you want," says Mr. Davis, as he guns his boat to swoop in on another one of his buoys. "But it seems our work is never finished. When we're not fishing, we're building or repairing our gear.

"What do we do for recreation?" grins Mr. Davis, "We work!"

Backyard rule abandoned   For decades the unwritten law among lobstermen has been: Fish your own backyard or lose your traps. But de facto regulation of the past is breaking down, as Maine fishermen begin to scramble for the few remaining lobsters.

Making matters worse from the fishermen's point of view was the court's recent striking down of the state's three-year residency requirement for lobster licenses. New license applications have been soaring. Yet it was the fishermen who were largely responsible for the defeat of a bill in the state Legislature last year that would have limited entry into lobstering. The fishermen said they feared such a limit might have kept their sons and grandsons out of the family business.

A strict trap limit for the entire coast seems unlikely because lobstermen up and down the coast have different styles of fishing. While the fishermen in Maine's Penobscot Bay might agree to a 600-trap limit, such a law would put the fishermen down the coast in Casco Bay out of business.

Toward regulation   But a small nucleus of Maine lobstermen is beginning to realize that proper regulations and fisheries management could protect lobstering as a livelihood and make work easier. "We could be catching nearly the same number of lobsters with much fewer traps if we could all agree to cut back," says Wesley Staples, head of the Swans Island Fishermen's Cooperative.

Many of Maine's troubled fishermen wistfully eye the lobstermen on Monhegan Island who since 1909 have agreed to fish only six months of the year January to July. While the lobsters surrounding Monhegan breed during the summer months, the fishermen supplement their income by painting summer cottages and performing other odd jobs tied to seasonal tourist trade.

According to Maine state Rep. Lawrence Greenlaw Jr. (D) who represents the coast from Rockland to Bar Harbor, the Legislature will have to pass specific trap limits for different geographical regions. The Legislature will also consider the more easily enforced method of time limits on fishing hours.

Talk of laws, however, won't fill his traps, says Kenny Davis, and he will continue to fish until something is done. He will continue to have days like that foggy summer morning.

After three hours on the fishing grounds he had hauled 83 traps, repaired four, and lost one. His total catch—11 lobsters weighing 16 pounds and worth about $32. Take away $10 for his lost trap, $5 for gasoline, $5 for bait, and $1 for a new pair of gloves, and he made about $11. It wasn't much for three hours work, but these days Maine lobstermen like Mr. Davis can't afford to stay in port.

Exercise 22: Mrs. Frisby and the Rats of NIMH

"I was reminded of a story I had read at the Boniface Estate when I was looking for things written about rats. It was about a woman in a small town who bought a vacuum cleaner. Her name was Mrs. Frisby, and up until then she, like all of her neighbors, had kept her house spotlessly clean by using a broom and a mop. But the vacuum cleaner did it faster and better, and soon Mrs. Frisby was the envy of all the other housewives in town—so they bought vacuum cleaners too.

"The vacuum cleaner business was so brisk, in fact, that the company that made them opened a branch factory in the town. The factory used a lot of electricity, of course, and so did the women with their vacuum cleaners, so the local electric power company had to put up a big new plant to keep them all running. In its furnaces, the power plant burned coal, and out of its chimneys black smoke poured day and night, blanketing the town with soot and making all the floors dirtier than ever. Still, by working twice as hard and twice as long, the women of the town were able to keep their floors almost as clean as they had before Mrs. Frisby ever bought a vacuum cleaner in the first place.

"The story was part of a book of essays, and the reason I had read it so eagerly was that it was called "The Rat Race"—which, I learned, means a race where, no matter how fast you run, you don't get anywhere. But there was nothing in the book about rats, and I felt bad about the title because it was a *people race,* and no sensible rats would ever do anything so foolish!"[4]

Draw the causal-loop diagram corresponding to the story of *Mrs. Frisby and the Rats of NIMH*.

Exercise 23: Merchants Make History

Read the following excerpt and draw the corresponding causal-loop diagram.

Once upon a time, when China was again in the grip of a famine, the Emperor sent his wisest official to a stricken province. "It's all the fault of the merchant Wang," the people told him. "He bought up all the grain and kept it in his vast storehouses so that he might sell it at a usurious price." Angrily, the Mandarin ordered the culprit to be brought before him in chains.

"How dare you keep back grain in order to sell it to the hungry people at an outrageous profit! You have been making money out of the misery of the starving!" shouted the Mandarin.

"Allow me to tell you the story of my actions," Wang the merchant replied. "Last year the harvest was exceptionally good. The grain remained in the fields; many would not even cut it because the price was so low that the work was hardly worth their while. The people were squandering their bread, which appeared to have lost all value. It was then that I began to buy grain. True, I paid a low price for it. But at least the peasants garnered it. Ought I to have paid more? Everybody was glad to see me buy grain at all.

"And then the harvest failed. All of a sudden there was no grain to be had. Apart from myself no one had laid in any reserves; everybody had felt sure that there would always be plenty. Then there was famine. People were beginning to come to me, saying, 'Your granaries are full; give us grain.' But they still had enough to eat, they still would not see that only with extremely careful management could disaster be averted. Prices were still not high enough to teach them to bear some measure of hunger. Ought I to have opened all my granaries then? My modest reserves would have been eaten up within a few weeks; the people would have lived again as in the previous year, the year of a good harvest. I therefore held back, no matter how they insulted me, maligned me, and threatened me.

"The famine got worse. Only when the price had risen again did I open my first granary. It was soon emptied—at a good profit, I admit. The people were beginning to get used to the idea that there was always a reserve. 'Wang has plenty of grain,' they would say. But soon the shortage was upon us again, and prices continued to rise. Once again I opened a granary. This time my profit was even greater. But ought I to have kept the price down? The people had to be made to realize that there was not enough grain available; they had to be taught to be even more economical, to restrict themselves even further. This they could learn only if everything was more expensive than it had been in the past.

"Eventually I opened my third and last granary. The price now was enormous. Everybody could see how scarce grain had now become. A lower price would have been self-deception.

"And now I was left without any reserves. But I kept this from the people to prevent panic and despair. Instead I sent out messengers to all provinces, to wherever

I had friends. To all of them I wrote: Send grain; I will pay exceedingly high prices—only send grain!

"Everybody thought that I was tremendously rich—that was the only reason why my business friends promised to send grain. No doubt I shall have to sell it at an even higher price than I did my own grain—but the people will pay the price."

The Mandarin was angry. "You have no heart for the hungry," he said. "You have profited from their hunger. You waited for prices to rise merely because you wanted to make greater profits. And now you are trying to tell me that you were serving the people, that you were trying to save them from starvation. You shall die for this!"

The merchant Wang blenched. He bowed down before the Mandarin and said, "My lord! From all the points of the compass the caravans are setting out with the promised grain. When the merchants hear that I am no longer alive they will turn back. There is no grain left in this province. Unless supplies arrive soon all the people here will die."

"I will have it proclaimed on all the roads that I shall accept the grain and pay for it on behalf of the Emperor. What do I need you for?"

"You would have to pay at least the same price as I had to promise—and that was higher than the price I was paid for the last of my own grain. Are you going to sanction the usurious prices which you say it was criminal of me to demand? Besides, how are you going to pay? In this entire province there is not enough ready money available."

"I shall use your fortune for payment!"

"I was not paid in ready cash. I was given land, mortgages on houses, promissory notes. I myself can use these things. The foreign merchants cannot. They trust only my word."

"They will have the same trust in the official of their Emperor."

"When they hear that you had me executed for demanding prices well below those which they must charge you they will take fright and turn back. There is famine in other provinces, too, even though not as great as in ours. They will have no difficulty in selling their grain elsewhere."

"Are you trying to say that you alone can save this province?"

Calmly the merchant regarded the Mandarin. "Yes, that is precisely what I am saying," he said with assurance. "They trust only me, they will send their grain to no one but me. I want to make a profit; I must make a profit. I am a merchant, not an official. If I lost through a miscalculation then I have failed as a merchant and that is the end of me. I want to make money—I must make money—by serving the community."

The Mandarin regarded the merchant in silence for a long time. Then he commanded: "Take off his chains. Let's hope he produces the grain he has promised."[5]

ENDNOTES

1. Adapted from Clifford C. Humphrey and Robert G. Evans. *What's Ecology?* (Northbrook, Ill.: Hibbard Press, 1971), pp. 8–10.
2. Story and map adapted from Claire Sterling, "The Making of the Sub-Saharan Wasteland," *Atlantic Monthly* 233 (May 1974): 98–105.
3. Adapted from an article by Arnold R. Chalfant, "Ecology Primer," *National Wildlife* 10 (February 1972): 42–43.
4. Robert C. O'Brien, *Mrs. Frisby and the Rats of NIMH* (New York: Atheneum, 1971), pp. 169–170.
5. E. Samhaber, *Merchants Make History* (New York: The John Day Co., 1964), pp. 13–15. Reprinted with permission.

PART III

GRAPHING AND ANALYZING THE BEHAVIOR OF FEEDBACK SYSTEMS*

OBJECTIVES

Focusing on gleaning information and understanding behavior from graphs, this part presents examples and exercises to aid the student in:

1. Inferring behavior patterns from graphs of data plotted against time;

2. Computing and graphing a cumulative level, given data on its rate of change over time;

3. Inferring a causal-loop diagram from a simple system, given graphs of key variables in that system plotted against time;

4. Identifying key turning points in the plot of a level against time and understanding the behavior of rates associated with that level;

5. Understanding systems involving delayed responses;

6. Inferring causal-loop diagrams from a variety of types of delays.

*The authors deeply appreciate the assistance of Tanette Nguyen McCarthy and Marian Steinberg in writing Part III.

CHAPTER 5

GRAPHING DATA
AND SEEING PATTERNS

Part II introduced several tools helpful in identifying the causal structure of a system. Identifying the causal structure is an important first step in defining a system's boundary, and it is a necessary prerequisite to constructing a formal simulation model. The focus now shifts from system structure to system behavior—especially behavior represented as data plotted against time. Formal simulation models, when run on the computer, show how a system's causal structure can lead to desirable or undesirable system behavior. Hence, knowing how to interpret data plotted against time is important, both in helping to construct formal simulation models and in analyzing and understanding model output.

The first example begins to show how graphs of numbers plotted against time can reveal insights into a system.

EXAMPLE I: THE WRIGHTS' DAILY USE OF ELECTRICITY

Figure 5.1 displays the amount of electricity the Wright family used in one day. By examining the data closely, it is possible to make some inferences about daily life in this household.

The horizontal axis of the graph displays time marked off in units of one hour, to represent the twenty-four hours in one day. It starts at midnight of the night before, progresses through the morning, past 12:00 noon, and ends at midnight. For each hour, the vertical axis displays the amount of electricity in kilowatts used during that hour. (For example, about 4 kilowatt hours were consumed between 11 a.m. and 12 noon.) Notice that the greatest amount of power was consumed between 4:00 and 6:00 P.M. The Wrights probably ate dinner cooked on an electric stove around 6:00 P.M. The spike (the pointed segment of the graph) in power usage at noon indicates that someone in the Wrights' house probably cooked something for lunch, rather than having only a cold lunch.

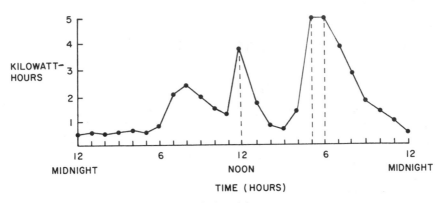

Figure 5.1 The Wrights' daily use of electricity

Because so much power was used between 6:00 and 9:00 A.M., the hours when everyone would most likely be getting up and getting ready to start the day, an inference can be made that the Wrights probably have an electric hot water heater and that most of the family showers or bathes in the early morning.

The Wrights probably get up at or about 6:00 A.M. (at least one of them does) and go to bed between 11:00 P.M. and 12:00 midnight, since the early morning power usage picks up markedly at 6:00 A.M. (probably due to the use of lights and hot water), and usage drops to a very low level between 11:00 and 12:00 P.M. Between the hours of midnight and 5:00 A.M., power remains at a uniformly low level. Everyone is asleep. The small amount of power is probably being used to run the refrigerator and the clocks.

Much of this information about the Wright family is, of course, speculation and inference. A visit with the family would be needed to determine how true it is. However, clearly something can be learned about the Wrights' living habits just by studying their patterns of electricity usage. Many observations can be made about most systems (the Wright family can be thought of as a system) by looking at plots of data over time, drawn from the system.

Exercise 1: People in a Room

Choose a room of interest. (A dining hall might be an interesting room to consider.) Graph the number of people in that room for every hour on a particular day, using the type of scales shown in Figure 5.2.

Exercise 2: Population of Boomtown, Colorado

The graph in Figure 5.3 plots the population of Boomtown, Colorado, from 1850 to 1890. Boomtown is located in a region of the Rocky Mountains known to be rich in deposits of precious minerals such as silver, gold, and uranium.

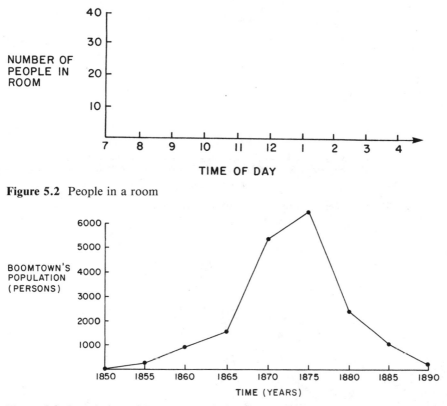

Figure 5.2 People in a room

Figure 5.3 Population of Boomtown, Colorado, 1850–1890

Write a paragraph describing what happened in Boomtown. Make as many reasonable inferences as are needed.

Exercise 3: Home Heating Oil Usage Rate

The Bertrands, who live in a suburb of Chicago, burn 1200 gallons of fuel oil each year.

Using a scale such as that in Figure 5.4, draw a graph that estimates how much fuel oil the Bertrands burn each month. (Be sure that the sum of the monthly amounts of oil consumed totals exactly 1200 gallons for the entire year.)

Exercise 4: Rainfall in Lagos

The graph plotted in Figure 5.5 shows the average precipitation by month in Lagos, a city on the Atlantic Ocean on the West Coast of Africa near the equator.

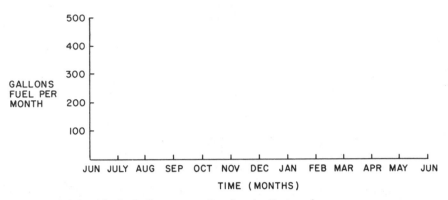

Figure 5.4 Monthly fuel-oil consumption for the Bertrands

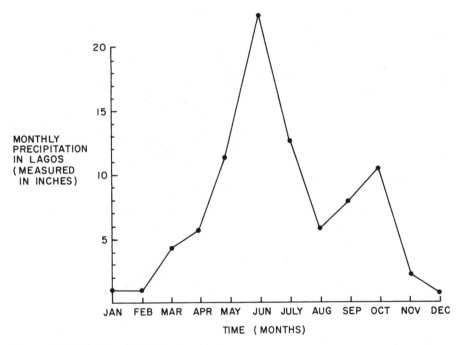

Figure 5.5 Monthly rainfall in Lagos

Write a short paragraph describing the climate in Lagos. When do you think would be the best time of year to visit Lagos and why?

PATTERNS

Numbers, especially numbers plotted against time (or "time series"), can convey important insights into the system under study. However, the trained eye can see patterns the untrained eye may miss. For example, if a curve comes to a peak and then declines, what causes the peaking behavior? If the slope of a curve increases dramatically, what does this indicate about the system generating the curve? These are some of the issues involved in the "eye-ball" analysis of time series data.

EXAMPLE II: MIDTOWN PARKING LOT—PART I

Dynamic structures are ubiquitous. Even very simple situations have some dynamic behavior built into them. For example, consider a parking lot near an office building in an urban area. Early in the morning, at about 6:00 A.M., very few cars are in the parking lot. As the morning rush-hour period approaches, however, more and more commuters arrive from the surrounding areas, and the parking lot activity builds up until no more space is available.

Table 5.1 shows the number of cars in a parking lot from 6:00 A.M. to 10:00 A.M. This table reveals that rush hour starts around 7:45 A.M., and that the parking lot is filled by 9:30. A more convenient way to display the data shown in Table 5.1 is to graph that data against time. Such a graph is shown in Figure 5.6.

Table 5.1 is a record of the number of cars in the parking lot between 6:00 A.M. and 10:00 A.M., updated every fifteen minutes. The time scale on the horizontal axis of the Figure 5.6 graph therefore begins at 6:00 and ends at 10:00, with each hour divided into four quarters.

Table 5.1 Cars parked in midtown lot, 6:00 A.M.–10:00 A.M.

Time	Number of cars
6:00	4
6:15	4
6:30	5
6:45	6
7:00	8
7:15	12
7:30	18
7:45	30
8:00	50
8:15	75
8:30	110
8:45	140
9:00	170
9:15	190
9:30	200
9:45	200
10:00	200

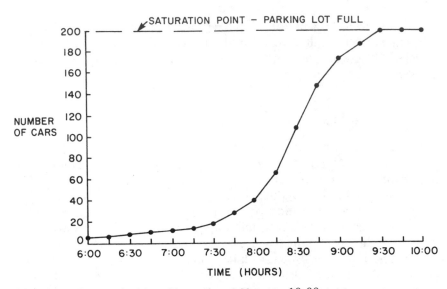

Figure 5.6 Cars parked in midtown lot, 6:00 A.M.–10:00 A.M.

Why are intervals of fifteen minutes chosen, instead of ten-minute intervals, or even sixty-minute intervals? Choosing sixty minutes would cause too much information to be lost. The beginning of the rush hour, or when the parking lot actually became full, might have been missed. On the other hand, not enough information is in the original data to graph ten-minute intervals. The table, for instance, does not indicate how many cars were in the parking lot at 6:10 A.M. or at 9:10 A.M. Therefore, the same intervals as the table are appropriate.

The vertical scale was determined by noting that the fewest number of cars in the lot at any time was 4 and the greatest number of cars in the lot was 200. The scale should include these minimum and maximum values, therefore the vertical axis begins with zero and ends with 200, in steps of twenty.

Each point in Figure 5.6 indicates the exact number of cars in the parking lot at a given time, with a line connecting those points reflecting a belief that cars are added linearly between the plotted data points.[1] The "number of cars" plotted in Figure 5.6 is called a *level* variable, a variable which shows the level (or accumulation) of cars present at any point in time.

EXAMPLE III: MIDTOWN PARKING LOT—PART II

The completed graph in Figure 5.6 can suggest some information about the behavior of the drivers of the cars. Example II illustrated how to tell the number of cars in the parking lot at a fixed point in time. What happens, however, over time? Notice that the slope of the curve in the diagram is flat from 6:00 to 7:45. Very few drivers are arriving for work. From 7:45 until about 9:00, the slope is very steep—cars are arriving at a much faster rate. After 9:00, the slope flattens until the lot is full. What does this tell about the rush-hour period? The graph suggests that the rush hour begins at about 7:45 in the morning and trails off at about 9:00. The data in Table 5.2 can provide this information more directly. By subtracting the number of cars at one point, say 7:00, from the number in the lot 15 minutes earlier, "net arrivals" can be determined.

Table 5.2 suggests a way to calculate "net arrivals" from data on the total number of cars in the lot. Notice, for example, that there were 50 cars in the lot at 8:00. At 8:15, fifteen minutes later, there were 75 cars in the lot. To calculate the net change in cars between 8:00 and 8:15 A.M., subtract the number of cars in the lot at 8:00 (50) from the number of cars in the lot at 8:15 (75) like this:

# of Cars at 8:15	minus	# of Cars at 8:00	=	Net Difference
75	−	50	=	25

There is a net difference of +25 cars from 8:00 until 8:15; 25 more cars were in the lot at 8:15 than at 8:00. Look at the net difference column in Table 5.2. Notice that there is a net difference of zero at 6:15 and after 9:45. Arrivals were equal to departures before 6:15 or after 9:30. What else does the net difference column suggest? Between 6:30 and 9:30, the lot was filling up. More cars were arriving than were leaving because the net difference is always positive during that time.

Table 5.2 Computation of net arrivals

Time	Number of cars	Difference	Net difference per 15 minutes	Net difference per hour
6:00	4	—	—	—
6:15	4	4 − 4	0	0
6:30	5	5 − 4	1	4
6:45	6	6 − 5	1	4
7:00	8	8 − 6	2	8
7:15	12	12 − 8	4	16
7:30	18	18 − 12	6	24
7:45	30	30 − 18	12	48
8:00	50	50 − 30	20	80
8:15	75	75 − 50	25	100
8:30	110	110 − 75	35	140
8:45	140	140 − 110	30	120
9:00	170	170 − 140	30	120
9:15	190	190 − 170	20	80
9:30	200	200 − 190	10	40
9:45	200	200 − 200	0	0
10:00	200	200 − 200	0	0

Notice that the net difference column is really the net difference *per fif-teen-minute period*. It is interesting to consider what would happen if the arrival rate for a particular fifteen-minute period were maintained for a full hour. How would the equivalent net difference per hour arrival rate be calculated? If 1 car arrives per fifteen-minute period, then 4 cars would arrive per hour. Two cars per fifteen minutes corresponds to 8 cars per hour, and so on. Using the simple rule:

$$\begin{array}{rcll} \text{Net Difference} & = & \text{Net Difference} & \times \quad 4 \\ \text{per Hour} & & \text{Fifteen-Minute Period} \end{array}$$

the net difference per hour column shown in Table 5.2 can be computed. Net difference per hour just standardizes net difference per fifteen-minute period into a unit that is more convenient to think about. (Most people think in terms of whole minutes or whole hours rather than fifteen-minute periods or three-hour periods, for example.)

Also notice that just because the net difference per hour between 8:00 and 8:15 is 100 cars per hour does not mean that 100 cars actually arrive between 8:00 and 9:00. This figure merely means that if cars continued to arrive at the same rate they had between 8:00 and 8:15, then 100 cars would arrive in a one-hour period.[2]

A graph of the net difference might show other patterns. Once again, time is on the horizontal axis, beginning at 6:15 A.M. (since there is not enough data to compute net arrivals between 5:45 and 6:00 A.M.) and ending with 10:00 A.M., using intervals of a quarter of an hour. The vertical axis scale is net difference, which in this case is the same as net arrivals of automobiles into the lot.[3] Look at Table 5.2. Notice that the lowest number, or minimum net arrivals, is zero, while the highest or maximum is 35. Choosing a vertical interval of 5 might cause a loss of too much information, so an interval of 2 is used instead. That means that the vertical scale will go from zero to 36, in intervals of two.

The graph of the net difference, or net arrivals (Figure 5.7), is a very different picture from Figure 5.6. The number of arrivals starts at zero at 6:15, increases slowly at first and then more rapidly, and declines by 9:45 to zero. The graph illustrates the *rate* at which cars arrive. In other words, the "net arrivals" represent rates of change. From the graph, when does the morning rush hour begin? Certainly, it seems under way by 7:30. Note that the maximum point on the graph, 35 cars arriving at 8:30, is also the *turning point*. Before 8:30 the rate at which cars arrive is continually increasing, while after 8:30 the rate of new arrivals begins to decline until the rate of new arrivals falls to zero at 9:45. Why must net arrivals to the parking lot go to zero? Look back at Figure 5.2. What happened by 9:30 to inhibit new arrivals?

Notice that Figure 5.7 could be rescaled quite easily to represent Net Arrivals of Cars per Hour by changing the vertical axis so that it ranges from 0 to 144 rather than from 0 to 36. This is so because, as noted earlier, an arrival rate of 36 cars per fifteen-minute period corresponds to an hourly arrival rate

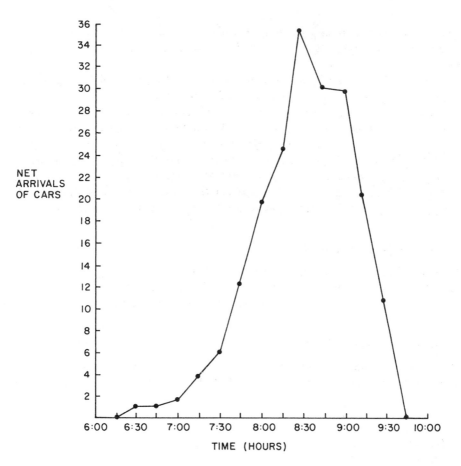

Figure 5.7 Net arrivals of cars into the parking lot

of 144 cars per hour. Deciding whether to express arrivals in terms of arrivals per fifteen-minute period or arrivals per hour (or even arrivals per half hour or per ten minutes) is, to a large degree, a matter of ease and convenience. Most people are accustomed to thinking in terms of rates per specific unit of time (per minute, per hour, per day, per week), so these measures are often preferred to other measures (such as the per fifteen-minute measure used here). In any case, care should be taken that the basic unit of time be defined consistently and clearly when analyzing plots of variables against time.

SUMMARY

The Midtown Parking Lot examples illustrate the transfer of numbers from a table to a meaningful graph. By carefully analyzing the graph, inferences may be made about the behavior of the cars over time. Two different types of variables have been identified: levels and rates. A level represents the number of cars in the parking lot at a given time, while a rate depicts the number of cars coming into the parking lot over a brief time interval; in this case, fifteen minutes.

Hence the number of cars present in the lot at exactly 8:00 (50 cars) is a level. It is represented by a point on the graph in Figure 5.2. The number of cars coming in between 8:00 and 8:15, which is represented by the slope between two points on Figure 5.2, is a rate. It is the rate of change or the amount of net arrivals (25 cars) between 8:00 and 8:15. The graph of the rates is plotted directly in Figure 5.7. Notice that when the same phenomenon (cars in the parking lot) is plotted first as a level (number of cars) and then from the perspective of the related rate (net arrivals per fifteen minutes), the two resulting graphs look quite different (compare the shapes of Figure 5.6 and Figure 5.7).

Exercise 5: Rate of Arrival

Compare the rate of arrival before and after 8:30. Is one steeper than the other? Can you explain why there is a slow, 2 1/4-hour increase of arrivals before 8:30 and a steeper 1 1/4-hour decline in arrivals after 8:30?

Exercise 6: Rate of Departure

In the evening, when offices close, the Midtown Parking Lot begins to empty out. Table 5.3 shows the number of cars in the lot from 4:00 P.M. until 7:30 P.M.

Table 5.3 Cars parked in midtown lot, 4:00 P.M.–7:30 P.M.

Time	Number of cars
4:00	200
4:15	200
4:30	195
4:45	185
5:00	170
5:15	120
5:30	65
5:45	30
6:00	20
6:15	10
6:30	5
6:45	4
7:00	3
7:15	2
7:30	2

a. Plot these data on a graph.

b. Which variable will you put on the horizontal axis? Which on the vertical axis? What scale will you use? (Consider the minimum and maximum number of cars. Remember to use constant intervals.)

c. Describe the behavior you have drawn. Why is the lot emptying out? Can you tell what time most people are leaving work?

d. Use Table 5.3 to generate a table of net departures.

e. Draw a graph plotting net departures against time.

f. Is there a maximum point? What is it and when does it occur? Is there a turning point? Where is it? Is it the same as the maximum point? Why?

g. Do people seem to leave quickly or slowly? Why do you think this is?

Exercise 7: The Yeast Experiment

Chapter Four discussed how yeast cells develop and multiply in a sugar solution. A student in a biology course has actually carried out the experiment in a laboratory setting. She gave the following report: "I emptied one package of dry yeast into a gallon of water. I added ten tablespoons of molasses. I then placed a drop of the yeast solution on a slide, colored the solution with a dab of iodine, and placed a cover slip over the slide. I looked at the slide under a microscope and counted the number of yeast cells on the slide."

   Table 5.4 comes from the student's daily log (count) of the yeast culture over a period of five days.

a. In preparation for drawing a graph of the yeast experiment, think about how you would label your axes. What would your scale be on each axis?

b. Plot the growth of the yeast colony on a graph.

c. What was the original number of cells in the culture? What was the maximum number of cells in the culture? When was the maximum reached?

d. Suppose you were to extend your experiment for another day. How do you think the yeast would behave during the sixth day? Would the culture

Table 5.4 Yeast culturing experiment results

Time (days)	Number of yeast cells
1	10
2	58
3	115
4	90
5	30

Table 5.5 Net change of yeast cells

Time (days)	Number of yeast cells	Increase (or decrease) in number of yeast cells per day
1	10	
2	58	
3		
4		
5		

grow or decay? Represent your guess about what would happen on the sixth day on the graph for Exercise 7(b).

e. Using a table such as that provided in Table 5.5, calculate the net increase (or decrease) in the number of yeast cells in the culture each day.

f. What is a good set of scales to use on each axis for plotting the net increase or decrease in yeast cells? Plot the net increase or decrease for the yeast colony on a graph.

g. When did the maximum increase rate occur? What do you suppose happened after that point? When was the net increase just equal to zero? How is this point related to the maximum point that you found in Exercise 7(c)?

Exercise 8: Littleton, N.H., Population History[4]

In 1764, Governor Wentworth granted a charter for the formation of the town of Littleton, New Hampshire. In 1769, a cabin was built on a meadow within the town bounds, but used by the builder for only one year. Knowing the availability of this cabin, Nathan Caswell and his family set out to take up residence there in 1770. Arriving at night, Caswell sensed there were Indians in the area. During the first night in their new home, Mrs. Caswell gave birth to her fifth son, while Mr. Caswell and his four other sons spent the night outside on guard. The next morning, sure that there were Indians menacing about, Mr. Caswell made a log raft and floated his family back to the nearest established town.

After resting for a few weeks, the family once again headed for their cabin in the meadow. They returned to find that the cabin had been burned to the ground by the Indians. Without delay, Mr. Caswell and his sons built a new cabin, which became the first lasting house in Littleton. Two years later, the Caswells were joined by the Hopkinsons, a family of six, and from then on, the town enjoyed a relatively steady population growth.

Table 5.6 shows the number of people living in Littleton from 1770 through 1970. Data for some years are unavailable.

Table 5.6
Population growth
of Littleton, N.H.

Year	Population
1770	7
1772	13
1773	14
1775	15
1790	96
1800	381
1810	873
1820	1,096
1830	
1840	1,778
1850	2,008
1860	2,292
1870	2,446
1880	2,936
1890	3,365
1900	4,066
1910	
1920	
1930	
1940	4,558
1950	4,817
1960	5,003
1970	5,290
1980	5,520

a. Construct a graph of the population of Littleton, New Hampshire, from 1770 through 1980. Select appropriate scales, plot the known data points, and sketch a smooth curve connecting the data points.

b. From the description of the events of the year 1770, construct a graph of the population of new settlers in Littleton for that year alone. The story provides enough information for you to be very accurate about the numbers of new settlers in the town and how they changed during the year, but you will not be able to be accurate about the times the settler population changed. You may pick the dates of population changes pretty much as you wish. After plotting known data points on your graph, connect them with straight horizontal and vertical lines to form a graph.

c. Why should the graph in Exercise 8(a) look rather smooth, while the graph in Exercise 8(b) looks jagged and full of right-angle turns made up of horizontal and vertical lines? Is the population time-graph for an area ever really smooth? Explain your thinking.

Exercise 9: Littleton, N.H., School Budget

As the town of Littleton began attracting more families, it was clear that some form of public education needed to be made available to the children. In 1787, the first schoolmaster, Robert Charlton, was hired. He received 8 bushels of wheat in payment for his services. By 1791, his compensation had risen to 16 bushels of wheat, then worth about $12.00. With this 16 bushels he had to purchase whatever was necessary for the school, as well as support himself. By 1795, 50 bushels of wheat were allocated from the town's budget for school needs. From 1796 to the present, school allocations have been made in dollars. Table 5.7 summarizes the available information.

a. Graph the history of Littleton's school budget based on the information from Table 5.7. (Be careful how you choose the vertical scale.)

b. Your graph should include approximate values for the budget between 1800 and 1880, for which data are missing in the table. What makes you think that your graph for these years is roughly correct?

c. Compare your Littleton population graph to your Littleton school budget graph. Describe the differences you see in the shapes of the two graphs.

d. If you were to graph "School dollars per person in Littleton" from 1800 through 1980, what would the graph look like? Describe it without graphing it, if you can.

Table 5.7 Littleton, N.H., school budget

Year	Budget allocation
1787	8 bushels of wheat
1791	16 bushels of wheat (worth about $12)
1795	50 bushels of wheat
1796	$40.00
1800	$100.00
. . .	. . .
1882	$2,935.50
1886	$3,813.25
1900	$12,386.53
1910	$18,500.00
1920	$30,900.00
1930	$70,277.51
1940	$69,225.13
1950	$117,438.76
1960	$442,247.59
1965	$642,200.00
1975	$1,480,589.04
1980	$2,414,513.90

e. Try to explain the differences you observed between the shapes of the population graph and the school budget graph. What might have caused what you observed in Exercise 9(a)?

Exercise 10: Lexington, Massachusetts, Population History[5]

The first "giving away" of land, which today is part of Lexington, was made to Richard Herlarkenden as an incentive for him to come to America from England. However, in order to hold the land grant, Richard had to improve the land during his first year of ownership; that is, begin a clearing and build a house. Richard Herlarkenden forfeited this land because he did not meet this condition. In 1638, the land was transferred to his brother, Roger, with the same condition. Unfortunately, Roger died that year, leaving a wife and two children.

The 600-acre land grant then went to Mr. Herbert Pelham, a widower, who the next year took Roger's widow for his second wife. The Pelhams lived in Cambridge but farmed the Lexington land. The land was left to Roger's son, Edward, who lived in Rhode Island and continued to hold the land in fields, pastures, and forests. After the land had remained for fifty years in the Pelham family, Edward finally broke it up into three parcels and sold them. The Lexington Historical Society records this transfer of land, in 1693, as a happy day for Lexington because the large farm was broken up and the area could then begin to grow into a prosperous village.

The first official estimate of population did not come until 1775, when there were about 700 people living in Lexington. The town continued to grow gradually until the twentieth century, when population began increasing more rapidly.

a. Graph Lexington's population data, given in Table 5.8. Use the same graph on which you have drawn Littleton, New Hampshire's, population. Notice that while the first population data for Littleton were specified for 1770, the first official estimate for Lexington was not available until 1775. Census data for some years are missing for both towns (1780, 1830, 1910, 1920, 1930 for Littleton; 1780, 1950, 1970 for Lexington). Another difficulty is the problem of scale. Population growth for Littleton goes from 7 people in 1770 to 5,290 people in 1970. The population scale for Lexington, however, is much larger, ranging from 700 in 1775 to 31,388 in 1965.

b. Despite these problems, compare the population growth curves for Littleton and Lexington. Discuss the similarities and differences of these two towns based on the graph. Hypothesize as to why the growth curves are much the same initially, but change around the twentieth century. You might want to make a similar comparison with data assembled from the town in which you live.

Table 5.8 Population growth
of Lexington, Massachusetts

Year	Population
1775 (estimate)	700
1790	941
1800	1,006
1810	1,052
1820	1,200
1830	1,543
1840	1,642
1850	1,893
1860	2,329
1870	2,270
1880	2,460
1890	3,197
1900	3,831
1910	4,918
1920	6,350
1930	9,467
1940	13,133
1965	31,338

DEFINING RATES AND LEVELS

In the preceding examples and exercises, an important distinction has been drawn between two types of variables: levels and rates. Levels are accumulations over time (called "stocks" in economics and "state variables" in engineering), and are usually measured in units such as number of cars in a parking lot, or the number of yeast cells in a culture. Rates (sometimes called "flows") are changes in some variable, measured in units per unit time, such as net arrivals of cars per fifteen-minute period, or the increase (or decrease) of yeast cells per day.[6]

In some of the exercises, levels have been graphed without any mention of the associated rates. For example, in the population and school budget cases, the unmentioned rates are the annual dollar increases in the budget and the annual number of persons added to the population either by net births (births minus deaths) or by in-migration. However, whenever a rate exists (that is, when something is changing), a level also exists (i.e., something is being changed). Conversely, levels that are changing over time must be associated with rates that are flowing.

The shapes of rates and levels curves often look very different when graphed against time. In the parking lot case, for example, the number of cars showed a steadily increasing "s-shaped" growth during the morning. However, the net arrivals graph was a somewhat jagged one that first increased and then suddenly fell to zero. Often in analyzing a system, data are available only for rates or only for levels within the system. By understanding how the shapes of rates and levels are the same or different, considerable insights can often be drawn concerning the behavior of the system under study. The following example will develop these points more fully.

EXAMPLE IV: THE FLU—PART I

Nearly every year the elementary and secondary schools in Hometown have been hit by a severe outbreak of influenza. The board of education is interested in taking action to lessen the annual impacts of the flu. Several policies have been proposed: closing the schools before the flu virus reaches a peak, providing free vaccinations to teachers in the school district, providing free vaccinations to children residing in the school district, or doing nothing. Any of these options could prove costly both in real dollars or in terms of teacher or student time missed from school. In addition, the vaccinations are known to have some adverse side effects for some portion of the population. Before making a decision on how to deal with their annual flu crises, the board decided to gather some additional data concerning the severity and timing of the problem.

Unfortunately, the board discovered that no systematic records of the severity of the flu have been kept on a system-wide basis. The only available records are those of Mrs. Stewart, the school nurse at the Maplewood School. Early in September she began to notice that many students were staying home from school. When they returned, they said that they had had the flu. Being a conscientious nurse, Mrs. Stewart had kept a careful record of recoveries, as shown in Table 5.9.

While it looks as if a large number of students recovered from the flu between September 11 and September 18, it is hard to see a clear pattern in these figures. Previously, converting data from a table into a graph made the pattern clearer. However, in this case, the graph in Figure 5.8 is not much easier to interpret than the table of numbers from which it came. The graph shows six peaks—September 11, 13, 18, and 25, and October 2 and 9. Why was the number of students returning to school on September 18 so high?

The number of students recorded on that date as having recovered from the flu is 74—a number significantly higher than the number recorded just before or after September 18th. Only 17 students are marked as having recovered on September 19. How many are listed as recovered for the day before the peak? The graph does not show a point for September 17th. In fact, the first point on the graph prior to September 18 occurs on September 15, indicating 34 recovered students. Table 5.9 gives no information for either September 16 or 17. That seems odd, but it does not help explain the peak at September 18.

Table 5.9 Mrs. Stewart's records of flu recoveries

Date		Number of recoveries
Sept.	4	0
	5	0
	6	1
	7	3
	8	6
	11	55
	12	37
	13	39
	14	38
	15	38
	18	74
	19	17
	20	14
	21	11
	22	9
	25	18
	26	4
	27	3
	28	3
	29	2
Oct.	2	5
	3	1
	4	1
	5	1
	6	1
	9	3
	10	0
	11	0
	12	0
	13	0

Perhaps the next peak, which occurs on September 25, can help shed some light on the problem. The number of recoveries shown on September 25 is 18. Again, this is quite a bit higher than the number of recoveries immediately before or after that date. Only four students are shown as having recovered on the following day, September 26, while the first data point before the 25th of September indicates 9 recovered students. But this occurs on September 22—two days worth of data have been lost again. Surely something unusual is happening.

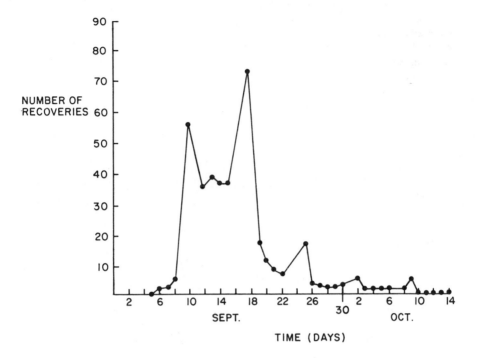

Figure 5.8 Mrs. Stewart's records of flu recoveries

Take a more careful look at Table 5.9. Two days of data are missing before September 18 and before September 25. Are there any other dates missing? Yes—September 9–10, September 30–October 1, and October 7–8. Notice that each of these two-day segments occurs with regularity every five days. Obviously what has happened is that since there is no school on weekends, all the students who recover on Saturday and Sunday return to school on Monday, so that the record for Monday includes the recoveries from Saturday, Sunday, and Monday.

All but one of the peaks (September 13) turn out to be Mondays. Averaging the Monday recoveries among Saturday, Sunday, and Monday, an "adjusted" table of recoveries which includes weekends is created. Consider Monday, September 18. Mrs. Stewart's log shows that 74 students have recovered from the flu on that date. By averaging the 74 recoveries over a period of three days (Saturday, September 16, Sunday, September 17, and Monday, September 18), there are approximately 24 recoveries per day ($74 \div 3 \cong 24$), give or take a few cases. One possible set of adjustments is indicated in Table 5.10 and graphed in Figure 5.9. The total number of recoveries from Saturday, September 16 to Monday, September 18 is equal to $29 + 24 + 21 = 74$ recoveries, which matches Mrs. Stewart's records.

Table 5.10 Adjusted
recovery data on
students at
Maplewood School

Date	Number of recoveries
Sept. 4	0
5	0
6	1
7	3
8	6
9	10
10	17
11	28
12	37
13	39
14	38
15	34
16	29
17	24
18	21
19	17
20	14
21	11
22	9
23	7
24	6
25	5
26	4
27	3
28	3
29	2
30	2
Oct. 1	2
2	1
3	1
4	1
5	1
6	1
7	1
8	1
9	1
10	0
11	0

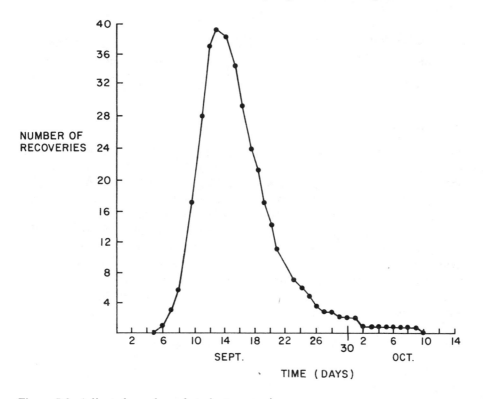

Figure 5.9 Adjusted number of student recoveries

Exercise 11: Finding Errors in Data Collection

The experiences with Mrs. Stewart's records indicate how data collected and graphed can sometimes hide a pattern.

Suggest other cases where the data collected might distort the pattern in the data.

EXAMPLE V: THE FLU EPIDEMIC—PART II

Mrs. Stewart's records allowed the reconstruction of data to show when Maplewood students recovered from the flu. The data show that the epidemic ended on October 9, the day when the last student who had been sick with the flu returned to school. But the data do not directly tell when the flu began, or how many people were affected.

In late October, the board of education asked Mrs. Stewart to provide some additional information on the recent flu epidemic. The board had three questions:

1. When did the students contract the flu?

2. What was the total number of students sick with the flu?

3. How many students were sick with the flu on a daily basis?

When Did the Students First Contract the Flu?

The data indicate how many students recovered from the flu on a daily basis. Knowing that this particular strain of flu lasts about five days, the data from Table 5.11 allow for an approximation of the date when the students were first infected. To determine when someone became infected with the flu, deduct five days from the day the person recovered. For example, Table 5.11 shows that on September 8, six students returned to school. Since the flu lasts five days, they must have become sick about five days earlier. Therefore, the students who returned to school on September 8 must have become infected on approximately September 3. Similarly, the 39 students who recovered on September 13 must have become infected on September 8, while the 21 students who returned to school on the 18th of September became ill on the 13th. In

Table 5.11 Number of infected and recovered students per day

Date	Number infected	Number recovered	Date	Number infected	Number recovered
Sept. 1	1		Sept. 23	3	7
2	3		24	2	6
3	6		25	2	5
4	10	0	26	2	4
5	17	0	27	1	3
6	28	1	28	1	3
7	37	3	29	1	2
8	39	6	30	1	2
9	38	10	Oct. 1	1	2
10	34	17	2	1	1
11	29	28	3	1	1
12	24	37	4	1	1
13	21	39	5	0	1
14	17	38	6	0	1
15	14	34	7	0	1
16	11	29	8	0	1
17	9	24	9	0	1
18	7	21	10	0	0
19	6	17	11	0	0
20	5	14			
21	4	11			
22	3	9			

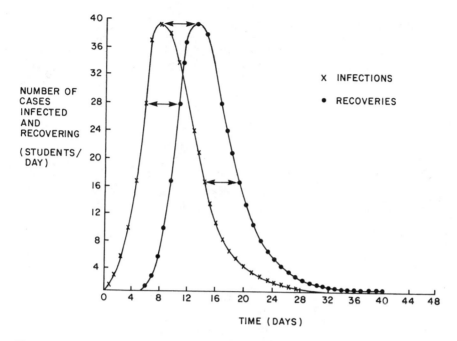

Figure 5.10 Number of new cases and recoveries by day

fact, by examining Table 5.11 closely, it is possible to know approximately how many students became infected on a day-by-day basis.

The data presented in Table 5.11 could also be plotted against time, as in Figure 5.10. Notice that the curve showing the daily number of infections is exactly the same as the curve showing the number of daily recoveries, except that the infections curve occurs five days earlier in time. The infections curve peaks on September 8, five days before the recoveries curve, which peaks on September 13. Of course, this five-day lag is due to the fact that this particular strain of flu lasted five days.

What Was the Total Number of Students Sick with the Flu?

There are two ways to determine the total number of students from Maplewood School who had the flu, using either the data on the number of infected students, or on the number of recovered students. One easy way is to keep a running total, also called a cumulative total, of the number of students who became infected, day by day. The results on this approach are shown in Table 5.12. The column on the left, labeled "Number of Infected Students," is the same as the data in Table 5.11. The second column, or cumulative total of students who became infected, shows the running total of infected students.[7] The cumulative total is therefore based on the number of students infected each day.

Table 5.12 Actual and cumulative flu data

Date		Number of infected students per day	Cumulative number of infected students	Number of recovered students per day	Cumulative number of recovered students
Sept.	1	1	1		
	2	3	4		
	3	6	10		
	4	10	20	0	0
	5	17	37	0	0
	6	28	65	1	1
	7	37	102	3	4
	8	39	141	6	10
	9	38	179	10	20
	10	34	213	17	37
	11	29	242	28	65
	12	24	266	37	102
	13	21	287	39	141
	14	17	304	38	179
	15	14	318	34	213
	16	11	319	29	242
	17	9	338	24	266
	18	7	345	21	287
	19	6	351	17	304
	20	5	356	14	318
	21	4	360	11	329
	22	3	363	9	338
	23	3	366	7	345
	24	2	368	6	351
	25	2	370	5	356
	26	2	372	4	360
	27	1	373	3	363
	28	1	374	3	366
	29	1	375	2	368
	30	1	376	2	370
Oct.	1	1	377	2	372
	2	1	378	1	373
	3	1	379	1	374
	4	1	(380)	1	375
	5	0	380	1	376
	6	0	380	1	377
	7	0	380	1	378
	8	0	380	1	379
	9	0	380	1	(380)
	10	0	380	0	380
	11	0	380	0	380

For example, notice the figures for September 2. The number of students infected that day was 3, but the running total is 4. That difference comes from adding the number of students infected on the very first day of the epidemic, September 1 (1 student) and the number infected on September 2 (3 more students), to get a cumulative total of four ($1 + 3 = 4$). A total of four students had been infected with the flu by September 2.

Since six more students became infected on September 3, a total of ($4 + 6$) or 10 students had contracted the flu by September 3. Keeping this running total through the epidemic produces a grand total of 380 students who had become infected by the time the epidemic died out. Notice that the cumulative total of 380 infected students was reached on October 4, and did not go any higher, even though the table does not end until October 11. This shows that the epidemic had ended on October 4, since no more students became infected after that.

When did the last student recover? It must be on October 9, since the cumulative total never gets any higher than it was on that day (meaning that there were no additional recoveries). Comparing the last date a student recovered, October 9, with the last day a student became infected, October 4, there is a five-day lag, or delay—the exact amount of time that a student would be sick with the flu.

To get a better picture of these data on cumulative numbers of infected and recovered students, Mrs. Stewart transferred the data from Table 5.12 to a graph (Figure 5.11). Notice how fast the number of infected students rises between September 4 and September 15, and how much slower the increase in new infections is after that time.

Another interesting thing can be seen by comparing the curves for infections and recoveries. Notice how the recovery curve is an exact image of the infected curve, only five days apart, showing that it takes five days after the date of infection until a person recovers.

How Many Students Were Sick with Flu on a Daily Basis?

To answer the board's last question, Mrs. Stewart reasoned that the number of students sick on any given day must equal to the total number who had been infected (cumulative number of students infected) less the total who had recovered (cumulative recoveries). For example, looking back to Figure 5.11, on September 5, the cumulative number of infected students was 37, and no students had yet recovered. Therefore, the total number sick on that day was 37 ($37 - 0 = 37$). By September 10, however, 213 students had become infected with the flu, of whom 37 recovered, leaving a total of 176 ($213 - 37 = 176$) sick with the flu. On September 25, when the epidemic was nearly over, 372 students had been infected since the epidemic began, of whom 360 had recovered. So, on September 25 only 12 ($372 - 360 = 12$) were sick. A more direct way to get this information would be to subtract the cumulative number recovered from the cumulative number sick for each day.

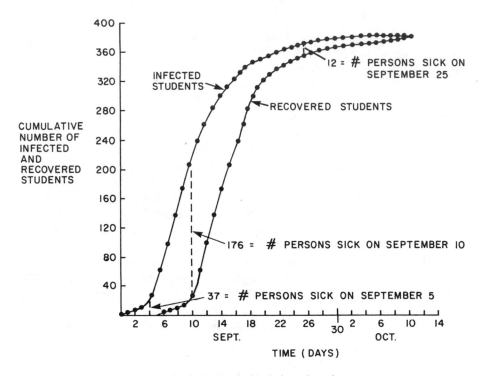

Figure 5.11 Cumulative recoveries/cumulative infected students

Exercise 12: Further Analysis of the Flu

a. In the preceding text, the number of persons sick on September 5 was 37, and the number of persons sick on September 10 was 176. Using Table 5.13 and the data presented in Table 5.12, compute the total number of persons sick for each day of the flu epidemic.

b. Plot the total number of persons sick for each day.

c. When did the maximum number of sick persons occur?

d. Look back to Figure 5.10. What else happened on the same day that the maximum number of persons was sick? Explain why this pattern is so.

e. Thus far we have looked at graphs for the number of persons recovered per day, the number of persons infected per day, the cumulative recoveries, the cumulative infections, and the number of persons sick each day. Draw a causal-loop diagram that relates these quantities causing the spread of the flu. (*Hint:* The key to this problem comes from thinking clearly about what causes people to get infected and hence become sick.)

Table 5.13 Total number of persons sick for each day

Date	Cumulative number of infected students	MINUS	Cumulative number of recovered students	EQUALS	Number of sick persons
Sept. 1	1		—		—
2	4		—		—
3	10		—		—
4	20		0		—
5	37		0		—
6	65		1		—
7	102		4		—
8	141		10		—
9	179		20		—
10	213		37		—
11	242		65		—
12	266		102		—
13	287		141		—
14	304		179		—
15	318		213		—
16	319		242		—
17	338		266		—
18	345		287		—
19	351		304		—
20	356		318		—
21	360		329		—
22	363		338		—
23	366		345		—
24	368		351		—
25	370		356		—
26	372		360		—
27	373		363		—
28	374		366		—
29	375		368		—
30	376		370		—
Oct. 1	377		372		—
2	378		373		—
3	379		374		—
4	380		375		—
5	380		376		—
6	380		377		—
7	380		378		—
8	380		379		—
9	380		380		—
10	380		380		—
11	380		380		—

Exercise 13: Estimating Slopes for the Flu Example

a. Turn back to Figure 5.11. Using a ruler or straight-edge, estimate on what day the cumulative number of infected students was rising most rapidly. Also estimate the day that the cumulative number of recovered students was rising most rapidly.

b. Return to Figure 5.10. What days show the maximum number of recoveries per day and the maximum number of infections per day? Is it a coincidence that the answers to this question and Exercise 13a turn out to be so close? Why or why not?

MORE ON RATES AND LEVELS

The flu example is an exercise in looking at the rates and levels that exist within a single system. Beginning with a set of data on one rate (the recovery rate), one additional rate (the infection rate) and three levels (the cumulative number of persons recovered, the cumulative number of persons infected, and the number of persons sick at any point in time) have been derived. Once complete data on a rate are known, then the associated levels can be computed by accumulating the changes due to the rates over time.[8]

Because levels can be computed from the accumulation over time of rates (and conversely rates can be computed at any point in time if enough is known about the levels at that point in time), certain strict relationships exist between graphs of rates and their associated levels. For example, when the graph of a level is completely horizontal (sloping neither upward nor down), then the sum of the rates associated with that level must be zero.[9] Also, when a level is increasing most rapidly (when its slope is steepest), then the rates associated with that level are at a maximum point.[10] For example, the maximum number of recoveries coincides exactly with the steepest sloping portion of the cumulative recovery curve. Mathematicians spend considerable time proving these relationships, but the most convenient and often the most powerful ways of understanding relationships between levels and their associated rates stem from graphing the variables under question and logically inspecting the patterns that appear.

In summary, then, rates cannot exist without levels and levels cannot exist without rates. This principle means merely that something cannot change unless there is a rate to change it.[11] Put another way, rates and levels are just the flip sides of the same coin—different ways of looking at the same dynamic process. This insight turns out to be of considerable practical importance in analyzing dynamic systems because, while levels are often of primary concern to policy makers (such as the level of homes crowding into a developing town or the level of pollutants in a river or lake), policies are usually directed at rates (the issuance of new building permits or the control of the rates of emission of pollutants, for example). Thus, it is important to understand how changes in system rates will eventually lead to changes in system levels.

In the chapters of this part that follow, the relationships between rates and levels are explored more fully, using graphs of variables plotted against time. Then, in Part V, levels and rates are analyzed in more depth, using computer simulation.

ENDNOTES

1. Note that this seemingly simple discussion masks several important assumptions about the behavior of the parking lot. For example, a reasonable guess would be that at 8:10 about 67 cars are in the lot. This is not an explicit fact; it is inferred by interpolating between one data point that shows 50 cars at 8:00 and another point that shows 75 cars at 8:15. The assumption is being made that nothing drastic has happened between 8:00 and 8:15. (For example, a trailer loaded with 100 cars could have entered the garage at 8:02 and left at 8:13, producing over 150 cars at 8:10.) Figure 5.6 implicitly assumes that the curve connecting 8:00 to 8:15 is in some sense well-behaved—smooth and continuous. A time interval should be selected so that this interval is small enough that the behavior of the system is "well-behaved" within a single time increment, yet large enough that it does not generate a lot of redundant data. Well-chosen axes are easy to recognize with the naked eye, but often surprisingly difficult to select.

2. The student with some additional mathematical background will recognize this discussion of net difference per hour as a basic definition of the rate of change with respect to time of the level of cars in the garage. Let $C(t)$ be the level of cars in the garage at any point in time. Then the net difference per fifteen minutes can be defined as:

$$\text{Net Difference} = \Delta C = C(t + \Delta t) - C(t)$$

where t is the given point in time and $t + \Delta t$ represents the point in time fifteen minutes later. Similarly, Net Difference per Hour is given by

$$\text{Net Difference} = \frac{\text{Difference in Cars}}{\text{Difference in Time}} = \frac{\Delta C}{\Delta t}$$
$$\text{per Hour}$$

where ΔC is as defined above and Δt is .25 (since 15 minutes is one quarter of an hour).

As the Differences in Time become small (as Δt approaches 0), $\Delta c / \Delta t$ approaches dc/dt.

3. The reader may wonder about the exact placement of points on the graph in Figure 5.7. For example, the arrival rate of 25 cars between 8:00 and 8:15 is plotted at 8:15. It might equally well have been plotted at 8:00. Net differences (rates) can be plotted at either the upper or lower ends of the time intervals used in calculating them—as long as the choice is consistent.

4. From James R. Jackson, *History of Littleton* (Cambridge, Mass.: The University Free Press, 1950).

5. Information adapted from *Proceedings of Lexington Historical Society,* volume II (Lexington, Mass.: The Historical Society, 1900).

6. Persons with some formal training in calculus will recognize that if some level, x, exists, then a change in that level may be denoted Δx. Furthermore, a rate associated with that level would be the change in x per some unit change in time Δt or $\Delta x / \Delta t$. As the unit of time becomes small (decreasing from hours to minutes to seconds), then the rate, $\Delta x, / \Delta t$, is identical to the time derivative of the level, dx/dt.

7. In more formal terms:

$$\text{Cumulative Number of Infected students} = \sum_{i=0}^{t} \text{Number of Infected Students } (i)$$

where all days have been numbered consecutively starting from 0 to t.

8. For students with some training in calculus, the major computations in this sections would be summarized as follows:

Cumulative No. of Infections (t) $= \int \text{Infection Rate x} dt$
Cumulative No. of Recoveries (t) $= \int \text{Recovery Rate x} dt$
No. of Persons Sick at Any Point (t) $= \int (\text{Infection Rate} - \text{Recovery Rate}) \text{x} dt$
$= \text{Cumulative Infections} - \text{Cumulative Recoveries}$

9. For some level x, its rate of change with respect to time, dx/dt, is just the sum of the rates adding to that level minus the rates subtracting from the level. Hence this statement reduces to the relationship in calculus:

$x(t)$ is at a Maximum or Minimum when $dx/dt = 0$

10. This must be so because by definition, the height of the net rate curve is the slope of the level, i.e., dx/dt is the slope of $x(t)$.

11. This is an applied rephrasing of the fundamental theorem of the calculus. A rate, dx/dt, is the time derivative of some level, x—rates cannot exist without levels. On the other hand, a level merely accumulates over time small changes in its rate. That is, in formal terms a level is the integral over time of its rate(s). Just as derivatives and integrals are differing manifestations of the same functional form, rates and levels are differing manifestations of a single process changing over time.

CHAPTER 6

LINKING CAUSAL LOOPS AND GRAPHS

A graph can often provide a fairly rich and detailed description of a system's behavior over time. This chapter will further pursue the analysis of graphs by considering ways of using graphs of system behavior to draw inferences about system causal structure.

EXAMPLE I: THE RABBIT POPULATION—PART I

Figure 6.1 presents a graph of the estimated number, over time, of snowshoe hares, a type of rabbit, along Hudson Bay in northern Canada. The graph shown here is taken from a computer simulation of the rabbit population in the Hudson Bay area, based on pelt collection data from the Hudson Bay Company. The number of pelts taken is used as a proxy for the number of rabbits in the region. (This assumes that trapping is directly proportional to rabbit density.) Although similar to the real data, the computer-generated curves are somewhat easier to work with in this example.

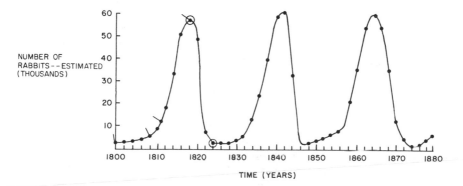

Figure 6.1 Behavior of the rabbit population on Hudson Bay

In 1800, the snowshoe hare population was very small. A total of 1,750 rabbits is accounted for at the company's various posts along Hudson Bay. In 1818, the rabbit population peaked, with 59,000 rabbits estimated. After that, it began a drastic decline, until it reached a low of 2,000 rabbits in 1824. Immediately afterward, the number of rabbits increased again, reaching a new peak in 1842, with 61,000 rabbits, followed by a new low in 1848, with about 1,800 rabbits. The successive high and low estimate points occurred in 1864 (58,000 rabbits) and in 1874 (1,500 rabbits), respectively.

What can be concluded about the behavior of the rabbit population from this graph? It appears that the snowshoe population goes up and down in regular cycles of about twenty-four years. This type of cyclical pattern is referred to as *oscillations*.

Three full oscillations are on the graph. Each oscillation corresponds to a complete cycle of growth and decline in the rabbit population. The first oscillation began in 1800, when the number of rabbits started to grow. It ended around 1824, when the population hit a record low of 2,000 rabbits. The second oscillation started right after 1824, when the rabbit population began to increase again, and lasted until 1846, when a new low point was reached. The third oscillation ran from 1846 to 1874. (Oscillations could also be measured from peak to peak!)

Look at the graph again. Any one of the three oscillations can be compared to a hill. Like a hill, it has two slopes, one slope going up and one slope going down. For the period of the first oscillation, for example, the two slopes can be distinguished very clearly. The first is an increasing slope, which lasted from 1800 to 1818. The second is a decreasing slope, and lasted from 1818 to 1824.

The time segment between 1800 and 1818 corresponds to the part of the oscillation where the slope is going up. This upward trend means that the estimated number of rabbits in the posts along Hudson Bay increased every year. The increases were very small at the beginning and became greater as the peak was approached. As a matter of fact, this increasing slope could be divided into smaller segments (as shown on the graph), with each segment depicting a different rate of population growth.

The first segment, for instance, goes from 1800 to 1808. It is almost flat and represents very small increases in the number of rabbits on Hudson Bay. The second segment, from 1808 to 1812, is steeper and indicates greater increases in the rabbit population. Finally, the third segment, which begins in 1812 and ends in 1818, is very steep. It shows very rapid increases in the rabbit population.

An easy way to compare the steepness of the slope of various parts of a curve involves using a ruler or straight-edge. Place the ruler along a segment of the curve so that the edge of the ruler approximates the slope of a small piece of the curve. For example, in Figure 6.1, a straight line can be drawn that represents the steepness of the rabbit population curve between 1800 and 1808.

Similar straight lines can be drawn for the segments of the curve representing 1808 to 1812 and 1812 through 1818. These straight lines, found by lining up a ruler with the rabbit population curve, would show that the steepest slope occurs in the period between 1812 and 1818.

Reading numbers from the graph allows even more precision about how steeply the rabbit population is increasing or decreasing at various times. For example, the total number of rabbits grew from about 1,750 in 1800 to 4,500 in 1808. This was a net increase of 2,750 rabbits (4,500 − 1,750) over a period of eight years (or approximately 344 rabbits per year). Within the next four years, from 1808 to 1812, the population expanded dramatically from 4,500 to 18,000 rabbits. This represented a net growth of 13,500 (3,375 rabbits per year), considerably greater than the increase over the previous period. The increase is even more drastic over the third segment (from 1812 to 1818), when the population reached a total of 59,000 rabbits in 1818. It means that 41,000 more rabbits (59,000 − 18,000) were counted along the Hudson Bay posts (at an annual growth rate of approximately 6,833 rabbits per year).

After 1818, however, the pattern oscillation is reversed, and the slope is actually going down. This downward slope, which lasts from 1818 to 1824, corresponds to a decrease over time in the number of rabbits present on Hudson Bay. Here again, this decreasing slope divides into several segments. The first segment lasts two years, from 1818 to 1820, and shows a net loss of 9,000 rabbits (59,000 − 50,000). The following segment covers another two-year span, from 1820 to 1822. It depicts a very sharp decrease in population. In fact, about 90 percent of the rabbits were lost over this very short period of time. In 1822, only 8,000 rabbits were registered by the personnel of Hudson Bay Company. Over the period covered by the next segment, between 1822 and 1824, the rabbit population kept decreasing until it reached a low point of about 2,000 rabbits in 1824. This represents a net loss of 6,000 rabbits (8,000 − 2,000), or three-fourths of the population registered in 1822.

Right after 1824, the rabbit population seemed to be increasing again. This new trend signals the beginning of the second oscillation on the graph. This oscillation behaves very much the same as the previous one. It also has an upward slope (from 1824 to 1842), and a downward slope (from 1842 to 1848). The upward slope corresponds to increases in population, and the downward trend to decreases in population.

EXAMPLE II: THE RABBIT POPULATION—PART II

Obviously the increases in the rabbit population over any given time period mean that the number of rabbits being born exceeds the number of rabbits dying. From 1824 to 1842, for instance, the number of rabbit births on Hudson Bay was greater than the number of deaths. On the other hand, a downward slope means that the rabbit population was actually decreasing. The number of deaths exceeded the number of births. Hence from 1842 to 1848

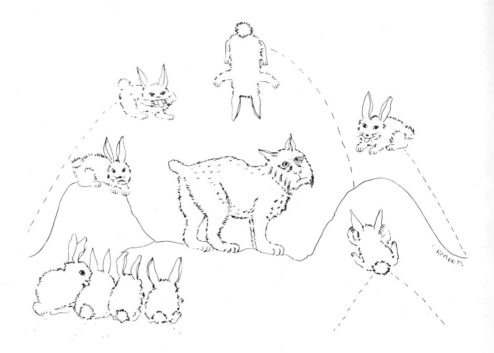

more rabbits were dying than were being born, and the population was decreasing.

The points in time at which the population reverses its pattern from increasing to decreasing, or from decreasing to increasing, represent key turning points in a system's behavior, and deeper analysis of those points can yield insights into why a system behaves as it does. As mentioned earlier, an oscillation can be compared to a hill. If we climb a hill, we will eventually reach a point beyond which we cannot climb any further. In fact, if we keep walking beyond this point, we will start going down the other side of the hill. This special point is the curve's maximum point. Another special point is the bottom of the hill, the minimum. Once we have reached this point, we cannot go down any further.

In Figure 6.1, the point that corresponds to the peak of a hill occurs at the time when the rabbit population reached its maximum, just before it started to decline. At this maximum point, the rabbit population was neither increasing nor decreasing. This means that the number of births of rabbits must equal to the number of deaths of rabbits. This special point represents the maximum number of rabbits along Hudson Bay, reached during the period of the first oscillation in 1818, with 59,000 rabbits (circled on the graph).

The other special point is reached when the rabbit population was at its lowest level, or minimum, and had not yet started to increase. Before this minimum point is reached, the number of rabbit deaths exceeded the number of births. Afterward, the number of births exceeded the number of deaths. Right at the minimum point, however, the number of births and death were again equal. The minimum population of rabbits for the first oscillation occurred in 1824 with about 2,000 rabbits. It is circled on the graph.

To be more specific about the behavior of the rabbit population over time, look at Table 6.1, which gives those data that have previously been put into graphic form in Figure 6.1. Here, as in the Midtown Parking Lot example, the net changes in population from one record to the next can be computed and plotted on a graph.

The positive changes in Table 6.2 correspond to increases in population where the number of rabbit births exceeded the number of deaths. These points are located above the horizontal line on Figure 6.2. Negative changes in the table correspond to decreases in population (with deaths exceeding births) and are located below the horizontal line. At the point where the graph crosses the horizontal line, the number of births equals the number of deaths.

When the net change rate is zero, it can be inferred that the level of rabbits is either at a maximum or a minimum point. The net change graph plotted in Figure 6.2 also provides some additional information concerning the slope of the rabbit level at any point in time. When the net change graph is positive, then the rabbit level is increasing. When the net change graph is negative, then the rabbit level is decreasing. This observation must be true because a mo-

Table 6.1 Population of snowshoe hares, 1800–1880

Time (years)	Rabbit population	Time (years)	Rabbit population
1800	1,750	1842	61,000
1802	2,000	1844	32,000
1804	2,500	1846	2,000
1806	3,000	1848	1,800
1808	4,500	1850	2,500
1810	8,000	1852	4,500
1812	18,000	1854	7,000
1814	32,000	1856	10,000
1816	50,000	1858	22,000
1818	59,000	1860	36,000
1820	50,000	1862	53,000
1822	8,000	1864	58,000
1824	2,000	1866	52,000
1826	2,500	1868	34,000
1828	3,000	1870	12,500
1830	4,000	1872	3,000
1832	7,000	1874	1,500
1834	14,000	1876	2,000
1836	23,000	1878	4,000
1838	40,000	1880	9,000
1840	58,000		

ment's reflection reveals that the definition of a positive net change rate is increasing rabbits and vice versa.

However, Figure 6.2 also suggests when the rabbit level was increasing *most rapidly*. The most rapid increase in the rabbit population (greatest positive slope) occurred when the net change rate was greatest. The peak in the net change rate for the first oscillation occurred at 1816. Referring to Figure 6.1, note that 1816 was also the time when the rabbit level was increasing most rapidly. Similarly, the most rapid decrease in the rabbit level occurred in 1822 when the net change rate was at its minimum. A quick examination of Figure 6.1 confirms that the most rapid decrease in the rabbit level occurred between 1820 and 1822.

Exercise 1: Net Changes

A close examination of Figure 6.1 and Figure 6.2 shows that each of the special points on Figure 6.1 occurs when the net change in the rabbit population plotted in Figure 6.2 is equal to zero. Can you find any way to distinguish between the peaks of the hills (maximum points) and the bottoms of the valleys (minimum points) as they occur on Figure 6.1? (Check whether the net change

Table 6.2 Computation of net changes in the rabbit population

Time (years)	Number of rabbits	Net changes in the rabbit population
1800	1,750	—
1802	2,000	+250
1804	2,500	+500
1806	3,000	+500
1808	4,500	+1,500
1810	8,000	+3,500
1812	18,000	+10,000
1814	32,000	+14,000
1816	50,000	+18,000
1818	59,000	+2,000
1820	50,000	−9,000
1822	8,000	−42,000
1824	2,000	−6,000
1826	2,500	+500
1828	3,000	+1,000
1830	4,000	+1,000
1832	7,000	+3,000
1834	14,000	+7,000
1836	23,000	+9,000
1838	40,000	+17,000
1840	58,000	+18,000
1842	61,000	+3,000
1844	32,000	−29,000
1846	2,000	−30,000
1848	1,800	−200
1850	2,500	+700
1852	4,500	+2,000
1854	7,000	+2,500
1856	10,000	+3,000
1858	22,000	+12,000
1860	36,000	+14,000
1862	53,000	+17,000
1864	58,000	+4,000
1866	52,000	−6,000
1868	34,000	−18,000
1870	12,500	−21,500
1872	3,000	−9,500
1874	1,500	−1,500
1876	2,000	+500
1878	4,000	+2,000
1880	9,000	+5,000

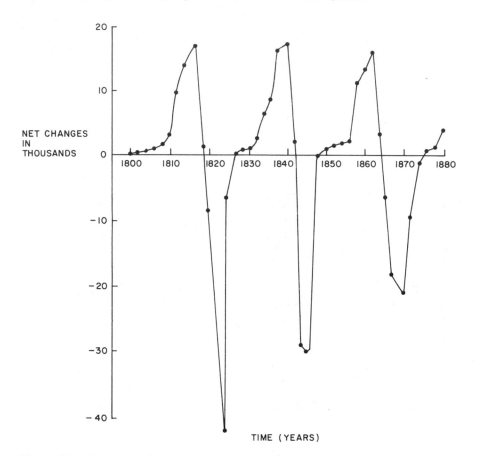

Figure 6.2 Net changes in the rabbit population over time

in rabbit population in the diagram is going from positive to negative or from negative to positive at each point where it is equal to zero.)

EXAMPLE III: THE RABBIT POPULATION—PART III

Only how the rabbit population behaves over time has been described, not why it behaves that way. What has been happening to cause these oscillations? One possible explanation would be a natural disaster. Northern Canada has a tundra ecosystem. As a result, it has an extremely sensitive ecological balance that can be upset easily by various natural phenomena. A severe drought, for instance, would cause the vegetation in the area to dry and die. The rabbits' food supply would then be depleted severely and starvation would follow, killing most of the rabbit population. Once the drought passed and the vegetation revived, the rabbits would start reproducing again. Increases in population would result until the next drought occurred.

This explanation seems plausible at first glance. There is, however, a flaw in the reasoning. According to the graph, a drought would have to occur regularly just about every quarter of a century, which is unlikely.

Another possibility would be to relate the cyclical decimation of the rabbit population to epidemics. In this case, population decline would correspond to periods of widespread disease, and population growth to periods of recovery from the epidemics. Here again, however, disease and epidemic are not enough to explain fully the behavior described on the graph. Clearly, the period of the oscillations, which is around twenty-four years, is much too long. The ordinary amount of time required for an epidemic cycle (which includes deaths from the disease and recovery) is about four to five years at most.

Still another interpretation of the behavior is possible. The oscillatory pattern of the rabbit population could be explained by a predator-prey relationship. As seen in Chapter 4, the rabbit population is being preyed upon by lynx. Figure 6.3 shows that the growth in the lynx population depends on the number of rabbits available at any given time. As the lynx population grows, however, it starts killing more and more rabbits. When the lynx's food supply (the rabbits) is depleted, the lynx population is deeply affected and starts dying, allowing the few rabbits that are left on Hudson Bay to reproduce.

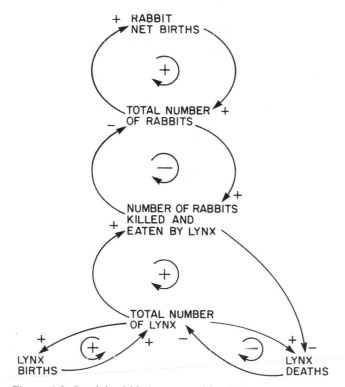

Figure 6.3 Partial rabbit-lynx causal-loop diagram

An examination of the simulated data pertaining to the lynx population in Hudson Bay lends support to the theory that the rabbits and lynx are related in a feedback system. The simulated data on lynx are based upon reports of actual lynx sighted in the Bay area by Hudson Bay Company representatives from 1800 through 1824. Those data are presented in Table 6.3.

The data in Table 6.3 were plotted in Figure 6.4 along with the available data on the rabbit population. Note that the scale for the lynx population differs from the scale for rabbits, since lynx sighted vary from 0 to 130, and the rabbit population goes as high as 60,000. If both sets of data are plotted on the rabbit scale, the variations in the lynx population would be too small to show up.

Figure 6.4 indicates that the lynx population was decreasing at the time when data were first being recorded. This is also the time when the rabbit population was at a minimum. There were probably so few rabbits in the Hudson Bay area that the lynx population was slowly starving. The lynx population hit a low in 1808. Interestingly, this is just about the same time that the rabbit population began to increase. The small number of lynx in the Hudson Bay area may have allowed the rabbit population to grow rapidly, with few rabbits being eaten by lynx. However, as the rabbit population started to increase very rapidly between 1810 and 1815, the lynx population also began to grow.

The presence of more and more lynx eventually must slow down the growth rate among the rabbits as more and more rabbits are eaten by the lynx.

Table 6.3 Active lynx population, 1800–1824

Time (years)	Lynx population
1800	125
1802	40
1804	10
1806	5
1808	3
1810	4
1812	6
1814	10
1816	15
1818	30
1820	52
1822	124
1824	50

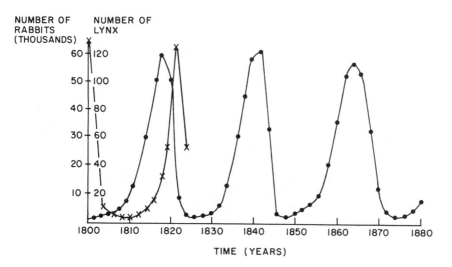

Figure 6.4 Behavior of the rabbit and lynx populations

This is, in fact, what happened. In 1818, the rabbit population peaked, in part at least due to the increasing number of lynx that were preying upon and eating the rabbits. Although the rabbit population began to decline in 1818, plenty of rabbits were still around to sustain growth in the lynx population for several more years. The lynx population continued to grow until 1822, when the sharp decline in the rabbit population ultimately led to starvation among the lynx. There were just too many lynx to be supported by the dwindling rabbit population, so two things happened: (1) the remaining lynx were devouring the rabbits, causing the rabbits to continue to decline at a rapid rate; and (2) the relative shortage in rabbits led to starvation and decline in the lynx population. By 1824, the entire rabbit-lynx system returned to the same situation that it was at in 1800. The rabbit population was near a minimum and the lynx population was rapidly declining. The system has progressed through one cycle of an oscillation and is ready to begin a second cycle.

Notice that the story about the rabbits and lynx that is so graphically displayed in the plots in Figure 6.4 is the same story that is told in the causal-loop diagram in Figure 6.3. The plots of data against time and causal-loop diagrams are just different ways of looking at the same system. The causal-loop diagrams are good for showing why things happened as they did. The plots of data show exactly when and how the events took place. These two ways of looking at a system are complementary. (Of course, sometimes, as with Mrs. Stewart's records, the data contain sources of error requiring adjustments. Similarly, the causal-loop diagrams always represent some degree of simplification and omissions.)

Exercise 2: Rabbits and Lynx

a. Study Figure 6.4, showing the behavior of the rabbits and lynx on the Hudson Bay in northern Canada. Attempt to predict when the next two peaks in the lynx population will occur. Explain why you think that the peaks will occur then.

b. Again referring to Figure 6.4, attempt to predict when the next minima (lowest points) in the lynx population will occur. Again, justify why the minima will occur where you predict.

c. Using your responses to Exercise 2a and 2b sketch what you believe to be the behavior of the lynx population in the Hudson Bay area between 1824 and 1880. Draw your sketch on a graph similar to Figure 6.4.

RABBIT-LYNX LEVELS AND RATES

The Rabbit-Lynx system, as discussed in the previous example, consists of two levels (the level of rabbits and the level of lynx) and four associated rates (rabbit births and deaths, lynx births and deaths). For purposes of analysis, one could limit attention to only two rates: net births (births minus deaths) for the rabbits and net births for the lynx. More complicated analyses might have divided the rabbit population into two levels (those capable of reproduction and those not capable, for example) or might have considered other rates (perhaps the rate of trapping of animals or the rate at which animals move into or out of a given geographical region).

The important point in the rabbit-lynx example is that the major rates depend critically on the levels. That is, the rate of change of rabbits depends on the level of lynx (because more lynx will eat more rabbits). Similarly, the rate of change of the lynx depends on the level of rabbits (many rabbits lead to rapid growth of the lynx and few rabbits lead to lynx starvation). Look back at the graphs of both rabbit and lynx population, and visualize how the decline in the level of rabbits had a dramatic impact on the rate of change (i.e., starvation) of the lynx.

The interrelated pattern of rates and levels observed in the rabbit-lynx example occurs much more widely. In fact, for dynamic systems characterized by feedback, this pattern holds as a general principle. Feedback could even be redefined in rate and level terms as follows: Any system level is the accumulation over time of its associated rates. In turn, all rates within a system depend on (are functions of) one or more system levels, thereby closing the feedback loop. Rates are functions of levels, and levels are functions (accumulations) of rates. This definition of feedback is an elaboration of the one found in Chapter 2. Part V will show how to take account of these principles while constructing computer simulation models.

EXAMPLE IV: POPULATION HISTORY OF THE BOSTON AREA—PART I

Figure 6.5 shows the population living within the city limits of Boston, Massachusetts, from 1810 through 1930. Notice that the population increases during this entire period, although not at an even rate. Placing a ruler against the population curve shows that sometimes the rate of population increase was faster and sometimes it was slower. Place one end of the ruler along the beginning of the graph, where the population in 1810 is shown to be about 25,000. Line the ruler up along the curve of the population. Notice that the line showing population parallels the ruler until about 1850. This means that the rate of increase in the population from 1810 to 1850 was nearly linear (falls on a straight line) over that period. At 1850, the rate of population growth suddenly increased. Move the ruler along to parallel the population at 1850, and see how much steeper the curve is from 1850 to 1910, compared to the first segment of the curve. If the rate of increase from 1810 to 1850 continued through 1930, the 1930 population would be only 300,000. But if the actual rate of change from 1850 to 1870 had continued, the 1930 population would be 650,000. In fact, the 1930 population was even higher—about 750,000. So the average slope of the population curve from 1870 to 1930 must be even steeper than it was from 1850 to 1870.

Exercise 3: Boston Population, 1810–1930

a. Line your ruler up with the population curve beginning in 1890. At what point does the population curve no longer parallel the ruler?

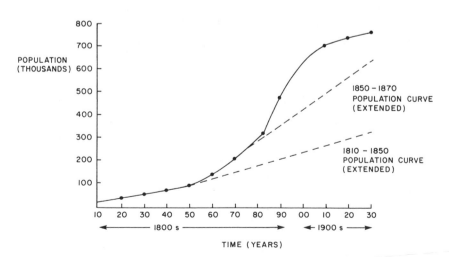

Figure 6.5 Population of Boston, 1810–1930

b. If the change in the population continued at the same rate that it was just around 1910, verify that the population in 1930 would be about 900,000. In fact, the actual 1930 population was only about 750,000. Why do you suppose that the population did not reach the 900,000 level?

c. While at some times in the period of 1810 through 1930 the population grew faster than others, it was always growing. During this period, what were the fastest periods and the slowest periods of growth? How can you tell?

d. Why do you think that the population grew faster at some times than at others?

EXAMPLE V: POPULATION HISTORY OF BOSTON—PART II

Has the population of Boston continued to grow? Figure 6.6 shows the population living within the Boston city limits through 1970. What has happened since 1930? Just after 1930, the population peaked, and then began to fall. However, it recovered and began to grow again about 1940, reaching a second peak in 1950. After 1950, the population suffered a net loss. In fact, by 1970, the population was under 700,000 for the first time since 1920.

Look at the population curve in 1950. Just before 1950, the population was increasing and the slope of the curve was positive. That means that more people were coming into Boston than were leaving. After 1850, the slope of the curve is negative. In other words, the rate of growth of the population is negative, which is the same as saying that the population was declining—or that more people were leaving the area than were entering. A turning point, then, is seen in 1950. The population was neither growing nor declining. It was simply staying even—the population was in balance. Increases in the population (from births and immigration) must equal decreases in the population (from deaths and emigration) if a balance or equilibrium is maintained.

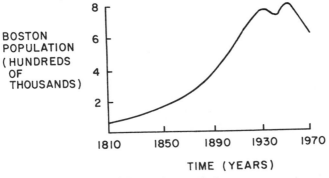

Figure 6.6 History of Boston's population

Exercise 4: Further Analysis of Boston's Population

a. Birth, death, immigration, and emigration are four natural processes that can work to change the population of an area. At different times in Boston's history only one or two of these processes were dominant. Which processes dominated from 1850 through 1900? From 1950 through 1970? How did the four processes compare during the period from 1930 to 1950?

b. Which of these ways of changing the size of a population do you think was primarily responsible for the decline in Boston's population from 1950 through 1970? Does the graph in Figure 6.7 show that you are correct, or are you making use of additional information?

c. Consider Figures 6.7 and 6.8. Figure 6.7 shows Boston and its surroundings divided into eight concentric circular regions, numbered 1 through 8.

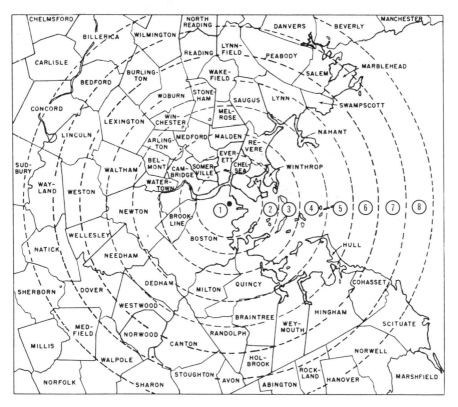

Figure 6.7* Boston and outlying concentric rings

*Reprinted from *Readings in Urban Dynamics: Volume I* by Nathaniel J. Mass, ed., by permission of The MIT Press, Cambridge, Massachusetts. © 1974, p. 107.

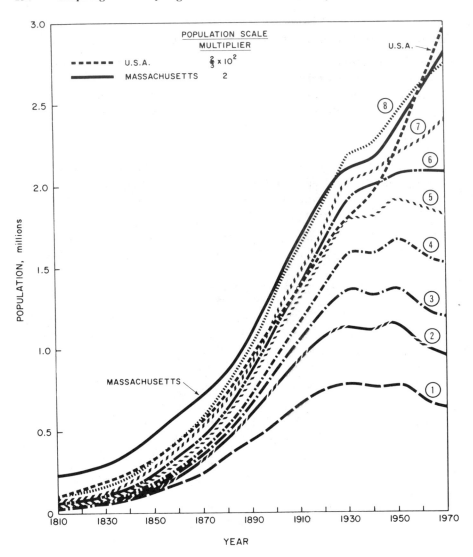

Figure 6.8* Number of persons in U.S., Massachusetts, and around Boston

Figure 6.8 shows on one graph the population histories of all eight of these regions. Curve number 1 shows the population of Boston and its immediate surroundings; curve number 2 shows the population history of the entire region within circle 2, that is, both regions 1 and 2; and so on, through curve number 8, which shows the population history of the entire region around Boston contained within circle 8. Also superimposed on the

*Reprinted from *Readings in Urban Dynamics: Volume I* by Nathaniel J. Mass, ed., by permission of The MIT Press, Cambridge, Massachusetts. © 1974, p. 108.

graph in Figure 6.8 are the population histories of Massachusetts and the entire United States for comparison. (The scales for the Massachusetts and United States curves are not shown.)

Most of the curves show a decline in population from 1950 through 1970, although not all the curves do. Answer Exercise 4(b) in the light of this new information. Explain what makes you think you are right. Why are you more certain now than when you first tried to answer Exercise 4(b)?

EXAMPLE VI: POPULATION HISTORY OF THE BOSTON AREA—PART III

Why do city populations, such as Boston, stop growing? First, what causes cities to grow? A population grows if the birth rate is greater than the death rate, but what is special about cities? The causal loop in Figure 6.9 suggests one reason cities grow, perhaps even faster than the population of the region in which the city lies.

This causal loop indicates that the greater the population and the larger the number of businesses in the city, the more goods and services are desired. The increase in the demand for goods and services, the market, draws new businesses and industries, and that creates new jobs. The increase in job availability attracts more people to the city, and thus increases immigration, especially if the city has a favorable location.

According to this causal loop, how does the size of the city's population behave over time? Since this is a positive or growth loop, the population should continue to grow, for there is no control on the growth. The larger the population is, the faster the city will grow.

The type of graph generated by positive feedback loops is often *exponential*. Exponentially growing quantities (such as population and business structures in the preceding example) not only continue to increase, but they increase at an ever faster rate.[1] These curves grow faster and faster because of the "snowballing" effect of a positive loop. More people and businesses lead to more labor, and goods and services desired, which in turn lead to more population and businesses.

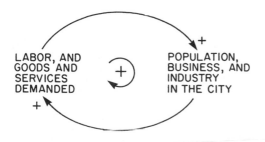

Figure 6.9 Growth of a city—Structure

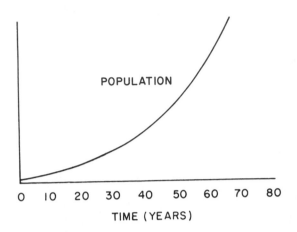

Figure 6.10 Growth of a city—Behavior

A sketch of the behavior from the preceding example might look like Figure 6.10. The graph shows that the growth of population is not linear, but exponential. In other words, the growth is not constant, but increases faster and faster. For the first twenty years, the population grows very slowly. It then begins to grow faster.

Exercise 5: Growth of a City

The causal loop shown in Figure 6.9 is a *positive* loop, which leads here to growth. Population, business, labor demanded—in fact, everything in the loop—grows as time passes. What causes that growth to stop? Nothing indicated in the positive loop stops the growth of the city. The reason is a *negative* loop, which also exists in the city system.

   a. Two causal loops are shown in Figure 6.11, one positive and one negative. The positive one produces growth, and the negative one by itself leads to-

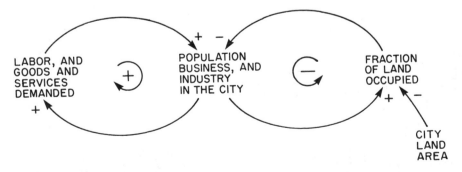

Figure 6.11 Effect of land on urban growth

ward equilibrium. Write a paragraph explaining how city land eventually comes into play to retard the growth of business and industry.

b. Which loop dominates the city system in the early history of the city? Which loop might eventually dominate, and halt the growth of the city? What brings about the change from dominance by one loop to dominance by the other?

DEVELOPING GRAPHS FROM CAUSAL LOOPS

The following exercises build upon the concepts and skills developed in the rabbit-lynx and urban growth examples. In the rabbit-lynx and urban growth cases, the information contained in graphs was used to formulate causal loops. This section starts with causal loops and investigates the sorts of graphs they might generate.

Exercise 6: Colds and Netsilik Eskimos

Reread the story of the Netsilik Eskimos (Exercise 17 in Chapter 4). The following tells a little more of the history of the Eskimos.

Until the European explorers encountered the Eskimos, the Eskimos had never had the ordinary colds that were common among the Europeans. Many of the Eskimos, especially the young children, got very sick from colds, and many of them died. Figure 6.12 shows what happened to the Eskimo population, births, and deaths after the explorers first met them in 1750.

a. Why were Eskimo births, deaths, and population time graphs just straight lines before the explorers came?

b. Write a paragraph that tells why the births, deaths, and population curves look the way they do.

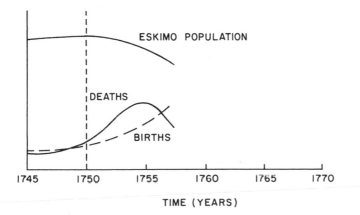

Figure 6.12 Time-graph of Eskimo population changes

c. Knowing what you know about the Eskimo population from the story of the Netsilik Eskimos and your causal-loop diagram of it, complete the time-graphs of Figure 6.12 with your guesses of what happened to Eskimos births, deaths, and total population over the next fifteen years or so.

Exercise 7: Regulation of Human Body Temperature

Reread "The Body Temperature System" (Exercise 12 in Chapter 4). This exercise is an extension of that description and is intended to show more completely how negative causal loops influence the behavior of a system. The following causal-loop diagram (Figure 6.13) shows two of the negative loops controlling human body temperature.

Shivering and perspiration are two of the body's automatic reactions to deviations in body temperature from 98.6°F.

a. Write a paragraph describing the temperature regulation system shown in the causal-loop diagram.

b. Figure 6.14 shows a person's temperature before, during, and after some physical exercise. The temperature rises because of the heat generated by muscle activity. On the same graph, sketch a curve showing your estimate of the body's perspiration rate during the same period.

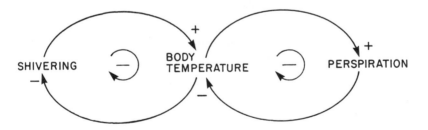

Figure 6.13 Body temperature controls

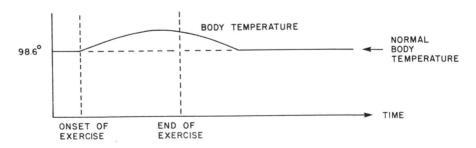

Figure 6.14 Graph of body temperature controls

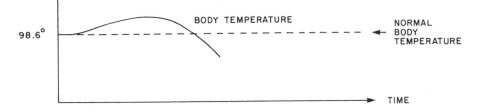

Figure 6.15 Body temperature affected by exercise

c. The body's temperature does not always return smoothly to normal as it did in Exercise 7(b). Describe a simple, real situation in which a person's temperature before, during, and after exercise could easily behave as shown in Figure 6.15.

d. Consider how the temperature regulation system given in the causal loop in Figure 6.13 would react in the situation graphed in Exercise 7(c). Draw on the same graph a curve showing perspiration rate and another curve showing shivering rate before, during, and after the exercise. Then complete the graph of body temperature. You should continue the graphs until the body has returned to a state of equilibrium at 98.6°F.

e. In Exercise 7(b) the body temperature returned smoothly to normal because of the controlled cooling effects of moderate perspiration. In Exercise 7(c), however, the body was apparently cooled too much by perspiration, given the situation, and the body temperature dropped below its equilibrium value of 98.6°F. In such situations we say there is "overshoot." Negative loops can thus produce two kinds of behavior: a smooth return to equilibrium, or overshoot behavior. It is not usually obvious from the causal-loop diagram exactly which kind of goal-seeking behavior is implied by a given negative loop.

Describe a real situation in which shivering brings a person's body temperature back to normal smoothly without overshoot. Can you think of a situation in which shivering could result in overshoot of body temperature over 98.6°F? What characteristic of perspiration makes it more likely that perspiration might produce overshoot than shivering?

Exercise 8: Graphing the Regulation of Body Sugar

Reread the story about insulin and glucose in the human body (Exercise 5 in Chapter 4). Make sure that you have a good causal loop describing that system. This exercise is an addition to that story.

If the body is at rest for some time, the glucose and insulin levels in the body come to rest at amounts that remain constant for a time, as shown in the first part of the time-graph in Figure 6.16. These constant levels of glucose and

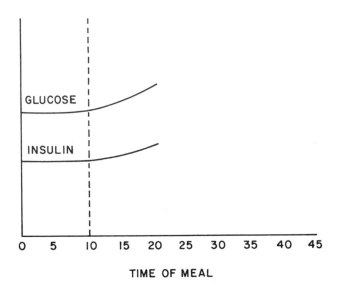

TIME OF MEAL

Figure 6.16 Glucose/insulin system

insulin are called the *equilibrium* levels. If one eats a big meal and the body starts digestion, then the glucose also causes a change in the insulin level. Suppose that Figure 6.16 represents what happens to the glucose and insulin levels immediately following a big meal.

Complete the time-graphs of the glucose/insulin system, and write a short paragraph explaining why things happen as you suppose. Do not be concerned about the exact numbers for the levels of glucose and insulin in the blood; just graph the general shapes of the time-graph.

SUPPLEMENTAL EXERCISES

The following exercises draw together some of the concepts developed in Chapters 5 and 6 concerning graphs, rates and levels, and causal loops.

Exercise 9:   The Growth of Rats in a Fixed Area

Figure 6.17 represents an experiment conducted by B. F. Calhoun on a population of wild Norway rats. Ten rats were confined in a quarter-acre enclosure with an abundance of food and room to live, and with predation and disease eliminated or minimized. As a result, only the animals' behavior toward one another remained as a factor affecting their growth. Figure 6.17 includes a monthly account of the rat population and the monthly rat birth rates and death rates.

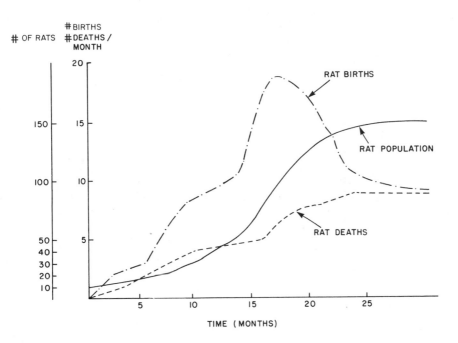

Figure 6.17 Rat population growth

a. Which curves represent rates? Which ones represent levels? Why?

b. Draw a causal-loop diagram explaining possible causes for the growth of the rat population observed in the experiment.

c. Looking at your causal-loop diagram and at the graph of the rat population growth, can you tell which loop(s) was/were dominant in the earlier months of the experiment?

d. What impact did those loops have on the behavior of the rat population?

e. When did these forces stop dominating the rat population growth? Why?

f. Look at Figure 6.17 more closely. What was the initial number of rats in the experiment? When did the rat population stabilize? At what level? What happens to birth rates and death rates at the same time?

g. What conclusions can you draw from the experiment?

h. Do you recall any other instances where a system behaves the same way as the rat population?

Exercise 10: Fishing in New Jersey

Figure 6.18 shows the annual landings of fish in New Jersey by commercial fishermen from 1880 to 1971.[2]

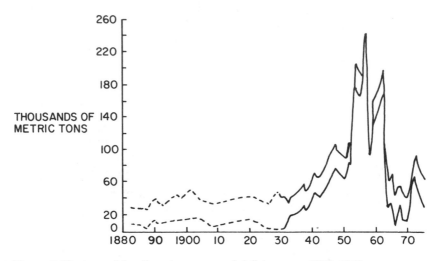

Figure 6.18 Annual landings by commercial fishermen, 1880–1971

a. Describe the behavior of the size of the total catch.

b. Suggest why the total harvest increased and decreased. What variables would affect the catch?

c. Draw a causal-loop diagram that might account for the behavior of the total catch.

Exercise 11: Finfish

Figure 6.19 shows the annual commercial landings of three species of finfish in New Jersey from 1880 to 1975: alewife, Atlantic herring, and menhaden.[3]

a. What do the graphs of all three species have in common?

b. What is the major difference in the behavior of the herring, alewife, and menhaden landings in the period 1880–1975?

c. What factors do you think affected the landings of these three species?

d. Write a few sentences to explain the behavior shown in these three graphs. Make sure you account for the shift in harvest from one species to another.

e. Draw a causal-loop diagram that explains the behavior shown in the three graphs. Make sure you account for the shift in harvest from one species to another.

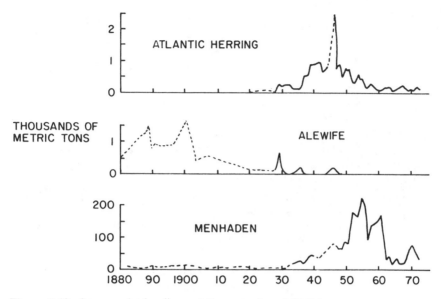

Figure 6.19 Commercial landings of three species of finfish

Exercise 12: The Fall of Mayan Civilization

Read the attached article entitled "Study Depicts Mayan Decline," by Harold M. Schmeck Jr.*

a. Translate the verbal description and analysis into graphs of several variables over time. Sketch your graphs on the same set of axes (or comparable axes if it gets too crowded). Identify coordinates on the (horizontal) time axis and any critical values on the y-axis.

b. In a paragraph or two, define your variables clearly, explain the time horizon you selected and showed for your graphs, and briefly justify (from the article) the dynamic patterns and interrelationships illustrated in your graphs.

Study Depicts Mayan Decline

By Harold M. Schmeck Jr.
October 23, 1979

One of the great mysteries of human history has been the sudden collapse of one of the main centers of Mayan civilization in Central America at a time when it was apparently at a peak of culture, architecture and population around A.D. 800.

---

*© 1979 by the New York Times Company. Reprinted by permission.

No one knows exactly why this society of several million people collapsed, but new research shows a gradually tightening squeeze between population and environment that may have been crucial to the fall.

Tropical environments are notoriously fragile. By understanding what the Mayas did to theirs, modern humans may get some useful guidance on how to treat tropical environments today—a realm of knowledge that could be particularly useful to the so-called Third World.

Just before the final cataclysm, the new research suggests, the population in one area ranged from about 200 to 500 persons per square kilometer (about four tenths of square mile). This population-density almost certainly required advanced agriculture or large-scale trade.

Within two to four Mayan generations, which probably meant less than 100 years, the population dropped back to what it had been almost 2,000 years before—20 or less per square kilometer and sometimes far below even that sparse population. Furthermore, after the collapse, whole areas remained almost uninhabited for a thousand years—virtually until the 1970's.

Some of the environmental changes appear to have been as long-lasting as the loss of population. Lakes that were apparently centers of settlement in the Maya time have not even today recovered the state of productivity that made their shores good living places more than 1,000 years ago.

Such clues to the past as these were found, in eight years of research, by scientists at Florida State University and University of Chicago. Their research, still continuing, showed there was an exponential growth in Mayan population during at least 1,700 years in the tropical lowlands of what is now Guatemala.

Human numbers doubled every 408 years, according to the new estimates. This trend may have caught the Maya in a strange trap. Their numbers grew at a steadily increasing pace, but, for many centuries, the growth was too slow for any single generation to see what was happening.

Over the centuries, the increasing pressure on the environment may have become impossible to maintain. Yet the squeeze could have been imperceptible until the final population spurt at the end. In more northerly regions, the quality of Mayan civilization may have deteriorated without so great a population drop, some specialists believe.

Based on Detailed Study

The new estimates for the southern lowlands are based largely on a detailed survey of traces of residential structures that were built, occupied and abandoned over the centuries.

The studies are focused on the region of two adjacent lakes, now called Yaxha and Sacnab, in the Peten lake district of northern Guatemala. The area was inhabited as early as 3,000 years ago and the first agricultural settlements appeared there about 1,000 B.C. The land was largely deforested by A.D. 250.

Gradually-intensified agriculture seems to have severe cumulative damage to an originally verdant environment. To this was added the impact of increases in human dwellings and other major architectural works on the land. Essential nutrients washed, slid and were moved downhill to be lost in the lakes, diminishing the fertility of agricultural land.

Increases in phosphorus in the lakes from agriculture and human wastes showed that pollution must have aggravated the environmental damage. Scientists even confirmed the population trend by estimates of the per capita increase in phosphorus going into the lakes as human numbers rose.

Authors of the research, part of which is summarized in the current issue of Science, believe theirs to be the first published report giving documented estimates for the growth pattern of Mayan population and correlating this growth with the damage to the environment that went with it.

The authors pointedly omit any claim that they have solved the longstanding mystery of the Maya collapse. They do say the kind of environmental pressure their studies suggest may have been one important factor among several. The scientists also believe the research may have some useful practical implications.

Exercise 13: Viewing Policy Problems as Graphs Over Time

Think of some policy that has failed to achieve the desirable results expected of it. It could be a government policy dealing with some problem in the public sector, a policy of a corporation, some "policy" of an individual concerned with a personal problem, or the like. View the policy you select, and the problem to which it was addressed, from the dynamic perspective; translate it or recast it from static to dynamic terms, if necessary. Sketch three graphs over time:

a. A graph of the pattern of dynamic behavior to which the policy was directed—the problem addressed by the policy-maker(s);

b. A graph of the pattern of behavior that would have resulted had the policy been deemed successful;

c. A graph of the behavior that actually resulted when the policy was implemented.

You may have to sketch the patterns of more than one variable on each graph to capture the essence of the problem and the policy.

Exercise 14: Find Your Own Dynamic Problem

Find a newspaper clipping or brief magazine article that may be interpreted dynamically. Follow the directions in Exercise 13 parts (a) and (b) for the article you find. (Be sure to hand in a copy of your article.)

ENDNOTES

1. Mathematically, an exponential curve may be characterized as:
   $X(t) = X(O)e^{at}$
   where $X(t)$ is the value of the quantity at time t; $X(O)$ is the initial value of that quantity, and a is a parameter directly related to the curve's doubling time.
2. From J. L. McHugh, *Fisheries and Fishery Resources of New York State* (Washington, D.C.: National Marine Fisheries Service, U. S. Department of Commerce, March 1977).
3. Ibid.

CHAPTER 7

GRAPHS OF SYSTEMS WITH DELAYED RESPONSES

Systems often respond sluggishly. For example, it takes time for consumers to react to higher gas prices by purchasing more fuel-efficient cars; it takes time for auto manufacturers to design and produce them. This chapter focuses on the behavior of time delays, and examines graphs that delays can produce. The examples that follow illustrate some of the main characteristics of delays. Chapter 17 introduces the use of simulation to more fully analyze delays and the behavior they generate.

EXAMPLE I: MARTAN CHEMICAL AND THE SPARKILL RIVER—PART I

Martan Chemical has been manufacturing a variety of chemical products, including insecticides, at its Amsterdam plant since the early 1960s. As a major source of employment, Martan has a significant impact on the town of Amsterdam, which is located in a rural area in the northern part of the country. Because of a short growing season and poor soil, Amsterdam is unable to support productive farming. The second major source of income for the town comes from tourism. The Sparkill River, which runs through the town of Amsterdam, is known throughout the state as a fine source of trout, salmon, and bass. During the fishing season, Amsterdam's population is swollen by an influx of fishermen who fill all available motel rooms and spend freely in the local restaurants and shops.

Located directly alongside the Sparkill River on the south side of town, Martan's waste products from manufacturing are released into the Sparkill through a buried outfall pipe. Martan's main product is Nobug, a biodegradable insecticide that breaks down rapidly into the environment after use or disposal. In accordance with state regulations, the town continuously tests the water downstream of the outfall pipe. Figure 7.1, giving the results of the tests over a typical month, shows a pattern that seems to repeat itself regularly. The

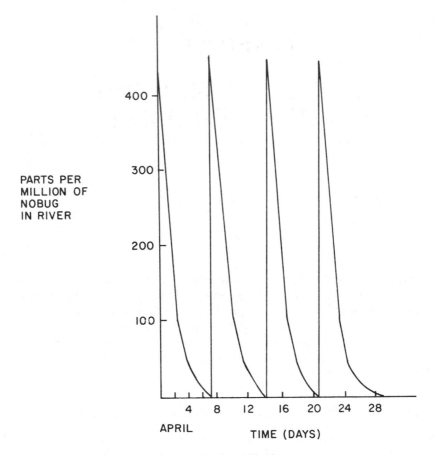

Figure 7.1 Amount of Nobug in the Sparkill River

level of Nobug measured on April 1 was 440 parts per million (ppm), which is the same as the level measured on April 8, 15, and 22. The interval between April 1 and April 8 is seven days, and the interval between April 8 and April 15 is also seven days.

Similarly, the graph shows approximately 100 ppm of Nobug in the river on April 3, 10, 17, and 24. Again, there is exactly a seven-day interval between each of these points. Not only is the shape of each segment similar, but the entire pattern repeats itself every seven days. Why? Look at the graph more closely. Notice that the first peak occurs on April 1. The amount of Nobug measured in the Sparkill on the following day, April 2, is considerably lower, and on April 3 it is again lower. In fact, the level of Nobug gets lower and lower daily, following the initial peak on April 3, reaching nearly zero by April 7. On April 8, however, the level of pollution shoots way up again, only to drop to near zero seven days later.

Three characteristics of the graph seem noteworthy. The levels of pollution begin with a sudden increase, which could be called a pulse, lasting for only a day, after which the level of pollution drops rapidly. Second, a pattern of pulses and declines occurs regularly, with each pulse followed by a decline in the level of pollution to near zero. Third, the pattern of a sudden rise in pollution followed by a relatively rapid decline repeats itself every seven days.

Something is happening at the Martan plant to account for this behavior. One explanation might be that the plant produces most of its Nobug on Monday, a smaller amount on Tuesday, and a still smaller amount on Wednesday, and almost none on Thursday through Sunday, disposing of the waste products daily. But that would mean having a large workforce on Monday, fewer people working on Tuesday, almost none on the production line on Wednesday, and none working from Thursday through Saturday, which seems unlikely. Martan would want to equalize the amount of work throughout the week.

If they do have the same number of workers on the production line all week, how else can this pattern of pollution be explained? Perhaps Martan puts all its waste products into a large holding vat, and dumps the entire week's supply into the river at one time. But why would they go to the expense of putting in a holding tank rather than simply dumping the waste products as they occur?

What actually happens is somewhat similar to this second theory. Once a week Martan flushes the entire production mechanism so that the machinery and pipes do not become clogged. The flushing is done automatically every Sunday night so that the plant will be ready to operate when the work crew arrives early Monday morning.

Figure 7.1 visualizes the impact of the flushing of the waste products from the manufacture of Nobug. A simple causal loop representing what is going on in the Sparkill River might look like Figure 7.2.

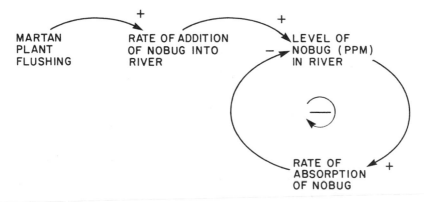

Figure 7.2 Simple causal loop for Nobug in Sparkill River

EXAMPLE II: MARTAN CHEMICAL—PART II

Figure 7.3 shows the effect of discharges from the Martan plant on the Sparkill River for April and May. The data for April are identical to the information given in Figure 7.1. Figure 7.3 shows several changes in the pattern of pollution beginning in May. A series of weekly pulses followed by a decline in the level of pollution in the river still occurs, but beginning in May each peak is higher than the last. In addition, the minimum weekly pollution levels are significantly higher than they were in April. Taken together, these two observations indicate that for some reason the total level of pollution in the river is rising.

Several possible explanations can be offered for this behavior. The first is that the company is increasing its production of Nobug. This would lead to more pollutants being released into the river. However, careful examination of Figure 7.3 shows this cannot be the reason. If Martan were disposing of larger quantities of the waste products from Nobug, the length of each pulse would increase, since the amount of added pollution is equal to the new maximum value less the previous level, but this is not the case.

Notice that the amount of pollution measured on May 6 is about 680 ppm (parts per million). The level the day before was about 250 ppm. This means an increase in the pollution level of about 430 ppm between May 5 and May 6. The next peak occurs on May 13, with a level of about 795 ppm. Subtracting the level on the previous day generates a difference of 420 ppm (795 − 375).

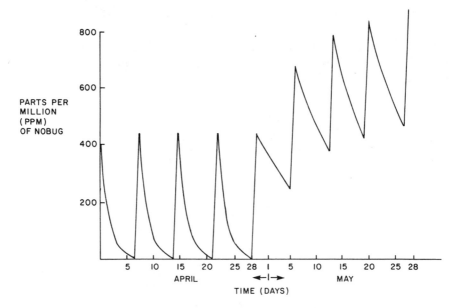

Figure 7.3 Amount of Nobug during April and May

Similarly, the difference between the reading on May 19 and 20 is 425 ppm (850 − 435), and between May 26 and 27 it is about 420 ppm (885 − 465). It seems, then, that the company is not discharging ever greater amounts of No-bug into the river. On the contrary, Martan seems to be discharging the same amount they had all through April, when the amount added weekly averaged about 430 ppm.

If Martan is not adding more Nobug, perhaps they are adding the same quantity, but more frequently. This might account for the increasing pollution in the river. This suggestion can be dismissed immediately, however, since the interval between peaks remains at seven days throughout both April and May.

Perhaps a clue has been overlooked. Look again at the minimum points and notice that the minimum levels of Nobug residue throughout April were nearly zero. That means that the river has dissipated, or absorbed, nearly all of the effluent pumped into it on Sunday by the following Saturday. But this is not the case in May. For some reason, the level of Nobug, even after a week, remains very high.

Why does the river absorb the 430 ppm of Nobug added weekly in April, but not in May? It turns out that beginning in May, Martan altered the formula for Nobug in response to customer requests for an insecticide that would remain active for a longer period of time. Gardeners liked Nobug because it was biodegradable—that is, it broke down completely and left no residue on fruits or vegetables. However, it broke down so rapidly that it had to be reapplied almost daily. Data for the first week in April on Figure 7.3 show that Nobug broke down very rapidly in the water, as well. On April 1, about 440 ppm of Nobug were in the river. By the next day, there was only half as much, just 210 ppm. And on April 3, there was only half as much Nobug as on the day before. Scientists refer to this type of behavior as the half-life of a chemical. The half-life of Nobug in April was just about one day, since the amount of Nobug measured in the river after each weekly addition halved each day.

When the chemists at Martan reformulated Nobug so that it did not need to be applied so often, they extended its half-life. The new half-life for Nobug-II can be estimated by determining the level of Nobug-II in the river after one of the weekly additions, such as May 6. Since the level of Nobug-II on May 6 was 680 ppm, the half-life is equal to the number of days it would take the level of pollution to be halved, that is, to reach about 340 ppm. That means that the halving time for Nobug-II must be somewhat more than seven days.

The lengthening of the halving time of Nobug explains the rise in the level of pollution shown in Figure 7.3. When the halving time was only one day, almost all the Nobug added on a Sunday would be broken down by Saturday. Therefore, when another dose of Nobug was piped into the river from the Martan plant on the following Sunday, the river was almost free from Nobug, and so there was no noticeable buildup.

However, when the halving time of Nobug was lengthened to seven days, the amount added on the first Sunday, April 29, only halved in the next week, to about 250 ppm on Saturday, May 5. Therefore, when an additional 430 ppm of Nobug was added on Sunday, May 6, the level rose from a base that began with the 250 ppm left from the previous week. And at the end of the next week, on Saturday, May 12, about half of this combined level of pollution remained, to be added to the weekly discharge the next day.

Exercise 1: The Martan Problem

    a. Draw a causal loop showing the concentration in the Sparkill River during May.

    b. In terms of the causal loop, how is the situation in May different from the situation in April?

    c. Obviously, the new formula for Nobug is creating a higher concentration of pollution in the Sparkill River with a potentially bad impact on the fishing industry and tourism. Recommend a plan for dealing with the potential pollution problem in the Sparkill River. Who would have to agree to this solution? How might this solution be implemented and enforced?

    d. Draw another causal loop showing some of the impacts of your proposed solution on both employment at Martan and tourism in the Amsterdam area.

DELAYS

The Martan Chemical Company example was a specific case of a very frequently occurring causal structure. The essential causal loop regulating the clean-up of pollution in the Sparkill River is shown in Figure 7.4. This causal loop shows that the higher the concentration of Nobug in the Sparkill, the higher would be the rate of dissipation of Nobug. This should make sense because if there were little or no Nobug in the river, there would be no rate of dissipation, and a lot of Nobug would lead to the river's dissipating a lot of the

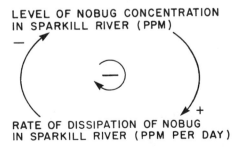

LEVEL OF NOBUG CONCENTRATION
IN SPARKILL RIVER (PPM)

RATE OF DISSIPATION OF NOBUG
IN SPARKILL RIVER (PPM PER DAY)

Figure 7.4 Basic negative loop controlling the clean-up of Sparkill River

chemical. The negative link between rate of dissipation and level of Nobug concentration means that dissipation of Nobug always leads to a decrease in Nobug concentration.

The particular piece of causal structure shown in Figure 7.4, involving one level and one rate with a negative loop between them, is characteristic of a "delay." Why? The rate of dissipation of Nobug can be looked at as a kind of delay of the rate of dumping into the river. Even though Martan does all of its dumping on Sunday, when it flushes its system, the dissipation of the chemical is often delayed by two, three, or more days as the river takes some time to "clean itself out." While this delay in chemical dissipation is going on, the level of Nobug concentration in the Sparkill is steadily decreasing.

An important factor in almost all delays is the delay time, or the amount of time that it takes the output rate (in this case, the dissipation rate) to react to the input rate (in this case, the Nobug dumping rate). In the Martan Chemical case, the delay time was directly related to the half-life of Nobug. In May, when Martan changed to a chemical formula that did not decompose so quickly, the delay in river clean-up was lengthened. An increase in the delay time can have dramatic effects on water quality and other variables that might depend on a delay.

Delays occur in many forms and many places. For example, if a batch of letters is sent through the mail (the "send rate," measured in letters per day), these letters will be delivered after some delay (the "arrival rate," also measured in letters per day). What accumulated during this delay (the "level in the delay") is the amount of mail actually being processed by the postal department. A causal loop for a mail delay is shown in Figure 7.5. Graphs of variables that go through delays can sometimes look quite different from the Martan Chemical example.

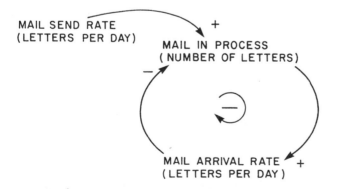

Figure 7.5 Causal loop representing a delay in mail delivery

EXAMPLE III: TREE HARVESTING

In 1930, Lester Splintz planted 10,000 fast-growing soft pine saplings on a piece of land that had some hills, some meadowland, and several shaded and rocky spots. He intended to harvest these trees and sell them to a paper pulp mill once they had reached a six-inch diameter. According to an agricultural bulletin that he received at the time of planting, the species of tree that he planted was intended to reach harvesting size after an average of twenty years.

Figure 7.6 shows the harvest rate (measured in trees per year) that Lester experienced between 1930 and 1980. Notice that no trees were harvested until after 1940 when the harvest rate began to peak. The peak harvest rate occurred in 1949 at almost 800 trees per year, and then greatly tapered off until the last tree was harvested around 1974 or 1975. Apparently what has happened is that some of the trees were planted in fertile meadowland with a sunny southern exposure. These trees grew very rapidly for harvest in slightly over ten years. Other trees may have been planted in rocky and shaded hillsides and did not mature until quite a few years later. However, the overall average time to harvest seems to have been somewhere close to twenty years. That is, the number of trees that was harvested before the twenty-year mark appears to just about balance the trees harvested after the twenty-year mark.

Since the trees were harvested some amount of time after they were planted, it might be possible to think of the planting and harvesting of trees as a delay process having an average delay of twenty years. A causal loop for such a delay might look like the one found in Figure 7.7.

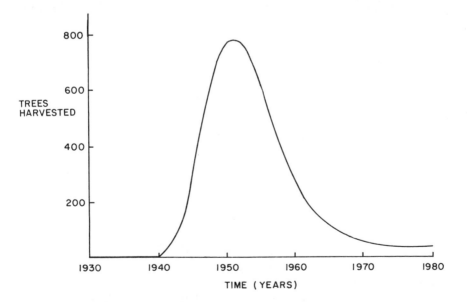

Figure 7.6 Harvest rates for 10,000 trees, 1930–1980

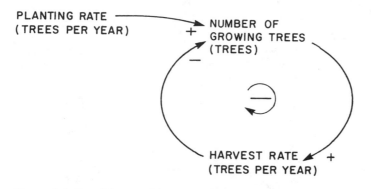

Figure 7.7 Possible causal loop showing harvest delay

The causal loop in Figure 7.7 looks quite similar to the causal loop for the delay in the Martan Chemical Company example. However, the shape of the parts per million concentration of Nobug in the Sparkill River looked quite different from the harvest rates of Lester's trees. Is there a reason why these graphs look so different? One good reason for the different appearance is that the graph of ppm concentration of Nobug is a level, and the graph of tree harvests is a rate. This is comparing apples with oranges, so to speak. To be truly comparable, the level of Nobug concentrate in the river should be compared with another level, the number of growing trees, not with the rate at which the trees are being harvested.

The data presented in Table 7.1 show the exact number of trees that Lester harvested on a year-by-year basis. Using these data (presented in the second column), it is possible to compute how many trees were still standing and growing at any point in time. (Since Lester's timberstand is hypothetical, it seems convenient to assume that all of the 10,000 trees survive to be finally harvested.)

From 1930 to 1939, the level of growing trees remaining stayed constant at 10,000, since none were harvested. The first 4 trees were harvested during the 1940 growing season, leaving 9,996 trees at the end of 1940. Then 14 more trees were harvested during 1941 leaving 9,982 at the end of the year, and so on. The result of all of the computations performed in Table 7.1 is plotted in Figure 7.8. Notice that each variable is plotted on a different scale. The number of trees growing is plotted on a vertical scale that varies from 0 to 10,000, while the rate of harvest is plotted on a vertical scale that varies from 0 to 800. The horizontal scale is measured in years for both variables.

This graph does show some interesting relationships. The number of growing trees is stable for the first ten years from 1930 to 1940 and shows its steepest rate of decline in 1949. However, the original puzzle about the differences in shape between the tree delay and the Martan Chemical Company delay still persists. In the case of Nobug concentration, the level reaches a sharp "spike" at the point when the chemical is dumped into the river. However, in

Table 7.1 Tree harvest

Year	Trees harvested	Trees remaining
1930	0	10,000
1931	0	10,000
1932	0	10,000
1933	0	10,000
1934	0	10,000
1935	0	10,000
1936	0	10,000
1937	0	10,000
1938	0	10,000
1939	0	10,000
1940	4	9,996
1941	14	9,982
1942	36	9,946
1943	78	9,868
1944	143	9,725
1945	230	9,495
1946	334	9,161
1947	446	8,715
1948	553	8,162
1949	645	7,517
1950	714	6,803
1951	753	6,050
1952	763	5,287
1953	745	4,542
1954	704	3,838
1955	647	3,191
1956	578	2,613
1957	504	2,109
1958	430	1,679
1959	360	1,319
1960	296	1,023
1961	239	784
1962	190	594
1963	148	446
1964	115	331
1965	87	244
1966	66	178
1967	49	129
1968	36	90
1969	26	64
1970	19	45
1971	14	31
1972	10	21
1973	7	14

Table 7.1 (continued)

Year	Trees harvested	Trees remaining
1974	5	9
1975	3	6
1976	2	4
1977	2	2
1978	1	1
1979	1	0

the case of the trees, the number of growing trees remains flat for ten years before it shows any decline—the prominent spike is nowhere to be found.

The graphs are not the problem. Rather, the causal loop to represent the delay in the growth of trees is imprecise. Reflection will suggest that all growing trees are not identical. For this purpose, it is useful to distinguish between saplings (trees that are 0 to 1 inch in diameter), small trees (1 to 3 inches in diameter), medium-sized trees (3 to 6 inches in diameter), and harvestable-trees (6 plus inches in diameter). In effect, by lumping all of these types of trees together, some important distinctions have been blurred (most notably saplings cannot be harvested, and it takes many years before a sapling grows through all of the stages until it can be harvested). To solve this requires a causal-loop diagram for the delays encountered in tree growth that is more complicated than the delay diagram involved in the clean-up of Nobug in the Sparkill

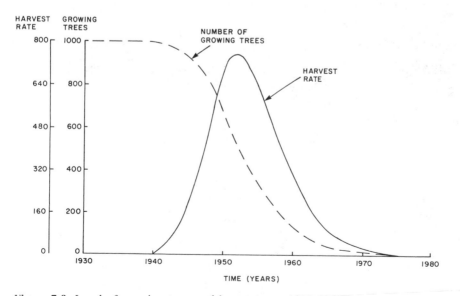

Figure 7.8 Level of growing trees and harvest rate, 1930–1980

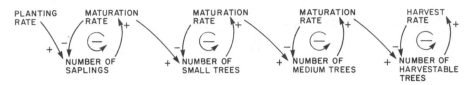

Figure 7.9 More complete causal loop showing levels of growth

River. This diagram has four levels, as outlined earlier, and five rates: the planting rate, the rate of maturation from sapling to small trees, the rate of maturation from small to medium trees, the rate of maturation from medium to harvestable trees, and, finally, the harvest rate itself. The complete causal loop needed to represent accurately the delay in the growth of trees is shown in Figure 7.9.

The causal-loop diagram of Figure 7.9 can show why the total number of trees does not change at all for at least ten years. During that period of time, the young trees are developing—and none is ready for harvest until it has passed through all stages of growth.

The Martan Chemical Company example showed how the length of the delay can be important in determining how some key variable (such as the amount of insecticide in a river) changes or fails to change over time. The tree example demonstrates how differing types of delay processes can be more or less complicated and can lead to different looking plots of variables over time (as in the "spiked" look of the Nobug concentration versus the gently sloping look of the number of growing trees).

Exercise 2: Tree Harvesting—Part II

In 1930, Lester's brother, Warren Splintz, also planted 10,000 trees. However, Warren continued to plant 10,000 trees every year for the fifty-year period between 1930 and 1980. Warren's timber harvest and planting rate are both plotted as hypothetical data in Figure 7.10.

   a. Would the causal-loop diagram representing the delay in harvesting of trees for Warren be different from the one developed for Lester in Figure 7.9?

   b. If you do not believe that the causal-loop diagram would be different, how do you explain the marked differences in the harvest rates between Warren's and Lester's crops?

   c. A system is considered to be in equilibrium when its inputs (in this case, Warren's tree-planting rate) are equal to its outputs (in this case, Warren's tree-harvesting rate). Does this system reach equilibrium? If so, when?

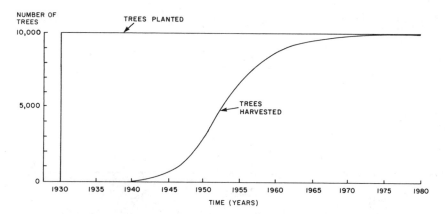

Figure 7.10 Planting and harvest rates for Warren's Lot, 1925–1980

d. Suppose that Warren stopped planting trees in 1980, yet continued his present harvesting policies. Would the system reach a new equilibrium? If so, what would that equilibrium be?

e. Sketch your guess about what Warren's tree harvest will look like between 1980 and 2030 if planting is stopped in 1980 and current harvesting policies are continued.

f. Explain why you think that the curved lines would behave as you have sketched them.

g. According to Figure 7.10, by 1970 Warren is harvesting 10,000 trees per year, each one of which has lived for an average of twenty years. Use these two facts to estimate how many trees Warren has standing on his farm in 1970. Explain your reasoning.

Exercise 3: The Growit Seed Company Delivery Problem

The Growit Seed Company specializes in Early Reds, a brand of early-bearing tomatoes that is quite popular with home gardeners. Early Reds must be planted on time or their growing advantage is lost. Hence, the Growit Company must assure the prompt arrival of the seeds to its customers throughout the United States. Due to differences in climate, Growit allows a three-week period in which to make its shipments. Typically, the Growit warehouse makes up a daily batch of shipments and sends them via parcel express. Because the Growit Company is interested in investigating the performance of the parcel express company in delivering its product, it devised the following experiment. On Monday, March 31, 1,000 packages were mailed. Each package carried a return slip that indicated the date and hour of the package's arrival. Figure 7.11 shows the arrival data (hypothetical) for the 1,000 shipments.

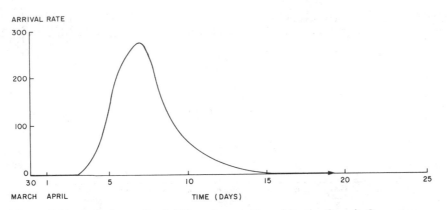

Figure 7.11 Arrival rate for 1,000 packages shipped by the Growit Company

a. Draw a causal-loop diagram showing the process that is determining the arrival rate for the Growit Company's packages.

b. What do you estimate to be the average delay time between shipping and arrival of packages sent by the Growit Company? How did you arrive at this estimate?

c. What might be a few of the reasons that some customers receive their packages as soon as four or five days after mailing and some other customers do not receive their packages for thirteen or fourteen days?

d. If Growit were to ship 1,000 packages per day for three weeks and then ship no packages after that, would the shipment and arrival rates ever reach equilibrium? Would more than one equilibrium point be reached? What would the equilibrium point(s) be?

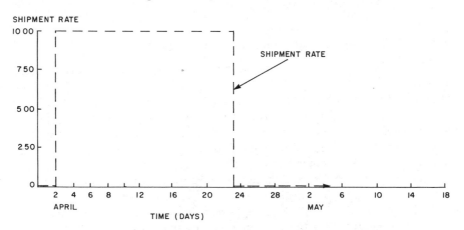

Figure 7.12 Growit shipments over a six-week period

  e. Figure 7.12 sketches a three-week period of constant shipments of 1,000
     packages per day with no shipments per day after that period. Sketch
     what the arrival rate would look like.

Exercise 4: The Overtraining of Teachers

The years immediately following the end of World War II were characterized
by an unprecedented baby boom. The sudden growth in population, combined
with an increasing awareness of the importance of education, led to very high
levels of student enrollment throughout the country. This phenomenon, in
turn, created a great demand for teachers. Figure 7.13 and Table 7.2 represent
what has been happening in the United States in the field of education at the
elementary and secondary levels since 1940.

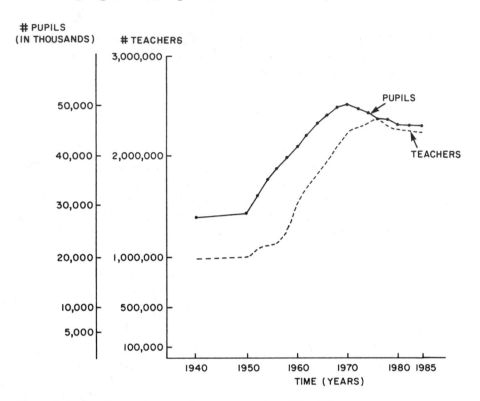

Figure 7.13 Pupil enrollment and teacher supply, 1940–1980

a. Draw a causal-loop diagram explaining what happened in the field of education from 1940 to 1980. (*Note:* Increased enrollment, which is determined exogenously (or independently), affects the demand for teachers. The capacity of teaching colleges will determine the supply of teachers for any given year.)

b. Looking at the graph of pupil enrollment in Figure 7.13, can you tell when the baby boom had an impact on pupil enrollment? When did the pupil population reach its maximum? What was that maximum? What happened afterwards?

c. Compute a table of changes in pupil enrollment. What would you call these changes: levels or rates?

d. Graph these changes. Compare your graph to Figure 7.13.

e. When did the teacher population react to the need for more teaching staff? Was there a delay in response? If yes, how long was it? Can you explain why there was such a delay? What was the maximum number of teachers? In what year did it occur? Did the teacher maximum occur later than the pupil enrollment maximum? If yes, can you explain why?

Table 7.2 Historical summary of pupil enrollment and teacher data*

Year	# Pupils (in thousands)	# Teachers
1940	28,057	995,366
1950	28,660	1,002,098
1952	32,064	1,073,198
1954	35,549	1,117,645
1956	37,919	1,154,715
1958	40,081	1,297,732
1960	42,112	1,559,995
1962	44,243	1,676,631
1964	47,016	1,819,373
1966	48,339	1,964,238
1968	50,042	2,124,131
1970	50,567	2,250,618
1972	49,780	2,292,083
1974	48,674	2,339,421
1976	47,750	2,352,276
1978	47,346	2,310,647
1980	46,692	2,365,692
1985	46,168	2,241,165

*Information adapted from W. Vance Grant and C. George Link, *Digest of Education Statistics* (Washington, D.C.: National Center for Education Statistics, U.S. Government Printing Office, 1979).

f. The ratio of the number of pupils enrolled to the number of teachers available on the job market for a given year is computed by dividing the number of pupils for that year by the number of teachers. For instance, in 1940, the student-teacher ratio was equal to

$$\frac{28,057,000}{999,000} = 29.19 \text{ pupils/teacher.}$$

Compute the table of pupil-teacher ratio using the data already provided in Table 7.2.

g. Assuming that the ratio required in order to provide adequate and efficient education is about twenty-five to twenty-eight pupils to one teacher, when did shortages of teachers occur? When did a surplus occur?

h. Given all the preceding data, and based on the answer you already gave, can you write an explanation describing the problems faced by elementary and secondary education over the years? What do you think will happen in the next decade?

PART IV

ANALYZING LESS-STRUCTURED PROBLEMS

OBJECTIVES

Part IV introduces, and then gives the reader practice applying, four critical components to defining an unstructured dynamic problem:

1. The *perspective* of the concerned person;

2. The *time horizon* over which the problem is considered;

3. A precise dynamic description of the *reference mode;*

4. The possible *policy choices* applicable to the problem.

CHAPTER 8

DEFINING A DYNAMIC PROBLEM

Questions are necessarily prior to answers, and no answers are conceivable that are not answers to questions. A "purely factual" study—observation of a segment of social reality with no preconceptions—is not possible; it could only lead to a chaotic accumulation of meaningless impressions. Even the savage has his selective preconceptions by which he can organize, interpret, and give meaning to his experiences.*

Gunnar Myrdal (1968)

The system dynamics approach to modeling and computer simulation is as a problem-solving tool. To evaluate possible solutions to a problem, the problem itself must first be clearly stated. That problem statement should answer four questions.

1. From whose point of view is the problem being seen?   (perspective)

2. Over what time period does the viewer perceive those changing aspects to be a problem?   (time horizon)

3. What is the pattern of change that constitutes the problem behavior? (reference mode)

4. What solution approaches are proposed to address the problem?   (policy choice)

Until now, causal-loop diagrams have been developed to fit specific problem situations in which pertinent aspects of the situation have been defined. Now, to define problems in less-structured situations, several specific dimensions of problem definition need to be explored. These dimensions are conveniently named perspective, time horizon, reference mode, and policy choice.

---

*Gunnar Myrdal, *Asian Drama: An Inquiry into the Poverty of Nations.* Copyright © 1968 by the Twentieth Century Fund, New York.

PERSPECTIVE

A given situation may appear quite different to different viewers. These differences arise from distinctions in roles within the situation. Since the "problem" is some system whose aspects are changing in time, it becomes a "problem" only to those whose fortunes are affected by those changing aspects. (Note the distinction between this kind of difference in perspective and that arising from limited information about the situation, as in the story of the blind men each viewing an elephant as a very different beast, based on the parts each had felt.)

EXAMPLE I: THE "OIL CRISIS"—PART I

The "oil crisis" looks very different when viewed from the perspective of a New York commuter, an oil company executive, and a Sierra Club member. For a commuter living on Long Island who must drive sixty miles to work and back five days each week, the primary impact of the oil crisis is the decreasing availability and the increasing cost of gasoline. The commuter might start to look for neighbors with whom to car pool into the city. He or she might become active in an organized effort to have rail service restored to his or her community. For the executive of a large oil company, the oil crisis becomes a time of extraordinary profits, with the price received for domestically produced oil far exceeding the actual cost of extracting, processing, and delivering the gasoline. For years, the Sierra Club has urged more judicious use of natural resources, but has had little impact on federal policy. Hence, a member of the Sierra Club might consider the oil crisis an opportunity to focus public attention on the depletion of natural resources.

Exercise 1: The "Oil Crisis"

a. Name at least two other perspectives for the oil crisis.

b. How does each perspective differ? State each perspective clearly and explain how it follows from the situations of the people having these perspectives.

TIME HORIZON

Over what time period does the viewer perceive changes to be a problem? Suppose a given situation contains periodic fluctuations that the problem solver would like to reduce. In that case, the appropriate time period for studying the behavior is several cycles. In addition to the fluctuations, there may be some undesirable steady increase in a system variable that takes hundreds of cycles before it is significant. The time scale for viewing the problem is then much longer.

EXAMPLE II: THE "OIL CRISIS"—PART II

How does John Scott, the owner of a small, oil delivery service in New Hampshire, view the oil crisis over one heating season? The irregular supply of oil leads to angry customers and a loss of income to John. In turn, his drivers have to be laid off because of fewer deliveries. When a shipment is finally received, John has to decide which customers most need the oil. Over a longer time period, say ten years, there will be a different set of problems. John will have to think about providing alternative fuels, perhaps selling and supplying wood for stoves. He might want to invest in a large storage tank so that he will be less affected by irregular deliveries. Since he offers oil on a fixed-price contract basis to some customers, guaranteeing delivery at a particular price for a fixed time period, he must consider the probable fluctuation of oil costs over the time period of the contract. If shortages are anticipated, a long-term contract may be feasible only if the price is high enough to allow John to purchase oil from a more distant, more expensive, source.

Exercise 2: The "Oil Crisis"—Part II

a. Suppose oil is now $1.20 a gallon and was $0.60 a gallon last year. What price would John Scott quote for oil to be delivered over the next heating season?

b. What further data on past oil prices would John need in order to quote a price for the next three years? (*Note:* Any contracted price represents some risk, even over one heating season.)

REFERENCE MODE

Problem behavior is undesirable behavior over time. This can be either past behavior or conjectured future behavior. A careful statement is needed of exactly which elements of the changing system's patterns are to be isolated as undesirable. Only with that behavior isolated and graphed can specific "remedies" be judged effective or ineffective. As was seen in Part III, a clear statement of this problem pattern is a time graph of each critical aspect. This graph is called a "reference mode" because it is the description of the problem that is used as a reference point by the modeler as he or she tests policies to correct the problem.

EXAMPLE III: EUTROPHICATION OF A LAKE

The literature in limnology (Anderson, 1973) abounds in descriptions of the process by which lakes, through the increasing concentrations of nutrients, become more and more oxygen-depleted, have higher and higher algae populations, and accumulate thicker and thicker layers of algal detritus, eventually becoming bogs, then dry land. While this is a natural process, it is greatly accelerated by a variety of human activities. Eutrophication is of considerable concern to many people, especially those who happen to gain aesthetic or economic benefit from a lake in its present state. Ironically, those who stand to lose most from the eutrophication of a lake are frequently the largest contributors to that process. The most obvious example is the lake-front property owner who fertilizes his or her lawn, producing runoff, which greatly adds to the nutrient load of the lake.

The best way to show the reference mode for this problem situation is by a graph over time of various constituents in a specific lake. In an article entitled, "The Effect of Changes in the Nutrient Income on the Condition of Lake Washington," W. T. Edmondson and John T. Lehman (1981) describe the response of the lake forming the eastern boundary of the city of Seattle, Washington. From 1941 to 1963, Lake Washington received more and more secondary sewage effluent from the growing city. The effect was an increase in the amount of nutrients in the water and in the kind and amount of phytoplankton. Between 1952 and 1959, eleven sewage treatment plants were built, discharging sewage treated by secondary methods into Lake Washington. The net effect was to increase the amount of nutrients and the amount of plant material growing in the lake, and to decrease the transparency of the lake. The data for phosphorus concentration, chlorophyll, and maximum transparency, during the years from 1957 to 1963, are given in Table 8.1.

Exercise 3: Graph of Eutrophication Reference Mode

Construct a graph by setting up scales and plotting the data from the three variables given in Table 8.1.

Table 8.1 Eutrophication data on Lake Washington*

Year	Phosphorus (micrograms/liter)	Chlorophyll (micrograms/liter)	Transparency (meters)
1957	3.1	12.9	3.2
1958	15.9	12.4	2.3
1959	34.5	. . .	2.0
1960	. . .	12.2	2.4
1961	. . .	. . .	1.3
1962	46.1	31.8	1.3
1963	55.3	34.8	1.0

*Adapted from W. T. Edmondson and J. T. Lehman, "The Effect of Changes in the Nutrient Income on the Condition of Lake Washington," *Limnology and Oceanography* 26 (January 1981): 1–29.

These data on the condition of the lake as it is changing in time define the *reference mode.* The problem has now been stated. A model of the system to be used for making plausible arguments about the system's behavior is needed next. The first model is a causal-loop diagram developed to explain the behavior of the reference mode. In the example of Lake Washington eutrophication, the causal-loop diagram must explain the rise in the levels of nutrients and plants. Similarly, any useful computer model of this system must first produce output plots or tables that fit reasonably well the historic data composing the reference mode.

Jay Martin Anderson (1973) has composed a causal-loop diagram from the situation with 10 feedback loops among 10 elements. Figure 8.1 contains the 10 elements ordered in the form of Anderson's final causal-loop diagram.

Definitions of these elements are needed to link them properly.

NUTRIENT

OXYGEN IN AIR
AND EPILIMNION

DECAY   GROWTH

SOLUTION   RESPIRATION

DETRITUS   BIOMASS

OXYGEN IN
HYPOLIMNION

DEATH

Figure 8.1* Eutrophication causal-loop skeleton

---

*Jay Martin Anderson, "The Eutrophication of Lakes," in *Toward Global Equilibrium: Collected Papers* (Cambridge, Mass.: Wright-Allen Press, 1973) p. 123.

Nutrient refers to the levels of phosphorus and nitrogen, which primarily serve as nutrients for the plant population of the lake.

Biomass can be taken to be the mass of all the plants in the lake.

Detritus is the dead plant material.

Epilimnion is the upper layer of the lake.

Hypolimnion is the remainder of the lake. The bulk of the water is in this layer.

Solution is the process of oxygen going into the hypolimnion.

Respiration is the consumption of oxygen and the release of carbon dioxide by the plants. (*Note:* The materials given off and absorbed are just the opposite of those in photosynthesis, which the model ignores —appropriately enough for a polluted lake where light penetrates only a thin layer at the top.) Respiration also includes other metabolic processes here, so that some nutrients are returned to the lake in respiration.

Growth and Death refer to rates of change of plants in the lake.

Exercise 4: Creating Causal Loops

a. Copy the incomplete diagram, Figure 8.1.

b. Using the definitions as given, and your own logic or knowledge, connect the following pairs of elements with causal arrows. Any pair may be connected in either or both directions. Place signs on each arrow.

Detritus	Decay
Nutrient	Growth
Growth	Biomass
Biomass	Death
Biomass	Respiration
Respiration	Oxygen in Hypolimnion
Solution	Oxygen in Hypolimnion
Decay	Oxygen in Hypolimnion
Growth	Oxygen in Air and Epilimnion
Solution	Oxygen in Air and Epilimnion

c. If you have created a loop between the two members of any listed pair, sign the loop. Write a sentence describing how each loop operates.

d. Search for larger loops. Did you have 10 loops?

e. Count the number of positive feedback loops. (There should be 3.)

f. Add to the causal-loop diagram an external source of nutrients added to the lake. Connect this element to the internal element *Nutrient*.

g. Consider the operation of the entire diagram. Of course with this many elements and this many loops, it is difficult to guess how the system will change in time. However, notice that there is no mechanism by which nutrients can leave the lake in the current diagram. You can therefore make reasonable guesses for the changing level of nutrients, of biomass or detritus, and of oxygen in the hypolimnion.

h. Check your guesses against the reference mode plots. Does the causal-loop model seem to fit the reference mode?

Anderson goes on in his analysis to trace the behavior of each loop in the system and argues for quantitative relationships for each arrow. These are then incorporated into a computer model, which is run and compared to the reference mode. The computer model is then expanded to include various possible policies designed to retard the eutrophication of the lake.

POLICY CHOICE

Policy choice is the identification of a particular set of actions as a possible means for altering the problem behavior. Proposed solutions will undoubtedly change as the system behavior is better understood. The journal article on the eutrophication of Lake Washington provides a real example of a policy choice. In 1963, the outflow from the eleven secondary sewage treatment plants was diverted into Puget Sound instead of Lake Washington. The article details the effect of this successful policy on the level of eutrophication of Lake Washington.

While it is not necessary to state a policy when defining a problem, having a policy in mind shapes the construction of the causal-loop diagram. Subsequent policy choices may require extensive redrawing of the causal-loop diagram.

A great deal of patience and care is needed to keep the problem definition clearly in mind as the diagram is built up, torn down, and rebuilt several times. The process of writing down the different dimensions for a problem provides a convenient description of the problem definition to use as a guide in building the causal-loop diagram.

EXAMPLE IV: RAINFALL

A farmer in California is experiencing a steady decrease in the amount of rain over the last few years. This reduced rainfall pattern causes reduction of the farmer's yield from one year to the next, and it appears likely to continue. His yield reduction causes a decrease in his profits, limiting his funds available for investing in farm improvements. The farmer's reference mode is an undesirable dynamic pattern of rainfall, crop yields, profits, and amount invested in farm improvements over several years. As policy choices for examination, the farmer might try out new crops that require less water or he might try seeding clouds. To seed clouds, he would have to hire a pilot to drop silver iodine crystals into clouds as they passed over his land.

Exercise 5: Rainfall: Policy Choices

a. Name another policy the farmer might try in order to solve his problem.

b. Draw a causal-loop diagram of the farmer's problem. Add to the diagram each policy choice, separately.

c. Describe the potential effect of each policy choice on the problem. Based on these discussions, is one policy clearly better than another?

CHAPTER 9

THE NUCLEAR POWER CONTROVERSY

BACKGROUND INFORMATION

A highly controversial subject for the last few years has been the use of nuclear energy to produce electricity. Nuclear power has powerful proponents and determined opponents. Whether or not the nuclear power industry will survive the next ten years is a serious question. The accident at Three Mile Island Reactor #2 near Harrisburg, Pennsylvania, has undermined the confidence of many Americans who are now, for the first time, listening carefully to the arguments against this technology.

The use of nuclear power has many advantages over its alternatives. These include the current cost of the electricity produced, the cleanliness of a plant that is operating normally, the reasonably large reserves of uranium ore, and the enormous available reserves of uranium and thorium ores if the breeder reactor is developed. A less well-known advantage is the relief from the increasing levels of carbon dioxide (CO_2) in the atmosphere resulting from the burning of fossil fuels such as coal and oil. As the levels of CO_2 in the atmosphere increase, the temperature of the earth rises. This process is known as the "greenhouse effect." Just as the windows of a greenhouse allow the visible light from the sun to come into the greenhouse, yet retard the loss of the heat inside, the layer of CO_2 allows visible light from the sun to come through to the earth's surface, but inhibits the passage of infrared radiation from the earth out into space. The Stanford Reseach Institute has predicted that at our present pattern of increasing use of fossil fuels, the greenhouse effect will cause the polar icecaps to melt, leading to the flooding of coastal cities by the year 2000.

The criticisms of nuclear power are directly related to its unique features. Most of the criticisms and fears related to nuclear power stem from the nature of its fuel. Natural uranium contains 3/4 of 1 percent of a vital isotope, U-235. This concentration is too low to be useful in the light water reactors used by U.S. public utilities to produce electricity. With the expenditure of considerable energy in huge federally owned plants, natural uranium can be enriched to raise the concentration of U-235 to 3 to 4 percent, appropriate for fuel in light water reactors. In a reactor, U-235 nuclei absorb neutrons and split (fission) into several fragments, releasing enormous amounts of energy per pound of fuel. This fission process must be controlled very carefully so that the rate of fissioning remains at the desired level, called criticality.

The products of the fission process are highly radioactive and would pose a severe threat to the public if they were not contained in the reactor for many half-lives. The half-life for a radioactive isotope is the time it takes for half of the isotope to undergo some disruptive process, which releases nuclear radiation and results in an isotope of a new element that is usually nonradioactive. For example, the radiation emitted by a sample of cesium-137, after one half-life (30.1 years), has dropped to one half its former intensity. The half-lives of the radioactive products of the fission process in a nuclear reactor vary from several hours for iodine-133, to 30 years for cesium-137 and strontium-90, to 24,000 years for plutonium-239. The problem of protecting the public from the fission products continues, then, far beyond the useful life of the fuel. The time it takes a sample of plutonium-239 to become one-half as dangerous is several times longer than the span of recorded history.

The heat produced from the fission process in the reactor is carried by circulating fluid to the boiler to produce steam. When water boils, it removes heat from the circulating fluid so that the fluid can again remove heat from the reactor core, thereby cooling the core. If a break occurs in a pipe carrying the circulating fluid cooling the reactor core, the accident is called a loss-of-cooling accident (LOCA). This breakdown would lead to a serious disaster if some back-up method of cooling the reactor core were not introduced. In this disaster, most of the fuel in the core would get hot enough to melt through the bottom of the reactor vessel, possibly through the bottom of the reactor building and into the ground below. This process is called a *meltdown*. Since the very hot molten mass is heading for the center of the earth, and, hence, in a sense to the opposite side of the earth, a reactor in a meltdown is sometimes said to exhibit a "China syndrome."

The core of a reactor contains the uranium fuel in vertical fuel rods that are surrounded by water. Neutrons emitted by the fissioning of a U-235 nucleus in one rod are slowed down in the water to where they have a high probability of causing a new fission event by interacting with a U-235 in another rod. As this process continues, fission products accumulate in the rod, many of which absorb neutrons without causing any fission event. As a fuel rod continues to produce neutrons and energy in a reactor core, the percentage of uranium in the form of U-235 drops to about 1 percent. The fuel rod must then be removed, since it absorbs too many neutrons for the number it generates. Such an expended fuel rod contains the long-lived radioactive products from its three years in the reactor core. The expended rods are stored in large pools of water in nuclear power plants where they must be cooled and kept apart to avoid any further fissioning. Eventually, these rods must be shipped elsewhere for reprocessing or long-term storage or disposal. The United States has not yet decided how to handle expended fuel rods. Nuclear power plants are currently having to expand their pool storage capacity, pending that decision.

As just described, the nuclear power "problem" is hardly defined as a problem at all. Partial lists of each of the four critical dimensions in problem definition are included next. To confirm an understanding of each dimension, the companion exercises request additional items to each list.

PERSPECTIVES

The first dimension is perspective. Here is a list of perspectives from which nuclear power might be viewed:

- An environmentalist

- A nuclear power plant builder

- An insurance executive asked to insure a plant

- A Navajo Indian working in a uranium mine

- A homeowner paying electric bills

- A manager of a nuclear power plant

- A member of the Nuclear Regulatory Commission

- An owner of a uranium mine

- A nuclear power plant control-room operator

Exercise 1: More Perspectives

Add at least two more perspectives to the preceding list by giving a role in society and a brief description of how nuclear power is viewed from that perspective.

TIME HORIZON

The pertinent span of time depends on which changing aspect is of interest. Some example time horizons include:

5 hours	Time after a loss-of-cooling accident during which a flow of cooling water must be maintained
2 days	Half-life of some dangerous gaseous isotopes

3 years	Time a fuel rod spends in the reactor core
10 years	Period during which one fourth of the U.S. uranium reserves will be exhausted, if present trends continue
30 years	Half-life of strontium-90 and cesium-137, major fission products
24,000 years	Half-life of plutonium-239

Exercise 2: Time Horizon

Add one or two time horizons and explain how they are pertinent to some aspect of nuclear power. At the same time, take one of the perspectives listed previously and indicate the time horizon pertinent to it.

PROBLEM BEHAVIOR

Among the many behaviors related to nuclear energy that might be considered "problems" from some viewpoint are: (1) accumulating expended fuel rods in nuclear power plants, (2) depletion of uranium reserves, and (3) behavior of a nuclear power plant after a loss-of-cooling accident.

Exercise 3: Other Undesirable Behaviors

Name a few more dynamic behaviors that could be considered problems. For each case, state the viewpoint and explain why the behavior is seen as undesirable from that viewpoint. Try to make that behavior more concrete by a rough graph of the values of important elements over the appropriate time horizon.

POLICY CHOICES

In general, for each unique combination of perspective, time horizon, and reference mode, a policy maker has several reasonable options. Performing the causal-loop analysis may help choose among these options. If not, the policy maker may wish to develop the model further into a computer model.

Here are a few policy options pertinent to nuclear power:

- Develop the breeder reactor.

- Call a moratorium on the operation of nuclear power plants.

- Call a moratorium on the licensing of any new plants.

- Establish a federal monopoly on nuclear power.

- Build all new plants in remote clusters ("power parks").

- Store spent fuel rods above ground.

- Develop solar power as rapidly as possible.

- Mount a public relations campaign.

Exercise 4: Policy Choices

Take one of the listed perspectives of a concerned person and write down a policy that such a person might adopt to deal with the nuclear energy "problem" as he or she sees it.

BUILDING A CAUSAL-LOOP DIAGRAM
OF THE NUCLEAR POWER CONTROVERSY

Consider the view of a long-range planner of a company that is contracted by public utilities to construct nuclear power plants. Looking back over the last few decades and ahead to the next few, the planner sees an eventual problem in the depletion of uranium reserves, despite the present lack of interest in nuclear power in the United States. Long-term growth of the nuclear power industry depends on continuing supplies of fuel. Since in our current technology this fuel must be made from uranium ore, the industry would eventually have to slow down if the ore ran out. One possible long-term solution to this problem is the breeder reactor. As the name implies, a breeder reactor actually breeds fuel in its core; that is, as the reactor operates, more fuel is generated than is used up. The "fuel" referred to here is any isotope that is easily fissioned, such as uranium-235. Such isotopes are said to be fissile. The French have a prototype breeder reactor in operation (Phoenix), and are building a full-scale breeder (Superphoenix). The U.S. breeder program is centered in the controversial Clinch River Breeder Reactor project in Oak Ridge, Tennessee. If and when the breeder reactor is developed, it will be able to supply nuclear fuel from unenriched uranium ore, thereby extending resources manyfold.

After receiving this information the company's planner decides to join a group lobbying for the rapid development of breeder reactors. Here are the four dimensions for a problem definition:

1. Perspective: Long-range planner of a company that constructs nuclear plants

2. Time frame: Fifty years

3. Problem behavior: Possible eventual depletion of uranium reserves

4. Policy choice: Lobby for development of breeders

This problem creates difficulties in diagram development because of its inherent complexity. Start with just the elements affecting supply and demand of nuclear fuel. Taking into account the planner's proposed policy of eventually developing a new way of producing fissile stocks (that is, the breeder), the diagram will use "fissile stocks" as including both uranium ore that has been mined and processed and fissile material "bred" in a breeder reactor. (See Figure 9.1. Compare this diagram with that developed for the meat industry, Figure 3.24, in Chapter 3.)

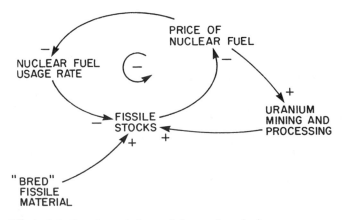

Figure 9.1 Supply and demand for nuclear fuel

Here, availability of fissile stocks (the inventory of nuclear fuel) is seen as the principal influence on fuel price. Usage rate is implicitly formulated as affected by price. The more direct representation of fuel usage by nuclear power plants will be developed later.

Exercise 5: Nuclear Fuel Supply and Demand

Temporarily assume that no change occurs in the overall demand for energy, and that a breeder reactor does not yet exist. Under these conditions, focusing on the contents of Figure 9.1, what type of dynamic behavior would you expect from this system? Over what time horizon?

Is this system consistent with the problem as defined by the long-range planner?

The basic behavior of a system's key variables over time is called the "reference mode" of the system. The reference mode serves as a basis for comparison with later behavior over time produced by policy changes. Look back at the list of perspectives given earlier in this chapter. Who on that list would be concerned with the situation generated by this system?

Broadening the system boundaries to include uranium ore reserves will extend the time horizon considerably. Uranium reserves include all ore in the ground that has been discovered or "proven." Mining depletes those reserves and exploration increases them. Reserves have a secondary influence on nuclear fuel price, as the reserves suggest the longer-term availability of fissile stocks.

Exercise 6: Uranium Reserves and Exploration

a. Add "uranium reserves" and "uranium exploration rate" to Figure 9.1, signing their connections with the rest of the diagram, and signing any newly formed loops. How should nuclear fuel price relate to uranium exploration rate in this causal diagram?

b. Describe the dynamic behavior this expanded system might exhibit, with a graph if possible. Before making such a graph, a question that must be answered is: "Over what time horizon?" The answer sets the time scale for the graph. Is the "reference mode" of this system consistent with the problem as defined by the long-range planner?

c. Whose perspective might be reflected in the system structure as now defined?

Beyond the issue of proven reserves is the potential exhaustion of all uranium ore in the ground, whether already proven or as yet undiscovered. That total amount is finite, at least within the defined time horizon of fifty years. As new discoveries reduce the uranium ore left to be discovered, the cost of exploration activities increase, directly discouraging further exploration but also contributing, to some extent, to further increases in the price of nuclear fuel.

Exercise 7: Undiscovered Ore and the Cost of Exploration

Add "Undiscovered Ore in Ground" and "Cost of Exploration" to the diagram developed in Exercise 6. Sign their connections to the rest of the diagram and sign any newly formed loops. Is it possible to predict which loops will be dominant over which time periods? Using a graph like Figure 9.2, sketch a fifty-year rough graph of "Nuclear Fuel Usage Rate" and "Undiscovered Ore in the Ground", a best guess at the reference mode of the system as now bounded. Assume that demand for energy grows over the period of interest.

Exercises 5, 6, and 7 provide a good opportunity to see the relationships between system boundaries, system dynamic behavior, and problem definition. At each expansion of the boundaries of a closed system to include new considerations, the focus of the problem definition changes, often with clear implications for the time horizon of relevance. The modeler's perspective on "the nuclear power problem" may cause her or him to define a problem situation quite differently from another modeler's problem. Each of the three system structures developed in the preceding exercises is "correct" for some problem.

The exercises also illustrate the increasing difficulty of anticipating system behavior from a causal-loop diagram as the system complexity grows. Clearly, by Exercise 7, confident analysis of system behavior and the effects of policy changes would require a computer model.

Further demonstration of the level of complexity encountered in real-world problems is provided in Exercises 8 through 10. Here, continuing with the perspective of a long-range planner of a company that constructs nuclear power plants, the necessary additional elements are added to produce an ap-

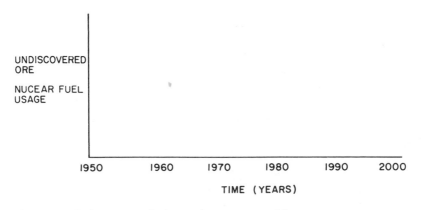

Figure 9.2 Reference mode for nuclear power problem

propriately bounded system for reference mode and policy analysis. It will not be difficult to build up the causal-loop diagram to take these new aspects into account. Unfortunately, the level of system complexity reached will make it extremely difficult to anticipate other than general patterns of likely behavior over time.

If the basis for the reference mode is an assumed growth of the nuclear power industry, rather than its current depressed state, the diagram should also include an exogenous element, "Rising Demand for Energy." In the absence of price increases and other factors such as environmental concerns, this demand would lead to a steadily rising number of nuclear power plants. The number of nuclear power plants in operation would grow exponentially, having about twice as many plants every eight years. As the number of operating nuclear power plants increases, the nuclear fuel usage rate increases.

Exercise 8: Changing the Perspective

a. Add the new elements "Rising Demand for Energy" and the number of nuclear power "Plants in Operation" to the diagram of Exercise 7.

b. How does this exogenous input operate? Describe a rough graph of the reference mode from 1950 to 2000 A.D.

The elements related to the perspective of the long-range planner now need to be added to the diagram. His company is most directly influenced by the orders for new power plants. As utilities decide whether to build new coal or new nuclear power plants, a crucial element is the price of nuclear fuel. There is no need to include the utilities directly in the diagram, but it does now seem reasonable to include the assumption that they place fewer orders for nuclear power plants when the price of nuclear fuel rises.

The planner must keep in mind three phases in the development of the nuclear power industry: the orders for new plants, the plants being assembled, and the actual number of operating nuclear power plants. The planner is only directly concerned with the first two stages, but the third stage determines the nuclear fuel usage rate: Fuel usage is proportionate to the number of nuclear power plants operating.

The three new elements, then, are "Orders for New Plants," "Plants in Assembly," and "Plants in Operation." Recognizing that in the diagram new nuclear power plants can be made, but there is no way of retiring them from service, the element "Decommissioning" could be added. Since the average lifetime for a nuclear power plant is forty years, the majority of those built in this century will still be operating in 2000 A.D. Hence, within the time horizon as now defined, the number that will be decommissioned will be small. For the sake of simplicity, "Decommissioning" can thus be omitted from the causal-loop diagram.

Exercise 9: Expanding the Diagram Still Further

a. Connect the three new elements to the old diagram, and give signs to arrows and any new loop symbols. (Note that the rising demand for energy affects the number of orders for new plants most directly, but that nuclear fuel price has an effect, too.)

b. Does the expansion of the diagram to include these connections to the nuclear power plant supplier change the behavior of that system over the fifty-year span?

The planner's proposed policy must now be taken into account in the diagram. Three new elements are required: "Lobbying Activities for Breeder," "Breeder Development Efforts," and "Number of Breeders" actually in commercial operation. Before adding the new elements, consider the time involved between successful lobbying and the existence of commercial breeders. A delay of about twenty years is probable in the development process, including such stages as the construction of small-scale demonstration plants, perfection of various aspects, and careful review of all safety aspects. As a reminder of these various steps, it helps to insert the word *delay* on one arrow to emphasize this significant lapse in time between the decision to go ahead in developing the breeder and the production of fissile material by commerical breeder reactors.

Exercise 10: Completing the Diagram and the Analysis

a. Complete the diagram by the addition of the new elements with signed arrows and any missing loop symbols. Decide whether "Orders" or "Plants in Assembly" or some other element has the most direct influence on the lobbying activity.

b. Now guess how the proposed policy works. Without the use of a computer to handle all the relationships quantitatively, one must guess how the future behavior will differ from the reference mode behavior. Plot on a graph an estimate of the reference mode (no breeder development before 2000), including curves for both undiscovered ore and price of nuclear fuel.

c. Plot how the price of nuclear fuel and the level of undiscovered ore will change if the decision to develop the breeder is made in 1985. Assume the delay time between the decision to develop and the first commercial breeder is twenty years. Extend the time horizon to 2020 to give the breeder policy a longer chance to have an effect.

ALTERNATIVE ISSUE: BUILD-UP OF RADIOACTIVE WASTE

Now consider a different question, though one clearly related to the prospects for growth of the nuclear power industry. Due to the accumulation of spent or expended fuel rods at nuclear power plants across the nation, the federal government is under increasing pressure to develop a safe way to dispose of radioactive waste. The proponents of nuclear power have claimed from the beginning that the disposal problem was of minor importance. However, both the nuclear power industry and the sponsoring federal agencies, first the Atomic Energy Commission (AEC) and now its successor, the Department of Energy (DOE), have as yet failed to resolve the problem.

Spent fuel rods contain plutonium-239, a fissile isotope that would be useful as a reactor fuel if it could be removed from the spent rod without too much expense. Several attempts at commercial reprocessing of nuclear fuel have failed. One of these, the Nuclear Fuel Services Plant at West Valley, New York, left the state of New York with millions of gallons of highly radioactive waste solution.

The federal government has been reluctant to develop radioactive disposal facilities because the absence of commercial reprocessing meant that the spent fuel rods would have to be disposed of, thereby destroying a potentially valuable resource. At this time, the federal government is being urged to adopt the policy of burying spent fuel rods in salt mines cut into what are known to be very stable formations. This is the safest disposal method now known.

Exercise 11: Problem Dimensions

a. List four possible dimensions of this problem:

Perspective:
Time frame: (200 years perhaps?)
Problem behavior:
Policy choice:

b. To simplify the problem somewhat, assume that no breeders are to be developed for 200 years. Among the various ways that the accumulation of spent rods in nuclear plants might affect the growth of the industry, assume that "public reaction" is the most important.

The situation, as given, assumes that to bury spent rods would be to destroy a source of fissile material, which might become important when the uranium ore begins to run out. Assume the reference mode has the continued accumulation of spent rods above ground, despite their threat to human health, and that in 2050 a way is developed to reprocess spent fuel rods economically.

ALTERNATIVE PROBLEM DEFINITION: URANIUM MINE OWNER

Now consider the owner of a uranium mine who is concerned about the increasingly negative attitude of the public toward nuclear power. This is a very clear threat to her business and may force her into bankruptcy in the next five years. She launches a campaign in her state to convince the public that nuclear power is clean, safe, and less expensive than alternative power sources.

A clear definition of the perspective and the other three dimensions of the problem can now be stated:

- Perspective: Owner of uranium mine

- Time horizon: Five years

- Problem behavior: Possible moratorium on nuclear power

- Policy choice: Launch a publicity campaign in support of nuclear power

The stage is set for a diagram-based analysis of how the number of nuclear power plants affects the uranium mine profits (Figure 9.3).

Exercise 12: Public Fear

a. Copy Figure 9.3, adding signs to the causal arrows.

b. What would cause a moratorium on new nuclear plants? How about public fear (of an accident)? What is the source of this public fear? Two possible sources for the fear are actual accidents, such as that at Three Mile Island, and increased public awareness of possible dangers of radioactivity. As the number of nuclear power plants is increased, the public fear coming from these two sources will increase, reducing the building of nuclear power plants. This closes a feedback loop. Expand the diagram to include these considerations. There is no need to list the two sources of public fear since they both elaborate a direct link between the number of nuclear power plants and the public fear. There is now a three-element loop in the feedback diagram.

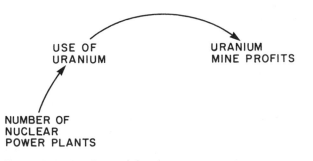

USE OF
URANIUM

URANIUM
MINE PROFITS

NUMBER OF
NUCLEAR
POWER PLANTS

Figure 9.3 Uranium mining

Exercise 13: Public Relations Campaign

a. On a copy of Figure 9.4, complete the loop and place signs on both arrows and the loop symbol. To study the effect of the proposed policy of a public relations campaign, the only new element needed is "Public Relations Campaign."

b. Add the new element to the diagram and connect it into the diagram with two arrows.

c. Place signs on the new arrows and the new loop.

d. Write a few sentences describing how the new loop affects the operation of the old loop.

e. Use the diagram to explain how the owner's policy will influence her profits.

From this analysis, it is not clear whether the campaign is cost-effective without quantitative data on the cost of the campaign, its effectiveness in altering public fear, and the effect of the reduction of public fear on the actual building rate for new plants.

Numerous different "nuclear power problems" have been described in this chapter. Each can be seen from one or more perspectives. In each situation, the determination of appropriate system boundaries depends on the problem defined over the relevant time horizon. Proper boundaries also take into account the likely policy approach to be adopted. Careful, focused thought is needed to develop each causally oriented analysis.

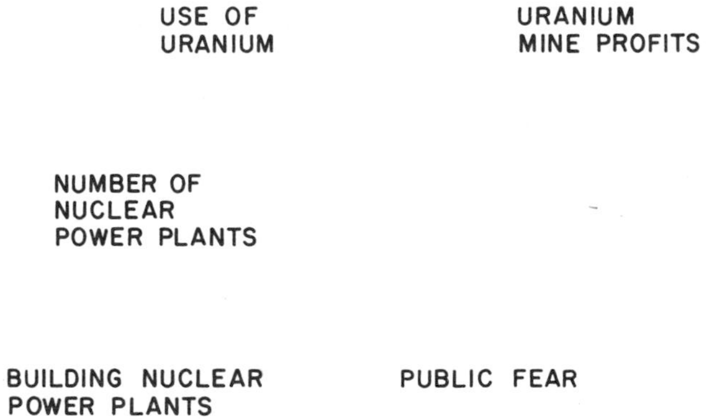

USE OF
URANIUM

URANIUM
MINE PROFITS

NUMBER OF
NUCLEAR
POWER PLANTS

BUILDING NUCLEAR
POWER PLANTS

PUBLIC FEAR

Figure 9.4 Expanded diagram of uranium mining

CHAPTER 10

THE DILEMMA OF SOLID-WASTE DISPOSAL

BACKGROUND INFORMATION

The following information represents the solid waste problem, as described by Randers and Meadows[1] in 1967 (Meadows and Meadows, 1973, pp. 165–212.) Since that time, the problem has grown along with population and industrial growth, despite the growing number of controls such as state "bottle bills." Regulations developed by the Environmental Protection Agency make the disposal of some industrial wastes, especially chemical, more expensive than before, and so may lead to increased recycling of waste materials. This chapter explores the solid waste problem and some proposed solutions.

Every person in the United States yearly discards 180 pounds of paper, 250 metal cans, 133 bottles and jars, and 388 caps and crowns. Seven million cars and 100 million tires are discarded in the United States each year. Of the 3,360 million tons of solid waste generated in the nation in 1967, 360 were urban waste, 100 mineral waste, and 2,000 were animal and agricultural waste. Urban waste now amounts to about 700 million tons per year, of which about 82 percent is hauled to dump sites. If this material were collected in one place and spread six feet deep, it would cover 80,000 acres. The nation's cities are overwhelmed by the solid refuse problem. Houston, whose garbage dump is now the highest point on the Texas coastal plain, has already passed the saturation point.

At the same time, we are running out of invaluable natural resources. The Council on Environmental Quality stated in its first annual report: "Even taking into account such economic factors as increased prices with decreasing availability, it would appear at present that the quantities of platinum, gold, zinc and lead are not sufficient to meet demands."

From 1870 to 1965, the consumption of metals in the United States increased by a factor of 9.6, corresponding to a doubling time of twenty-nine years, or a 2.5-percent annual increase in consumption. Assuming a continuing growth rate of 2.5 percent, the number of years the reserves of various metals would last has been calculated to be:[2]

Metal	Years
Chromium	88
Iron	68
Nickel	52
Copper	16
Lead	10
Zinc	8

The twin goals of slowing down both the generation of solid waste and the depletion of natural resources can best be satisfied by reducing the flow of materials from natural resources to solid waste. Two things might be reduced: the waste produced when each product is discarded, "Waste per Product"; and the number of products in use, "Products." The lifetime of the products, "Lifetime," might also be increased.[3] Reducing the number of products, however, implies a decrease in the standard of living, which might make such a policy unsatisfactory.

PERSPECTIVES

Some of the viewpoints on this situation are:

- A Detroit auto designer

- A manager of an industrial waste center

- A member of a city council responsible for urban solid waste disposal

- A volunteer running a local recycling center

- A manufacturer of pop bottles

Exercise 1: Perspectives on Solid-Waste Generation

a. State at least two perspectives on solid-waste generation that are different from those just listed. For each case, write a sentence explaining how that "concerned person" will view solid-waste generation.

b. How is the problem structure affected by the new perspectives? That is, are some aspects of the situation now more important than others? Would you expect different elements included in the causal-loop diagram for each case?

TIME FRAMES

The lifetimes of various products and the world reserve indices—the times that reserves are expected to last—are the main time variables for solid-waste generation. For example:

Time	Significance
2 years	Life of an automobile tire
5 years	Life of a car
10 years	Reserve index for lead

Exercise 2: Time Horizon

a. Name a familiar product and give an estimate of the average length of service for that product.

b. Take one of the metals listed earlier and explain what the reserve index means for that resource. Write an equation for the reserve index in terms of U, the amount used each year, and R, the present reserves for that resource.

PROBLEM BEHAVIORS

While the principal undesirable dynamic behaviors are quite explicit—the depletion of natural resources and the growth of garbage piles—there are some other pertinent problem behaviors.

Exercise 3: Problem Behavior

a. What would a bottle manufacturer consider a problem in the area of recycling?

b. What problems would a refrigerator manufacturer see in a bill to tax manufacturers so as to reward production of more durable items?

POLICY CHOICES

Some options that might reduce solid-waste generation are:

- Remove depletion allowances in mining industries. (Depletion allowances are tax deductions allowed to investors in exhaustible mineral deposits for the depletion of the deposits.)

- Remove tax deductions for cost of exploration for minerals.

- Make freight rates as low for scrap as for virgin material.

- Remove federal government stipulations that prohibit use of anything but virgin material.

- Make people sort their own wastes in their homes.

- Prohibit nonreturnable containers.

- Reduce packaging.
- Impose a high tax on all but the first car owned by a family.

Exercise 4: Policy Choices

a. Take any policy suggested above. How would it affect either solid-waste generation or natural resource depletion?

b. Note next to each proposed policy the appropriate variable or combination of variables that might be affected by the policy, putting next to each policy w (the mass per product), P (the number of products), or L (the average lifetime of a product or of the material used in the product).

Exercise 5: Defining the Problem

a. From the previous list of perspectives, choose one viewpoint and an appropriate possible policy choice for that concerned person.

b. Decide on the problem behavior and the time horizon to fit the viewpoint and policy. Develop a reference mode. Now the problem is defined.

c. Draw a causal-loop diagram of the problem.

d. Estimate how the loops in the diagram will make the system behave.

e. If the patterns of change in time predicted in (d) do not correspond to the reference mode in (b), redraw the causal-loop diagram and study the problem more carefully. (Perform this last exercise several times for other problems in the area of solid-waste generation, choosing a different perspective and time horizon each time.)

EXAMPLE I: SECRETARY OF THE INTERIOR'S VIEWPOINT

Consider the point of view of the Secretary of the Interior of the federal government toward a bill that would encourage homeowners to sort their own wastes at home. Under this law, the homeowner would have separate waste containers for garbage, paper items, tin cans, glass bottles, and plastics.

Since the bill would operate for five years on an experimental basis, five years is an appropriate time scale. The bill would encourage recycling since the separated waste can be recycled more cheaply. Hence, the problem behavior leading to the bill is the combination of resource depletion and solid-waste accumulation. Now that all four dimensions of the problem have been defined, causal-loop boundary analysis can be developed.

Development of the causal-loop diagram begins with the flows of raw, finished, waste, and recycled materials. Product demand is satisfied by two sources: production from virgin raw materials and production from recycled materials. Only the "Virgin Production" depletes "Natural Resources." Both virgin material and recycled material are used in part in producing "Disposables," which are the sources of the "Solid Waste" materials and items that can, in part, be recycled. Figure 10.1 shows a causal-loop diagram at this stage:

Exercise 6: Interpreting the Causal-Loop Diagram

a. Describe the generation of solid waste in the absence of any recycling.

b. Given present and foreseeable technology (as well as social behavior), the percent of solid waste that will be recycled is less than 100 percent. Under these conditions and constant product demand, describe the dynamics of Figure 10.1. What is the behavior of the positive feedback loop?[4]

c. The "Cost of Resources" is affected by the availability of "Natural Resources," and the "Cost of Recycling" is influenced by "Home Sorting." Add these elements to the diagram at appropriate places, recognizing that the percent of waste that is recycled is strongly influenced by the relative costs of natural versus recycled materials. Connect the new ele-

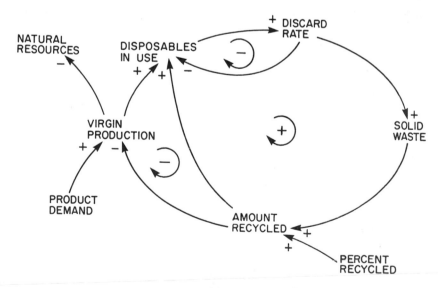

Figure 10.1 Materials flows in production and recycling

ments to the rest of the diagram with signed arrows, also signing any new feedback loops. Comment on the likely dynamic behavior of any newly formed loops.

d. Describe the change in generation of solid waste due to the new policy of home sorting.

e. Now suppose that product prices reflect their materials costs (both natural and recycled), and that product demand tends to fall as price rises. Add the element "Price of Product" to the diagram, connecting it as appropriate with signed arrows, also signing any new feedback loops. Comment on the likely dynamic behavior of any newly formed loops.

ENDNOTES

1. This article presents one set of views on the overall process of solid-waste generation. Randers and Meadows develop a rather complex causal-loop diagram by adding more and more small loops, and present a computer simulation based on that diagram.

2. Based on assumptions in D. H. Meadows et al. (1972, p. 56). These assumptions are not uniformly accepted.

3. A simple equation expresses the solid-waste generation rate, s. If P products are in use, each with an average lifetime of L years, the number of products discarded per year is P/L. If w kilograms of solid waste are produced when each product is discarded, the total mass of solid waste being generated per year is $s = (wP)/L$.

   Since the goal is to reduce the mass of solid waste being generated each year, any suggested policy must work to make s smaller. s becomes smaller if the mass per product (w) is reduced, or the number of products (P) is reduced, or the lifetime per product (L) is increased.

4. With a loop gain < 1, a positive feedback loop generates saturating rather than explosive behavior.

CHAPTER 11

FAMILY DYNAMICS FROM A SYSTEM PERSPECTIVE

Most people live within a web of social relationships called a family. This web controls much of their lives. Through their active influence in the web, they exercise considerable control over their lives. This is a feedback system with which the reader is intimately acquainted.

This chapter analyzes some familiar situations with causal-loop analysis. This area of analysis has immediate application in exploring how effectively each member of a living group deals with the problems he or she perceives.

PERSPECTIVES

Each family member has a unique perspective on the family situation, being intimately involved in the shifting relationships between members of the family. In addition, others have a real interest in and, hence, a perspective on a family, such as:

- A friend of any member of the family

- A minister, priest, rabbi, or family counselor

- An employer of any member of the family

- An insurance agent insuring a family member

- A sociologist

TIME HORIZONS

The period over which dynamic problems develop and persist varies from a childish tantrum of a few minutes to a long-standing jealously between siblings lasting for decades.

PROBLEM BEHAVIORS

In addition to those mentioned already, there are:

- Infant care strain

- Parental difficulty in adjusting to the changing needs of maturing children

- Care of incapacitated family members

- Romantic involvement of any member with a nonmember

- Financial strain associated with too much or too little money

- Complications of both adult members working

- Sharing responsibilities of running a house

POLICY CHOICES

This list will include hasty responses by family members that hardly deserve the label "policy" with its implications of due deliberation. In any case, they are surely not based on the kind of careful problem definition developed here! With that warning, here are a few familiar possibilities:

- Running away

- Screaming

- Physical abuse

- Presenting a desirable distraction

- Calling a family conference

- Breaking chairs and dishes

- Scrubbing floors

- Retreating with snacks to one's own room

EXAMPLE I: SIBLING INTERACTION

Mark and Eileen are twelve and eight years old, respectively, and frequently engage in skirmishes, which their parents terminate as quickly as they can. A common pattern in their relationship is Eileen becoming bored and picking on Mark to get some kind of excitement. Mark usually hits her back and so ends up being punished for what was originally Eileen's act of aggression. This string of events can be represented in the sequence diagram shown in Figure 11.1.

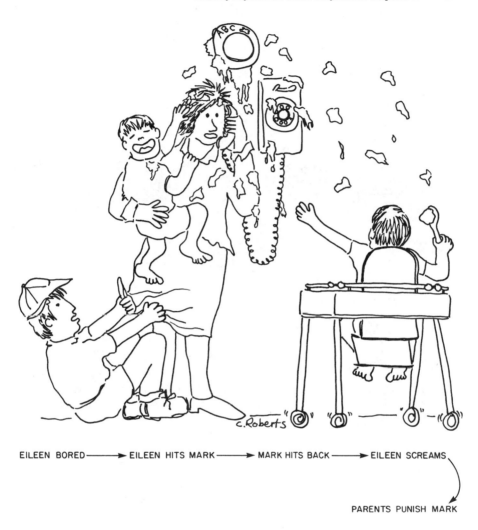

EILEEN BORED ⟶ EILEEN HITS MARK ⟶ MARK HITS BACK ⟶ EILEEN SCREAMS

PARENTS PUNISH MARK

Figure 11.1 Sibling interaction

In translating these observed events into a causal-loop diagram, two different sets of feedback phenomena can be identified. The first is the aggressive interaction between Eileen and Mark, with commonplace positive feedback escalation. The second is the negative feedback intervention by their parents, initially aimed at controlling Mark's aggressiveness. The two loops are pictured in Figure 11.2.

The loops suggest a dynamic scenario of Eileen's boredom initiating a spiraling growth of sibling aggression, leading to Eileen's complaints, which generate parental punishment of Mark, eliminating his aggression and ending the positive feedback loop escalation. Renewed boredom by Eileen can retrigger the same system dynamics.

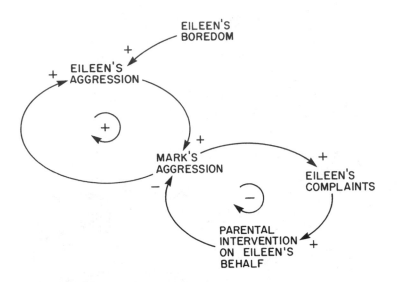

Figure 11.2 Sibling aggression and parental response

In Exercise 1 that follows, the perspective is that of Mark, who does not like this cycle. The time horizon is a few minutes for one cycle, perhaps months for the pattern to persist. The problem pattern is clear from the diagram. The policy Mark chooses to change this undesirable pattern is to complain to his parents as soon as Eileen screams.

Exercise 1: Mark's Current Strategy

Add new elements to Figure 11.2 corresponding to Mark's current strategy. Eileen's complaints now influence Mark's complaints. It is reasonable to assume that when Mark complains, the parents punish Eileen.

What is the new loop structure? Will Mark be happier as a result of his policy?

Exercise 2: Mark's Successful Policy

Mark eventually works out another policy. Instead of hitting Eileen when she hits him, Mark complains to his parents immediately. What does the feedback system become under this policy?

It seems that Eileen might now think twice before hitting Mark. Draw the causal-loop diagram for the situation with Mark's successful policy.

EXAMPLE II: SECOND INCOME

Carl and Marilyn are parents of three children. Carl is now the only breadwinner and he and his wife are unhappy with his income. They are dissatisfied with the quality of food, housing, transportation, and entertainment they can afford on Carl's income. Marilyn is considering looking for a job. Carl and Marilyn have talked this over and agreed that her job would affect the quality of their meals, home cleanliness, and child care, as well as their income. The four dimensions defining the problem are:

- Perspective: Carl's

- Time horizon: Four months

- Problem behavior: Inadequate family income

- Policy choice: Marilyn getting a job

The policy seems to solve directly the problem as initially defined. (See Figure 11.3.) If Marilyn works, their income will rise, reducing the discrepancy and, hence, the dissatisfaction. Note the use of the term "discrepancy." To be clear, it needs to be stated that discrepancy is defined here as (ideal income) − (actual income). This type of comparative element is frequently useful in causal-loop diagrams and later in equation writing. Some modelers use the word "*gap*" instead of discrepancy for the difference between the goal (ideal income) and the actual level (income).

Including the effects of Marilyn's working on the quality of meals, housekeeping, and child care, and Carl's share in those tasks, leads to a more complex causal-loop diagram, shown in Figure 11.4. With the inclusion of the new elements within the boundary, the impact of the proposed policy can be seen more clearly. Carl needs to analyze the situation carefully. Will Carl's net dissatisfaction level be higher or lower after Marilyn starts working? Only Carl can answer that. The feedback diagram makes the trade-off clear.

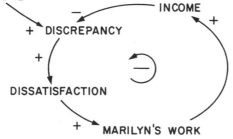

Figure 11.3 Carl and Marilyn, I

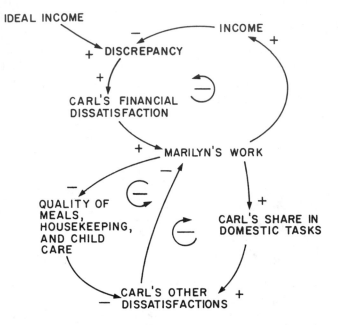

Figure 11.4 Carl and Marilyn, II

Exercise 3: Carl's Dissatisfaction

a. Do the two lower loops of Figure 11.4 work to increase or decrease Carl's net dissatisfaction if Marilyn works? Explain.

b. Does the upper loop of Figure 11.4 operate to increase or decrease Carl's net dissatisfaction when Marilyn works? Explain.

When the description of the same situation as the last example is changed slightly, the problem changes greatly. Suppose Marilyn is bored at home and is looking for an interesting job.

Exercise 4: Marilyn's Perspective

a. List the four dimensions of the new problem. Take the perspective to be Marilyn's.

b. Draw a causal-loop diagram with the elements "Boredom," Time Working," and "Time Marilyn Spends at Home." (Assume she gets an interesting job.)

c. How many loops are there?

d. Describe how the policy works or does not work in changing Marilyn's level of boredom.

e. Expand the causal-loop diagram to include the elements "Marilyn's Dissatisfaction with Income," "Income," and "Discrepancy."

f. Describe the working of the enlarged causal-loop diagram.

g. Suppose the job is not interesting. Mark this change in the diagram and reanalyze the behavior of the feedback diagram. How is it different from your description in Exercise 4(f)?

EXAMPLE III: A COMMUNE

Roy is a very pleasant, charming member of a small commune where each person is assigned tasks every day. Roy frequently forgets to do his tasks and has been able to get away with this behavior since his peers are placated by his charming banter. However, his peers have slowly built up resentment over several weeks. They show this resentment by avoiding Roy when possible. Roy is hurt by this rejection, but takes comments about his work pattern as a joke, as he always has.

Roy asks Mildred, a close friend, to help him to understand his predicament. She recognizes that Roy has the tendency to feel less responsible the more he feels accepted by the group. Having read these chapters, Mildred produced a causal-loop diagram for Roy and, through it, helps him formulate a policy to attack his problem.

Based on Mildred's analysis, Roy decides to keep a detailed diary of his own work habits and to let others know he is doing so.

- Perspective: Roy's

- Time horizon: several weeks

- Problem behavior: Roy's rejection by fellow members

- Policy choice: A "public" diary

Mildred's initial causal-loop diagram (without Roy's policy choice) is shown in Figure 11.5.

Exercise 5: Diagram for a Commune

a. Draw arrows with signs to show causal relationships.

b. Identify any loops by a signed loop symbol.

You may have drawn an arrow from Roy's task performance to acceptance of Roy. While this is appropriate, it is a weak link compared with the link through the group's perception of Roy's carrying his share.

ROY'S TASK
PERFORMANCE

ROY'S SENSE
OF RESPONSIBILITY

ROY'S SENSE
OF BELONGING

ROY'S CUMULATIVE
PERFORMANCE

ACCEPTANCE OF
ROY

DELAY

PERCEPTION OF
ROY'S CARRYING
HIS SHARE

ROY'S CHARM

Figure 11.5 Roy's performance

The positive loop involving Roy's charm can be properly limited by a negative loop involving a comparison of Roy's charm with the maximum charm anyone could have. As Roy's charm increases through the positive loop, it approaches a limit; the closer it gets, the slower it increases. This element might be called the comparison element "Charm Discrepancy," which would be how far Roy's charm falls below the maximum. The negative loop has the form shown in Figure 11.6.

Roy maintains his charm as long as he continues to be accepted. As his cumulative performance builds up a poor perception of his carrying his share, his acceptance actually drops enough to push the (+) charm cycle into a positive downward spiral, leading to his being dejected and rejected. Roy understands his problem behavior as analyzed by Mildred, including the (−) link from acceptance to his own sense of responsibility. He decides to work on building up his share-carrying reputation by demonstrating in a convincing way his increased sense of responsibility. By openly recording in his diary when he is supposed to do his tasks and writing next to each entry the actual time he started working on each task, Roy can actually improve his own sense

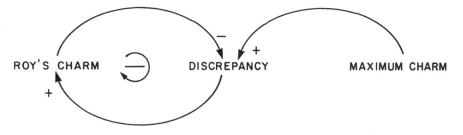

Figure 11.6 Limit on Roy's charm

of responsibility and improve more rapidly the community's perception of his carrying his share.

Exercise 6: Expanding the Diagram

a. Add to the original causal-loop diagram the new element "Use of Diary of Performance" and link it appropriately into the diagram.

b. How many new loops are there now?

c. Describe the way each of those loops operates.

d. How likely is Roy's strategy to succeed? Explain.

Exercise 7: Fraternal Interaction

George and Frank Parker have planned for weeks to drive two friends to Paragon Park for most of a Saturday. When they ask their mother on the preceding Saturday for the use of her car for their trip, she agrees on the condition that George clean the garage and Frank wash the car—before the trip. Frank washes the car on Thursday, but George puts off the garage job until Friday, when he has to practice basketball. When Frank asks his mother for the keys on Saturday morning, she checks the cleanliness of the car and the garage and maintains her condition that both tasks be complete before the trip.

Frank is angry since he had done his job. At the same time, Frank is embarrassed by not being able to leave when he had promised. George apologizes for holding them up and asks for Frank's help in cleaning the garage. Frank knows he will be even more angry if he has to help George, especially since this same type of situation has occurred before.

a. Describe this repeated situation from Frank's point of view.

b. List the dimensions of perspective, time horizon, and problem behavior for Frank.

c. Copy and add arrow and loop signs to Figure 11.7, the diagram of Frank's dilemma.

d. Suppose Frank sees a way out of the dilemma by asking George to pay him for the time he spends cleaning the garage. Add this policy choice to the causal-loop diagram as the element, "George's Pay to Frank."

e. Analyze the effect of the payment on the completion of the task and on Frank and George's relationship.

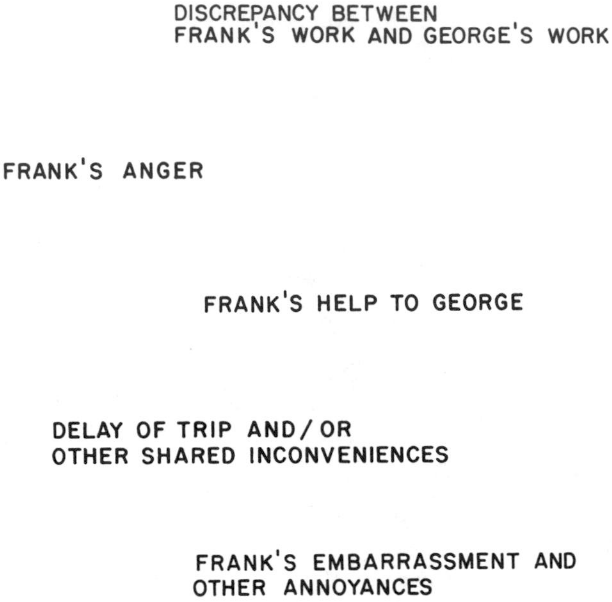

DISCREPANCY BETWEEN
FRANK'S WORK AND GEORGE'S WORK

FRANK'S ANGER

FRANK'S HELP TO GEORGE

DELAY OF TRIP AND/OR
OTHER SHARED INCONVENIENCES

FRANK'S EMBARRASSMENT AND
OTHER ANNOYANCES

Figure 11.7 Frank's dilemma

Exercise 8: The Baby

Margaret, six months old, is just getting her top teeth and is occasionally fussy. Her mother, Lois, does not like to hear Margaret cry, and works hard to distract her by playing with her, giving her toys, or feeding her. Lois cannot carry on a conversation with her husband, Sam, when she is comforting Margaret. After a tooth has come through and Margaret should not be fussy, Sam notes that Margaret still cries frequently. He guesses that Lois has been reinforcing the crying by rewarding Margaret when she cries.

  a. List the four dimensions of this situation from Sam's point of view. Choose some policy that might work to lessen Margaret's crying. Assume Sam is correct in his analysis of Margaret's excessive crying.

  b. Draw a causal-loop diagram for the problem.

  c. Argue from the diagram the way Margaret's behavior is changed over time.

Exercise 9: Lois's Perspective

Now suppose Lois disagrees with Sam and feels that it is important that Margaret be soothed when she is uncomfortable. Lois feels that this response will develop Margaret's confidence that she can have an impact on her own condition, affecting Margaret's way of dealing with life as an adult.

a. Reanalyze the situation from Lois's perspective. Write out the four dimensions, all of which may be different from Exercise 8.

b. Draw the causal-loop diagram.

c. Use the causal-loop diagram to trace the changes in attitudes of Sam, Margaret, and Lois over the next few decades.

Exercise 10: A Personal Example

a. Consider a recent personal family cooperation or competition experience. Write out the four dimensions of the problem definition.

b. Construct a causal-loop diagram.

c. Argue the dynamics of the situation from the causal-loop diagram.

d. If the scenario in (c) docs not correspond to what happened, go back to (a) or (b).

Exercise 11: Another Example from Personal Experiences

Do the same as in the previous exercise, but use a different pattern of interactions among family members.

CHAPTER 12

DESCRIPTIONS OF UNSTRUCTURED PROBLEMS

Exercise 1: BURROS*

The following article appeared in the Saturday–Sunday, March 29–30, 1980, issue of the *International Herald Tribune.*

BURROS: Beasts of Burden Have Become
Burden of Beasts in U.S. Deserts and Parks.

John Barbour

Los Alamos Lake, Ariz. (AP)—They keep turning them out the way they used to—one at a time.

It's always the same model—the same perky, impudent ears, the same big, innocent eyes, the same cuddly, oversized head. Elsewhere they call them ass or donkey or equus asinus. Here they call them burro, but they mean pest.

In a world where the snail darter challenges the dam, this little creature which once roamed North America before there was man, this trusty little beast which followed Don Quixote and Sancho Panza to their battles and carried Mary to Bethlehem, which Congress protected along with the mustang in 1971, this hardy, gregarious, sure-footed, obstinate, stupid, wise and patient ass is in trouble again, caught between man and nature and history.

Hereabouts, they round them up and ship them out to foster homes around the country. West of here, some say, they are causing trouble for the U.S. Navy at China Lake, Calif. South of here, some say, they compete unfairly with bighorn sheep. North of here, in the Grand Canyon National Park, there are plans to kill off several hundred if ways cannot be found to economically remove them from the remote and fragile canyons that run to the Colorado River.

Everyone is sad about the whole state of affairs. No one wants to shoot the little critters, but they probably will be shot.

---

*Reprinted with permission from The Associated Press.

The highways that sprint through this bittersweet desert accommodate the thousands of rubber-footed beasts-of-little-burden turned out by auto plants the world over.

But the desert just can't handle the lowly burro, which has been carting man and his belongings over a thousand landscapes since before the Old Testament.

Yet you rarely see one. By day they retreat into the gullies and arroyos. By night they venture toward the manmade lake, leaving clumps of manure along the road and occasionally breaking the dark stillness with an adenoidal agony inadequately called a bray.

Park rangers fence in their trailer-homes to protect what trees the meager desert allows from both burros and ranch cattle. The burros eat almost anything, and there is little to eat.

Which is the problem. On less than four inches of rain a year, desert plant life grows slowly. The largest plants are the palos verdes trees, which rarely grow much taller than a man, and the sparse mesquite.

The browsing burro grabs the tender growing tips of the palos verdes branches in his teeth, gives it a jerk with a twist of his massive head. The brittle palos verdes cracks and breaks well into the shrub. The burro consumes the tender bark and leaves the broken branch to the desert sun.

It takes the desert a long time to grow another branch. Officials estimate that half of the palos verdes trees within a mile of the lake are dead from animal grazing. Within three miles, the trees have been stripped of 85 percent of their vegetation.

Furthermore, the small bands of burros leave the desert hillsides scarred by their tracks. The desert is so frozen in time that tank ruts still survive from World War II training.

One might think back not too long ago when the notion about deserts was to make them bloom, not preserve their sterility. However, land managers who have to deal with deserts today feel obligated or are mandated to preserve their natural state. The Los Alamos Lake region consists of a state park where people come to swim, fish, and boat in a desert wonderland, surrounded by federal land under the Bureau of Land Management. The bureau is also mandated to maintain its lands for multiple uses—everything from mining to wildlife preservation, from range management to recreation.

Adopt a Burro Program

The burro gets in the way. So, since 1975, the bureau has rounded up 1,600 of the critters and sent them out on its Adopt a Burro program for people who want them for pets. There are many more applications than there are burros.

But the program is generally humane and keeps the herd down to the 175 or so that 212,000 acres can support.

Not so lucky is Merle Stitt, superintendent of the Grand Canyon National Park. His burro herd is scattered over some of the most remote and inaccessible land in the West. Old solutions won't work.

The park had its first burro problems in the early 1920s. In 1932, Chief Ranger J. P. Brooks reported: "Overgrazed conditions existed on all areas ranged over by burros. In many places herbage growth was cropped to the roots and some species of shrubbery were totally destroyed. Soil erosion was greater in burro-infested areas."

In the next seven years, nearly 1,500 burros were shot and left to decompose, reducing the herd to about 50 head. Park officials thought they could live with that, but the burro came back and 370 more were removed from 1932 to 1956.

Since the control program began, until 1969 when it was ended, almost 2,900 burros were killed or removed from the park. Publicized burro hunts of the late 1960s were responsible for a public outcry against the killing.

But the burro, which reproduces a fifth of its number each year, has come back again. Park Service people, looking at the damage done, figured they had a herd of 2,500. A sampling indicates there are only about 400, although Stitt says, "they do an awful lot of damage for 400."

Dean Durfee, who manages the Los Alamos Lake burro program for the BLM, explains that the burros have little else to do. "Jenny will drop a colt and when she's still nursing Jack will come along and get her ready for another one."

It used to be, in the Los Alamos Lake area, that casual hunting held the numbers down. And some folks actually relished burro meat.

But Congress put an end to that, and since the burro has no natural predators in the desert and is remarkably immune to most disease Jack and Jenny keep multiplying and multiplying.

Stitt says the burro problem in the park became intolerable, and in 1978 the Park Service proposed a control plan. There was no opposition at first, but then various groups came forward to say there must be a better way.

The Park Service did an Environmental Impact Statement on its plan. It put two cowboys with dogs to work on the most accessible area—the Tonto Plateau—to see how much it would cost to bring the burros out alive. It came to $440 apiece.

Next, they sent in park rangers with tranquilizer guns and helicopters to see how much that would cost. It came to $1,200 per burro. "Shooting is more certain and it only costs $60 apiece," Stitt says.

Now some conservationists are trying to collect money to bring the animals out alive—and the Park Service has given them two months to see if it is feasible. If it is, they can have longer.

Meanwhile the burros keep chomping away in their secret places, and they claim natural grasses like the dropseed and Indian rice grass, mesquite, black brush, brittle bush and a crust-forming lichen that helps hold the land together. Some of the damage is irreparable.

"It takes 50 to 100 years for a desert to repair itself," Stitt says.

But it only takes an average of 18 months between colts for every Jenny—and burros sometimes live to the age of 20.

Domesticated by the Spanish from African herds, the burro came here with the Conquistadors. The hunt for gold and silver in remote mountains gave the burro a special status. It could go anywhere and carry anything.

But when prospecting dried up, the burros were turned loose and adapted to the wild again. Therefore they are not considered native wild creatures, although there is evidence that asses roamed the North American continent in prehistoric times.

Congress protected the burro against slaughter when it protected the wild horse—but it also told the National Park Service even earlier to keep the national parks in their natural state. The two objectives seem to be incompatible.

a. What is the problematic behavior described in the article?

b. From whose point of view?

c. Is the time frame for the problem clear? If so, what is it?

d. Describe a federal policy that has definitely made the burro problem worse.

e. List the policies that have been adopted since 1972 to deal with the problem. List the four dimensions for the burro problem as tackled by those policies.

f. Draw a causal-loop diagram for the burro problem.

g. Predict the reference mode behavior for burros from your causal-loop diagram. Pay attention to the relation between reproduction time and the desert ecology recovery time.

h. Draw a new causal-loop diagram for each of the policies you listed in d. Again make a graph of number of burros over time before and after the applications of each policy.

i. Present short arguments for and against a control program in the Grand Canyon National Park of shooting burros to keep their population in the park down to about 1/10 its present size.

For a situation that may be somewhat similar to the burro problem, see the discussion of deer on the Kaibab Plateau in Chapter 18.

Exercise 2: The Oil Price Spiral*

The following article appeared in the February 29, 1980, issue of *Science*.

The Oil Price Spiral

Recent events, coupled with those of the last several years, point toward three conclusions:

---

*© Copyright 1980 by The American Association for the Advancement of Science. Philip A. Abelson, "The Oil Price Spiral," *Science* 207 (February 29, 1980).

- Supplies of Middle Eastern oil are subject to sudden interruption.

- Excessive dependence on such oil invites World War III.

- The oil cartel could easily further sharply increase its revenues while cutting production.

Any one of these considerations should be sufficiently persuasive to induce the consuming nations to seek to limit dependence on imported oil. In practice, the most effective goal is likely to be high prices. Past experience indicates that the limit on what OPEC can charge has not yet been reached. A small shortfall of supplies can lead to a great increase in price. In 1973 and 1974, production of oil in the free world was cut by 10 percent. A quadrupling of the price of oil followed quickly. The revolution in Iran led to a decrease in production there, but increases elsewhere held the drop to about 5 percent. This shortfall gave rise to a doubling of the price of oil. Imports by the developed countries have been little affected by the doubling, although at the moment there is a softening of prices on the spot market.

It is obvious that OPEC could extract much more money from the consumers while extracting less oil from the earth. The questions become: When will the next major squeeze occur, and how high will the price go? Any estimate is a wild guess, but a further doubling could occur within a year.

Price increases might be avoided if demand for oil were curtailed substantially. For the short term, this could be achieved by drastic conservation in the developed countries—for example, by gasoline rationing—but at the moment meaningful conservation seems politically unfeasible. For the longer term, prospects for cutting the use of oil are better, and one can visualize how the price spiral might eventually be brought under control through conservation and by the development of renewable energy sources. For the intermediate term, the most feasible solution is enhanced substitution of coal for oil and natural gas.

The energy potentially available in the form of coal is more than an order of magnitude greater than in oil. Important amounts of coal are present in many countries, including all the continents. Most important, the cost of thermal energy from coal is already substantially less than that from oil. In some parts of the world, the contrast is a factor of ten or more. Prospects for steadiness in the price of coal are good, and the large number of potential sources frees coal from the kind of political instability that now characterizes oil.

Quick substitution of coal is feasible in only a limited number of situations where oil had previously replaced coal. But the current contrasts in costs and uncertainties are serving as powerful incentives for exercise of ingenuity in adapting to coal. Action or lack of action by the United States will be an important factor in determining how fast substitution of coal will occur. More coal could readily be produced for both domestic and foreign consumption, but actions to implement the switch to coal have been slow.

Many foreign countries would like to obtain coal here, and delegations from France, West Germany, Japan, Spain, and Denmark have come to the United States during the last 2 months. However, concern has been expressed about the unreliability of supplies due to sudden domestic political moves and about the lack of infrastructure for exports. To make a really significant impact on world energy would require the existence of better rail transport, enlarged port facilities, and larger coal-carrying ships.

Switching toward use of coal will not be easy. However, new technology is being developed to improve the convenience and versatility of coal as a source of energy and chemicals. The United States can make many contributions to such developments. By moving resolutely this country could be crucial in helping to bring energy prices under control and in reducing dangerous tensions.

a. Is there a clear perspective presented in the editorial? If so, describe it.

b. Suggest an appropriate time scale. Explain why you chose that time period.

c. Describe a pattern of problematic time behavior for the oil supply/price situation that will serve as the reference mode for your causal-loop analysis.

d. Several policies are suggested to reduce the impact of the spiraling oil prices. List three.

e. Make a list of elements from which to construct a causal-loop diagram for the conversion to coal policy. Complete the causal-loop diagram from those elements.

f. Describe how you think the price of oil will change over time as implied by the causal-loop diagram with the policy implemented. Be sure to contrast this behavior with the reference mode description you gave in Exercise 2(c).

g. Contrast the time scales for the three policy choices you listed in Exercise 2(d).

Exercise 3: Amory Lovins' Soft Energy Path*

This article by John Popham summarizes concisely the economic perspective of the influential economist, Amory Lovins.

A Hard Look at "Soft" Energy Path

Lookout Mountain, Tenn.—The most vigorous debate on the nation's energy future currently involves the scientific community and the White House under the rather alluring title "The Soft Energy Path."

At this point President Carter, at his own request, along with his energy advisers, has been briefed on the novel "soft path" thesis. Intensive discussion has been

---

*Reprinted with permission from the author, John N. Popham (*Birmingham News,* Birmingham, Alabama, September 26, 1978).

made public in such prestigious publications as the weekly "Science" magazine of the American Association for the Advancement of Science, and the Center for the Study of Democratic Institutions' bimonthly "The Center Magazine."

Friends of the Earth

The foremost disciple of the proposal is Amory Lovins, a 31-year-old consultant physicist with the University of California's Energy and Resources Program. He is the British representative of the Friends of the Earth, Inc., an American nonprofit conservation group.

Lovins in 1976 initiated quite a debate on America's energy future in an article published by the distinguished quarterly "Foreign Affairs." It brought the quarterly's readers to their feet with such enthusiasm that the topic spilled over into other areas of the intellectual and scientific communities.

Simply put, Lovins sees our present large-scale power system as "The Hard Path" method because it is a policy of energy strength through exhaustion of fuels.

In this way we build ever-larger plants and our capital investments become astronomical as we seek offshore and arctic oil and synthetic fuels for such giant-size facilities. This starves other sectors of capital and we turn to selling military weapons which is both inflationary and immoral. We wind up with too much unemployment, alienation and crime.

"The Soft Path," as Lovins describes it, stresses diversity and smaller-scale components, each doing what it does best. It operates on renewable resources such as sun, wind, water, biomass residues. It does not rely totally on depletable fuels.

For this reason, Lovins says, the chance of technical failure is a lot lower because you spread the risk among a wide range of relatively simple things known to work, rather than put all your bets on a few rather adventurous technologies like big coal gas plants and breeder reactors. And this approach, he adds, is at the heart of the needs of the developing Third World.

Better Cash Flow

He says the soft path is cheaper in capital costs and has a better cash flow in delivered-energy price. It is faster in the sense that the invested dollar gives you more energy, money and jobs because the things you are doing in the soft path are so relatively simple that it takes only days, weeks, or months to do each one, not ten years.

Lovins says there is now a wide range of soft technologies that are technically mature and useful if we shop around. He cites solar heating, conversion of farm and forestry residues to liquid fuels, wind pumps and such.

He warns there is little time to get started. In 50 years we can revolutionize the supply of energy needs for mankind, he adds. You and your children will be debating many facets of "The Soft Path" for some years to come.

Lovins subsequently was invited to the White House to spell out his ideas on energy. It must have been a delightful session as Lovins is quite brilliant at turning a phrase and coming up with metaphors and similes that bowl you over.

For example, in discussing what he calls this country's "hard path" energy program he pointed with alarm to the economic dislocations of large scale energy systems such as support our vast grid systems and huge thermal power stations.

As such thermal power stations got bigger, he said, the fraction of the time they did not work also got bigger, increasing from about 10 to 35 percent for very good technical reasons. He then wrote:

"The picture gets worse. If one of these thousand-megawatt stations dies on you, it is like having an elephant die in the drawing room—you simply have to have another elephant standing nearby to haul the carcass away. You need a thousand-megawatt reserve margin to back you up. That costs a lot of money."

If, instead, Lovins contends, you build several stations of a few hundred megawatts each, they probably would not all fail at the same time, so you would not need as much reserve margin.

"In practice," he wrote, "that kind of change can, in most cases, let you do the same job for about a third less new capacity. If you went to, say, ten-megawatt units at the substation, you could do the same job with something like 60 or 70 percent less new capacity. These numbers were discovered only a year ago. Everyone had assumed they were negligible."

That is just a peep-hole view of the broad canvas on which Lovins has painted his view of the needs of the future. This column can only touch a few highlights to assist the general reader. But there seems to be little doubt that "The Soft Path" controversy is going to have a great impact on your political and economic life and this is an effort to alert you to its grand contours.

(If you wish to pursue Amory Lovins' analysis further, try *Soft Energy Paths: Towards a Durable Peace* (Cambridge, Mass.: Ballinger Publishing Co., 1977).

a. State as precisely as you can the problematic behavior characterized by our "hard" energy-dominated economy. Make a graph of the reference behavior over time. (Make time plots of "Alienation" and "Large Power Plants.")

b. Taking Lovins' perspective over the time frame you used in your graph, construct a causal-loop diagram including the current "hard" energy mechanisms. Include the element "Hard Energy Policy." Consider a long perception delay between alienation and an effect on hard energy policy.

c. Make a new causal-loop diagram for Lovins' "soft" energy policies by replacing "Hard Energy Policy" by "Soft Energy Policy." Again plot the same variable you plotted in a over the same time span, assuming that Lovins' suggestions have been taken seriously.

d. Try to build a diagram allowing soft and hard energy policies to compete. Try to include some feedback to influence which policy dominates.

e. What aspect of your analysis do you think would be considered by a hard energy enthusiast to be the most questionable?

f. If you were a supporter of soft energy policies, would your causal loop help you defend Lovins' policy against such an attack?

Exercise 4: Insect Strategies*

The following review of *Bumble Economics* by Bernd Heinrich (Cambridge, Mass.: Harvard University Press, 1979) appeared in the March 28, 1980, issue of *Science.*

This small book (whose information content and provocative nature make it seem much larger) is a remarkable exercise in anthropomorphics, starting with the title, which does not refer to the cost of keeping bumblebees (although the reader will find a fascinating essay on how to do so in the appendix). It is the reader, not the author, who will introduce most of the human analogies. The bumblebee behavior described here has been shaped by 80 million years of evolution into forms that are familiar in the human marketplace.

The title refers to how bumblebees produce, distribute, and consume wealth. Bumblebee wealth, nectar and pollen, is produced by foraging at flowers and is consumed to propagate a new bumblebee colony next year. This economic system has its counterpart in flowering plants. Their wealth of nectar and pollen is used to pay bumblebees for pollinating other plants. Bumblebees and their kin use various tactics to wrest wealth in excess of costs from flowering plants, which have their own tactics for obtaining the services of bumblebees without paying excessively.

Heinrich organizes the complex co-evolution of bumblebees and flowering plants first with a description of the annual cycle of a bumblebee colony followed by a presentation of experimental results that describe the energetic costs of foraging. Then he presents field observations and experiments of his own and others that illustrate how bees and flowers interact. The bumblebees he studied keep their flight muscles at a temperature of 30 degrees C or more while flying, although they may fly at air temperatures of 0 degrees C. A high metabolic rate and a hairy, insulated thorax are responsible for the large temperature difference. Insulation is adjusted by controlling the amount and timing of blood flow to the nearly naked abdomen. The queen bumblebee can produce heat at a high rate without flying and keeps her brood of eggs and larvae warm by incubating them. All this requires nectar for fuel plus fat and protein from pollen for the growing larvae. To bring home adequate supplies, the bees must juggle a number of variables: fuel aboard at takeoff, flight time to destination, flowers selected relative to nectar content and ease of gathering, air temperature, whether to hover or perch, whether to cool off or keep warm, and more.

The flowers advertise their nectar contents by their shape, color, arrangement, and odor. The author's observations on the last illustrate his straightforward and effective experimental technique. He covered a patch of clover flowers with bridal veil to exclude foraging bees, then lay back on the lawn with his eyes closed while a student held clover flowers for him to sniff. He could with 88 percent accuracy determine whether a flower had been visited by a bumblebee. Flowers may "cheat" by producing no nectar but looking like other flowers that do reward bees. Some orchid flowers resemble female insects closely and achieve cross-pollination by luring male insects into attempted copulation. On the other hand, bees may rob flowers by biting into the nectar cup rather than struggling through the pollen apparatus. Bees and plants have obviously reached a mutually satisfactory arrangement. One cannot help but admire a transport system where the fuel is nearly pure carbohydrate made on the spot from air, water, and sunshine.

Heinrich sets out to tie his research in with everyday experience so that both laypersons and professional biologists may share the fascinating continuity between physiology, behavior, and ecology. The overall aim of the book is to use the bumblebee as a model to explore biological energy costs and payoffs. All this Heinrich has achieved, with good science, pleasing style, and obvious linkages to the human condition.

Vance A. Tucker, Department of Zoology,
Duke University, Durham, North Carolina 27706

Since we must assume that individual bumblebees or groups of bumblebees do not "formulate policies" in the same sense as do humans, we must proceed carefully in giving the four dimensions defining the problem. The first sentence in the review refers to anthropomorphics, acts of projection of human characteristics onto nonhuman species. We shall do the same in our analysis. Take the description of how different bees have organized their social tasks and compare strategies. The payoff on any strategy for a bee colony is survival.

Different strategies for gathering food are compared on page 148 of *Bumblebee Economics:* "There is no set formula for the best foraging behavior. For example, the optimum response changes drastically when food resources become less compact. The wider resources are scattered, the less efficient it is to recruit and defend specific items, and the more difficult it is to patrol and defend an area. Competitors then appear to work peacefully (without contact) side by side, but they may still compete relentlessly by trying to remove resources faster than the next individual. Aggressive encounters then become a liability, for even the winners lose—they have only expended time and energy that could have been used for foraging. The nonaggressors, which do not interrupt their foraging, reap more food energy and are competitively superior. Such competition, called scramble or exploitation competition, generally results in the depletion of resources to the very minimum of economic profitability. In turn, it selects for energy economy and foraging efficiency in the contestants."

a. List the four dimensions for the problem analysis for a bee colony that has adopted the scrambling strategy.

b. Select appropriate elements and try out several causal-loop diagrams until you find one that works well. Some suggested elements are "Size of Colony," "Number of Sources Tapped," "Food Gathered," and "Supply of Food."

c. Describe how your diagram works, emphasizing the relative roles of positive and negative feedback loops.

d. Repeat steps a, b, c for a bee colony adopting an aggressive foraging strategy.

Exercise 5: Taping Albums From the Air*

The following article appeared in the *Boston Globe* on April 3, 1980:

To Play or Not to Play—
That is the question record companies and radio stations face.

By Jim Sullivan

The relationship between record companies and album-oriented rock FM radio stations is no longer the perfect friendship although they still need each other. For the record company, radio airplay gives artists valuable exposure, leading to greater record sales; for the radio station, record companies supply them with new music—its lifeblood—arrange interviews with artists, develop promotional give-aways, set up concert broadcasts and, of course, buy advertising.

But a monkey wrench has been thrown into the system. It has prompted verbal warfare in radio and record industry trade magazines and strained the relations between record and radio personnel. The issue is radio stations playing new albums in their entirety without commercial interruption, which many record company executives view as an invitation to listeners to tape albums off the air instead of buying them in a store. (Although older albums are played at times, it's the playing of new releases that has strained the friendship.)

. . . While no one believes home taping single-handedly put . . . many people out of work, record industry executives believe that it has substantially contributed to the decrease in sales. Blank recording tape sales have risen dramatically over the past six years.

. . . It seemed painfully ironic to record companies that their radio "partners" were making things so easy for listeners to tape their favorite albums by paying only for blank tapes. (A 90-minute cassette generally costs about $4.50 and would allow the taping of two albums. The average list price on an album is $7.98, though most stores discount albums to $5.99).

. . . The controversy puts the record company promotion personnel in the strange and ironic position of requesting stations not to air their records.

---

*By Jim Sullivan, *Boston Globe*. Reprinted with permission from the author.

"It's like being caught between a rock and a hard place." says Mike Bone, Arista Records vice president of promotion. "It's like you want it, but you want it your way. It's a very difficult position to be put in. We want exposure but we want it spread out over a period of time."

Ed Hynes, Columbia Records vice president of promotion, says he directed his 47 local representatives to ask radio stations not to feature Pink Floyd's double album, "The Wall." "I've had incidents where our people around the country have threatened us off radio if they tracked the album," says Hynes.

. . . WAAF program director Dabe Lee Austin says that the album features (one each night, six nights a week at midnight and six in a row on Saturday) are his station's most popular feature, and he has no plans to alter it.

. . . The record industry has no easy solution to the situation—though the suggestions put forth have ranged from the ludicrous and reactionary to the more rational.

Ben Bartel, a major stockholder in two West Coast retail chains, takes the prize in the former category. He suggested in a letter published in Billboard, ". . . an immediate approach to Congress to foster legislation prohibiting media from playing more than five minutes of any recorded work in any given two-hour period."

. . . A more moderate approach (though still sure to rest uneasily with some program directors) is a compromise, according to Arista's Bone, . . . to substitute "artist's features" for album features. With these the station would still play tracks from a new album but would intersperse them with older songs and commercials so a listener couldn't simply turn on his tape recorder and have a copy of the album.

. . . The real long term problem, Bone says, may be that as sales dim, record companies would be less able and willing to invest and work with new talent—thus depriving radio and public of new music. "If the record companies continue to do badly," he says, "they are not going to be able to continue to sign and develop new artists like the Dwight Twilleys, the Sports and the Graham Parkers."

"Then," Bone continues, "radio stations that used to play records will go all news and talk and everybody will end up watching television."

He laughs, and adds, "that's being sort of flip, but it's a possibility."

a. View the situation from the perspective of Arista Records' Mike Bone. Describe the ultimate problematic behavior for the record industry and the radio stations from his point of view. Be explicit about the time span (a reasonable guess is the best you can do). Express the problematic behavior as a plot of listening audience and record sales over time.

b. Draw a causal-loop diagram for the problematic behavior. (Suggestion: Assume that a decrease in the listening audience leads to an increase in stations playing full albums in competition for that audience.)

c. Suggest a policy that Bone might adopt, together with other record company executives, to combat the problem.

d. Make a causal-loop diagram with the proposed policy and predict how well your policy will work by arguing from the causal-loop diagram. Plot those predictions over the same time span as before.

PART V

INTRODUCTION TO SIMULATION

OBJECTIVES

The causal-loop methods discussed in Parts II, III, and IV can provide much insight into a system's structure. But as has been suggested, it is often difficult to infer the behavior of a system from its causal-loop representation. Part V introduces the use of computer simulation as a more precise tool to analyze the behavior of complex social and economic systems. Objectives of this section include:

1. Moving the student from a causal-loop representation of a system to a computer model by introducing flow diagraming and equation writing;

2. Demonstrating hand-calculated simulation as a forerunner of computer simulation;

3. Describing how to use computer simulation to analyze simple positive and negative feedback loops;

4. Discussing the representation of more complex causal relationships;

5. Introducing the simulation of delays.

CHAPTER 13

LEVELS AND RATES

DIAGRAMING LEVELS AND RATES

The first step in moving from a causal-loop representation to a computer simulation model is the identification of system levels and rates. Recall from Part III that a level is a quantity that accumulates over time, and a rate is an activity, or movement, or flow that contributes to the change per unit of time in a level. For example, the number of cars in the Midtown Parking Lot is a level, and the number of cars arriving per hour is a rate. Similarly, the number of children at Hometown Elementary School sick with the flu is a level, while the number of children recovering per day is a rate. Population is a level, and the number of babies born per year is a rate.

In identifying a system's levels and rates, it is generally helpful to represent the system in flow diagram form. Figure 13.1 depicts the symbols that are used to represent levels and rates in flow diagrams. A level is depicted by a rectangle (which is supposed to resemble a box or a bathtub), and a rate is depicted by a symbol that looks somewhat like a valve. (A rate might be thought of as a faucet, controlling the flow of water into the bathtub.) A complete set of flow diagram symbols is included at the end of Chapter 15.

Figure 13.2 shows that the number of cars in the Midtown Parking Lot is influenced by the number of cars arriving per hour (the arrival rate). The flow of cars arriving at the parking lot increases the level of cars in the lot, much as the flow of water into a bathtub increases the level of water in the tub.

Figure 13.3 is a somewhat more complicated diagram, indicating that the number of children at the Hometown Elementary School sick with the flu is influenced by both the number of children catching the flu each day and the number recovering. The flow of children catching the flu—the infection rate—adds to the number of children sick with the flu; and the flow of children recovering—the recovery rate—subtracts from the number who are sick. (The situation is somewhat similar to a bathtub with both a faucet and a drain. The flow of water into a bathtub adds to the amount of water in the tub, and the flow of water out of the drain subtracts from the water in the tub.)

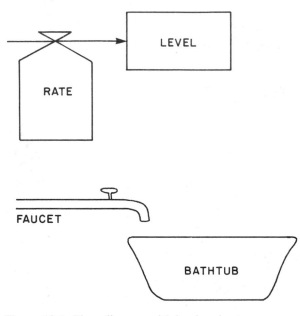

Figure 13.1 Flow diagram with level and rate

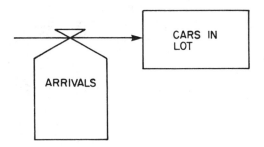

Figure 13.2 Midtown Parking Lot

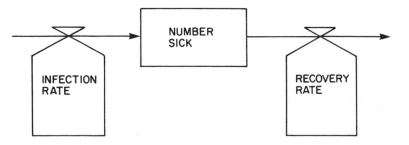

Figure 13.3 Hometown Elementary School

Exercise 1: Flow Diagrams

a. Draw flow diagrams for the following situations:

1. The population of rabbits is influenced by the number of rabbit births per year.

2. The number of yeast cells in a sugar solution is influenced by the number of buds formed per minute.

3. A child's knowledge is influenced by his or her learning rate.

b. What rates influence the population of Boston? (Draw a flow diagram including whatever rates you think might be appropriate.)

c. What rates might influence the number of students enrolled in an urban high school? (Draw a flow diagram.)

d. Add the number of students susceptible to the flu and the number of children who have recovered from the flu to the Hometown Elementary School flow diagram, shown in Figure 13.3.

FROM CAUSAL LOOPS TO FLOW DIAGRAMS

Moving from a causal loop diagram to a flow diagram requires a few additional symbols. Figure 13.4 depicts a causal-loop diagram and a corresponding

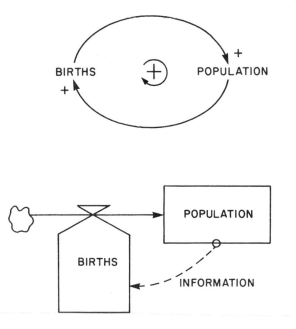

Figure 13.4 Causal-loop and flow diagrams of population and births

flow diagram of the interaction of population and births. The level in this instance is population, as indicated by the rectangle, and births is a rate, as indicated by the valve symbol. The positive link from births to population in the causal-loop diagram is depicted in the flow diagram as the flow of births into population. The direction of the solid arrow in the flow diagram indicates that births add to the population. The positive link from population to births, in the causal-loop diagram, is shown as a dotted line in the flow diagram, indicating that the size of the population influences the birth rate.

The "cloud" at the tail of the solid arrow represents the "source" of people. (Sources represent systems of levels and rates outside the boundary of the model. In this case, the source allows bypassing the issue of where babies come from!) Although not shown on this diagram, "clouds" can also be used to show "sinks," where flows terminate outside the system.

The flow diagram is a more detailed representation of the positive feedback loop than is the causal-loop diagram. It identifies population as a quantity that accumulates, and it identifies births as a quantity that influences how rapidly the population accumulates. The solid arrow shows the flow of people into population. The dotted arrow shows that the size of the population affects births, or that there is a cause-and-effect link from population to births. The causal-loop diagram ignores the distinction between a rate of flow and a cause-and-effect link not involving a rate of flow, but the flow diagram calls explicit attention to this distinction.[1]

Exercise 2: Population and Deaths

Figure 13.5 shows a causal-loop diagram of the interaction of population and deaths.

Identify the level and rate in the system, and draw a flow diagram.

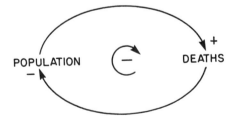

Figure 13.5 Causal interaction of population and deaths

Exercise 3: Natural Resources

Figure 13.6 depicts a causal-loop diagram of an interaction of natural resources and usage. The diagram depicts how a decreasing supply of a particu-

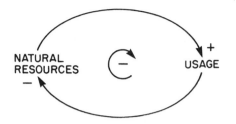

Figure 13.6 Causal-loop diagram of natural resources and usage

lar natural resource can result in less use of the resource, because it is harder to find.

Draw a flow diagram of this interaction. Begin by identifying which quantity is a level and which is a rate. Assume that no new quantities of the resource are created, so that there will be no source and no inflow. Treat the place where the used resource goes as a sink.

EXAMPLE I: CHILDREN AND ADULTS

Figure 13.7 depicts another causal-loop diagram of the growth of population through births. However, in this diagram population is separated into adults (individuals mature enough to bear children) and children (individuals too young to bear children).

To draw a flow diagram based on this causal-loop diagram, it is easiest to begin by identifying the levels and rates. Children and adults are levels, since they are quantities that accumulate over time; while births and children maturing are rates. (Note that the rates have units "People per Year," whereas the levels have units "People.") A partial flow diagram is shown in Figure 13.8. Births flow into the population of children, as indicated by the positive link from births to children in the causal-loop diagram. Furthermore, the flow of children maturing decreases the level of children and increases the level of adults. (Thus the flow of children maturing incorporates two links from the

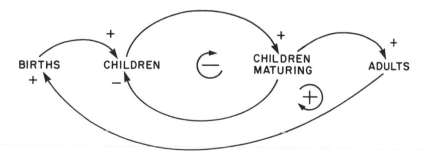

Figure 13.7 Causal-loop diagram of children, adults, and births

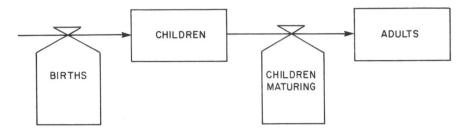

Figure 13.8 Levels and rates for children and adults

causal loop diagram: the negative link from children maturing to children, and the positive link from children maturing to adults.)

The next task in completing the diagram is to add the cause-and-effect links. For example, a positive link connects adults to births, since the more adults there are, the more births there will be (other factors remaining equal). This link is shown in the flow diagram as a dotted line running from the level of adults to the rate of births. One link in the causal-loop diagram remains to be inserted in the flow diagram—the positive link connecting children to children maturing. As this link indicates, the more children there are, the more children will mature. This link is represented in the flow diagram by a dotted line connecting the level of children to the rate of children maturing. Finally, the flow diagram is completed by adding a source symbol to the left of the births, as shown in Figure 13.9.

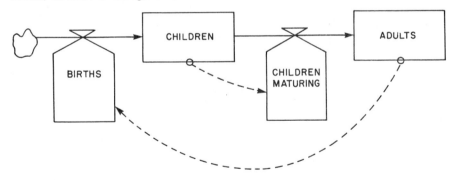

Figure 13.9 Completed flow diagram for children and adults

Exercise 4: Resource Processing

Figure 13.10 depicts a causal-loop diagram of the life cycle of aluminum used in cans. As aluminum is refined, it passes from the stage of being ore to being aluminum in process. The metal is then made into cans. The cans have an average life, after which they become solid waste. At each stage, the flow into the next stage depends on how many cans are at the current stage.

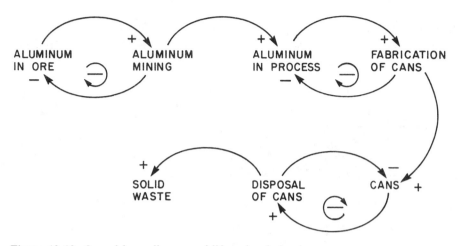

Figure 13.10 Causal-loop diagram of lifecycle of aluminum

Draw a flow diagram based on the causal-loop diagram. What would you add to your flow diagram to represent a recycling program?

SIMULATION, STRUCTURE, AND BEHAVIOR

The main reason for moving from a causal-loop representation of system structure to a flow diagram is to provide additional insight into the behavior a proposed model generates over time. For example, does the hypothesized model generate continued growth? If so, how rapid is the growth? Or, does the model generate decline? If so, how precipitous? Does the model exhibit goal-seeking behavior? If so, do model variables approach equilibrium smoothly, or do they oscillate? If the model produces oscillations, what is the period from peak to peak? How dramatic are the cycles? And so on.

In order to provide full answers to these questions, it is necessary to move one final step and express each model relationship in equation form. Of course, the translation from a verbal description of each model relationship to a statement as an equation often requires a good deal of ingenuity. However, in drawing out the implications of a model, equations are essential.

The strategy generally followed in formulating a model is to begin with a causal-loop diagram, then formulate a flow diagram, then write equations, and finally, use the equations to simulate the model on the computer. Once a "running" model has been developed, it can then be used to explore the consequences of alternative model assumptions and proposed policy interventions. Indeed, one of the main advantages of simulation is the opportunity it provides to move quickly and easily from one set of assumptions to another.

In the discussion of equation writing and simulation in the next few chapters, a good deal of attention is given to ways of using simulation to draw out the implications of hypothesized system relationships. Much less attention is given to methods of using empirical evidence to choose numerical values for model parameters. Nor is much attention given to methods of assessing the "match" between the behavior generated by a simulation model and the historical behavior of the actual system under study. This is not because these questions are unimportant or easy—they are not. Often, however, a good deal can be learned about a system by exploring the implications of alternative hypothetical models. In addition, in estimating model parameters and assessing the match between model behavior and historical evidence, it is worth paying a fair amount of attention to the relationship between the structure of a proposed model and the behavior it generates.

EQUATIONS FOR LEVELS AND RATES

Once a flow diagram has been developed, the next step in building a model is to write equations. The following examples introduce the general ideas involved. Chapter Fourteen then provides more detailed information on equation-writing using the DYNAMO computer simulation language.

EXAMPLE II: THE KINGDOM OF XANADU

In the mythical kingdom of Xanadu, exactly 100 babies are born every year, and no one ever dies. In last year's census (the year 2020, according to the Xanaduian calendar), the population was found to be 5510 people. Everyone in Xanadu believes that births will continue in the future as they have in the past.

The king of Xanadu wishes to have a model that will estimate the population of the kingdom for the next twenty years (the years 2020 through 2040). What will such a model contain? First, the model will contain variables, things whose numerical values change over time. As a notational convention, we will always refer to model variables using names written in ALL CAPITAL LETTERS. In the model for the king of Xanadu, population and births are variables, and for convenience they can be called POP and BIRTHS. POP, of course, is a level, and BIRTHS is a rate, as indicated in Figure 13.11.

The second thing a model must contain is a set of rules for computing the values of variables. For example, from the preceding description it is clear that the rule for births in Xanadu would be:

Set births equal to 100 people per year

An equation is a concise way of specifying a rule for computing a variable. For example, the rule for births in Xanadu could also be written in equation form:

BIRTHS = 100 people per year

This is called a "rate equation," naturally enough, since it is the equation for BIRTHS, which is a rate.

Now, how can an equation be written for the level of population over the twenty-year period from 2020 to 2040? The simplest approach is to break up the twenty-year period into one-year intervals, and then calculate the population year by year. In the year 2020, according to the Xanaduian calendar, the population was 5510. So, the first year that needs to be calculated is the year 2021. Recall that, in Xanadu, no one ever dies, nor does anyone enter or leave the Kingdom. Hence, the only change in the population from one year to the next is the number of new babies born—which is exactly 100. Thus the population in 2021 is just the population in 2020 plus 100.

$$POP(2021) = POP(2020) + 100 = 5510 + 100 = 5610$$

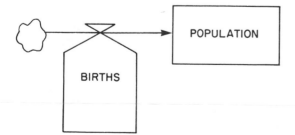

BIRTHS

POPULATION

Figure 13.11 Flow diagram for Xanadu

Using this same procedure, it is easy to calculate what the population will be in the year 2022. The population in the year 2022 is simply the population in 2021 plus 100.

$$POP(2022) = POP(2021) + 100 = 5610 + 100 = 5710$$

The same idea can be used to simulate the level of population over time, breaking up time into one-half year intervals. In this case, the first time at which population must be calculated is half-way through the year 2020. The population half-way through the year 2020 is just the population at the beginning of 2020, plus the number of babies born during the half-year interval. And, since 100 babies are born each year, one-half a hundred, or fifty are born in a half-year.

$$POP(2020.5) = POP(2020) + 0.5*100 = 5510 + 50 = 5560$$

By the same token, the population at the beginning of year 2021 can be calculated on the basis of the population in the middle of year 2020; the population in the middle of year 2021 can be calculated on the basis of the population at the beginning of year 2021; and so forth.

This procedure suggests a way to write a general equation that can be used to calculate the population at any moment in time, based on the population one time interval earlier. To clarify the development of the equation, it is helpful to refer to the moment in time at which the population is currently being calculated as the "present time," and it is helpful to refer to the interval between calculations as "one time interval."

The equation to be developed combines two fundamental ideas. First, the population at the present time (i.e., the time currently being calculated) equals the population one time interval earlier, plus the births that occurred over the interval. Second, the number of births occurring over one time interval equals the length of the interval, multiplied by the number of births per year. Combining these two ideas produces the following equation:

POP(present time) = POP(one time interval earlier)
        + (length of time interval)*BIRTHS(per year)

As equations go, this one appears somewhat cumbersome. One way to improve matters is to use symbols for the terms "present time," "one time interval earlier," and "length of time interval." Although many symbols are possible, the following are used throughout the text because they are consistent with the notation used by the DYNAMO simulation language to be introduced in Chapter Fourteen.

LEVEL.K   a level calculated at the present time

LEVEL.J   a level calculated one time interval earlier

DT   the length of the time interval between J and K

Figure 13.12 displays these symbols in graphic form. (The symbol L appearing in the figure will be discussed later in the chapter.)

Using this notation, POP(present time) is written POP.K, and POP(one time interval earlier) is written POP.J. Thus the level equation for population can be written:

POP.K = POP.J + DT*BIRTHS

For clarity, it is conventional to express the product of DT and BIRTHS as (DT)(BIRTHS). Hence the equation for population would generally be written:

POP.K = POP.J + (DT)(BIRTHS)

This equation can be read, "The population at time K equals the population at time J plus DT multiplied by BIRTHS."

Altogether, our model of the Xanadu population includes two equations. The first is a simple rate equation, indicating that the number of births per year is 100. The second equation is a level equation, indicating that the change in population over one time interval equals the number of births per year times the length of the time interval.

BIRTHS = 100
POP.K = POP.J + (DT)(BIRTHS)

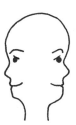

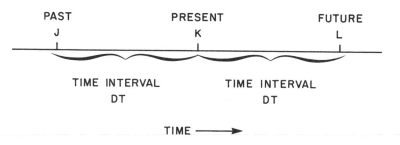

Figure 13.12 Definition of timescripts

In general, to simulate a model, it is necessary to write an equation for each level and each rate in the model, much as in the Xanadu case. The Xanadu model is a bit unusual in one respect, in that the rate (the number of births per year) is a constant. Often, rate equations are more difficult to formulate. But level equations are generally formulated exactly as the level of population in Xanadu. The value of a level at the present time *must* equal its value one time interval earlier, plus whatever flowed into the level over the time interval (minus whatever flowed out).

Consider, for example, the parking lot illustration discussed at the beginning of the chapter. (See Figure 13.1.) According to the example, the number of cars in the parking lot is a level, and the number of cars arriving per hour is a rate. If the name CARS is used to represent the number of cars in the lot, and ARRIV is used to represent the arrival rate (in cars per hour), then the level equation for CARS can be written:

$$\text{CARS.K} = \text{CARS.J} + (\text{DT})(\text{ARRIV})$$

The flu example shown in Figure 13.2 provides a somewhat more complex illustration. The number of children sick is a level, and the number of children who become infected per day, as well as the number who recover per day, are rates. If the name INFEC is used to represent the infection rate (in children per day), RECOV is used to represent the recovery rate (in children per day), and NSICK is used to represent the number of children sick, then the level equation can be written:

$$\text{NSICK.K} = \text{NSICK.J} + (\text{DT})(\text{INFEC-RECOV})$$

Exercise 5: Writing Level Equations

Review Exercise 1, and then write *level* equations for each of the levels in parts (a) through (d). (Choose whatever variable names you wish, and write them in ALL CAPS. Try to pick names that will aid you in remembering the subject of the equation!)

EXAMPLE III. CALCULATING THE POPULATION OF XANADU

The equations for POP and BIRTHS developed in Example II can be used to hand-simulate the population of Xanadu over time. As the year-by-year calculations are carried out, it is convenient to record the results in a form similar to Table 13.1.

Table 13.1 Table for computing population of Xanadu

Time	Change in population	Population	Births
2020	-----	5510	100
2021	100	5610	100
2022	100	5710	100
2023	100	5810	100
2024	100	5910	100
2025	100	6010	100
2026	100	6110	100
2027	100	6210	100
2028	100	6310	100
2029	100	6410	100
2030	100	6510	100
2031			
2032			
2033			
2034			
2035			
2036			
2037			
2038			
2039			
2040			

Population in 2020 = 5510 people
Births = 100 people/year
Time Interval = 1 year

Year 2020.   At the beginning of the simulation, the present time or time K is the year 2020. To get the simulation going, an "initial value" for the population in 2020 must be selected. For Xanadu, the population is known to be 5510 people in the year 2020. Thus, 5510 is entered in the *Population* column of Table 13.1, for the year 2020. The table then looks as follows:

Time	Change in population	Population	Births
2020	-----	5510	

The next thing that must be calculated is the birth rate for the year 2020. In this case, births are calculated according to the equation

$$BIRTHS = 100$$

Thus 100 is entered under *Births* for the year 2020, yielding an entry like this:

Time	Change in population	Population	Births
2020	-----	5510	100

Year 2021.   At this point, the calculations for the year 2020 are complete. To carry out the calculations for the year 2021, it is necessary to "advance the calendar" one year. Thus the year 2021 becomes the "present time" (time K) and the year 2020 becomes time J. The population in the year 2021 can then be calculated using the formula:

$$POP.K = POP.J + (DT)(BIRTHS)$$

The calculation is easiest if (DT)(BIRTHS) is computed first and written down. The column *Change in population* in Table 13.1 is reserved for this purpose.

Time	Change in population	Population	Births
2020	-----	5510	100
2021	100		

The population in the year 2021 (time K) can then be computed by adding the population in year 2020 (time J) to the *Change in population* column.

Time	Change in population	Population	Births
2020	-----	5510	100
2021	100	5610	

The birth rate for the year 2021 can then be calculated as before, using the rate equation BIRTHS = 100.

Time	Change in population	Population	Births
2020	-----	5510	100
2021	100	5610	100

Year 2022. The calculations for the year 2021 are now complete, and, once again, it is necessary to advance the calendar another year. Thus the year 2022 becomes the "present time" (time K), and the year 2021 becomes time J. Then, computations can be carried out exactly as in the year 2021, producing the following results.

Time	Change in population	Population	Births
2020	-----	5510	100
2021	100	5610	100
2022	100	5710	100

The simulation can be continued for as long as needed, with each advance of the calendar producing another iteration.

Exercise 6: Computing Additional Values for Population

Following the procedure in Example III, compute the population of Xanadu through the year 2040. Graph the results.

Exercise 7: Population at Other Times

If the population of Xanadu is 5510 at the beginning of year 2020, what is the population after the first month of 2020? What is the population in the year

2520? Is there a way to compute the answer without iteratively calculating the numbers as in Example III?

EXAMPLE IV: WORLD POPULATION GROWTH

Much the same approach used in writing the equations for the Xanadu model can be used in writing equations for a model of the growth of world population. Consider the causal loop and flow diagram in Figure 13.13. According to the diagram, the number of net births each year is influenced by the size of the population, and the size of the population is influenced by the number of net births. (The number of net births each year is the difference between the number of births and the number of deaths.)

The main problem involved in formulating a model of world population growth is formulating the rate equation for net births. In Xanadu, the rate equation was simple, since the number of births each year was constant, but for the world population, the number of net births each year is not constant. It increases as the size of the population increases.

What sort of equation should be written to express the relationship between the size of the world population and the number of net births each year? Many alternative formulations are possible, but the simplest assumption is that the number of net births each year is a *constant percentage* of the world population. In fact, over the recent past, the world population has grown at about 2 percent per year. This means that 0.02 net births are generated each year for every member of the population. Or, perhaps more sensibly, two net

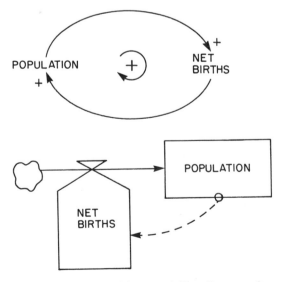

Figure 13.13 Causal-loop and flow diagrams for world population

births are generated each year for every 100 members of the population. (These 2 net births might correspond to a combination of 3 deaths and 5 new babies born.)

In analyzing population growth, it is essential to distinguish between the net birth rate, measured in *people per year,* and the annual percentage growth in the population, measured in *percent per year.* To call attention to this distinction, call the percentage growth in the population the growth fraction, or GF.

Using the growth fraction GF, the rate equation for net births NBIRTH (in people per year) can be written:

$$NBIRTH = POP.K*GF$$

Precisely speaking, the growth fraction GF equals 0.02, and it is measured in units (persons/year)/person. That is, 0.02 persons per year are added to the population, for each person in the population. The expression (persons/year)/person can be reduced algebraically to the expression (1/year), which in words is simply "per year." The expression (1/year) may seem odd at first glance, but after some reflection, it should be clear that a percentage growth of 2 percent per year amounts to a growth fraction GF = 0.02 "per year."

Once the rate equation for net births has been formulated, the level equation for population can be written rather easily. It has the usual form:

$$POP.K = POP.J + (DT)(NBIRTH)$$

Thus the complete model for world population growth consists of two equations:

$$NBIRTH = POP.K*GF$$
$$POP.K = POP.J + (DT)(NBIRTH)$$

One small technical matter needs to be taken care of. The equation for net births indicates that the number of births per year depends on the size of the population—and, of course, the size of the population varies over time. In the equation for net births, the variable POP has a subscript K to indicate the time, but net births NBIRTH so far does not. What subscript should be used?

The easiest way to determine the answer is to carry out a hand-simulation. For simplicity, carry out the simulation using a time interval DT equal to one year, and begin the simulation in 1975, when the world population was roughly 4 billion. According to the rate equation for net births, the number of net births per year over the period 1975 to 1976 is:

$$NBIRTH = POP.K*GF$$
$$= 4*(0.02)$$
$$= 0.08 \text{ billion persons per year}$$

Furthermore, the size of the population in 1976 is simply the size in 1975, plus the number of net births that occurred over the year interval from 1975 to 1976.

$$POP.K = POP.J + (DT)(NBIRTH)$$
$$= 4 + (1)*(0.08)$$
$$= 4.08 \quad \text{billion persons}$$

Since the net birth rate used in the calculation of the population in 1976 is, by definition, the birth rate that persists over the interval 1975 to 1976, it is plausible to give the net birth rate two subscripts: one for 1975 and one for 1976. In assigning these subscripts, however, it is necessary to pay strict attention to the "calendar time" at which the calculations occur. The net birth rate for the period 1975 to 1976 was calculated on the basis of the population in 1975, when the "present time" (time K) was 1975. Thus the rate equation should be written:

$$NBIRTH.KL = POP.K*GF$$

This indicates that the net births per year during the period from time K (1975) through time L (1976) is equal to the population in 1975, times the growth fraction GF. (Recall from Figure 13.12 that time L is one time interval DT following time K.)

The population in 1976 is calculated when the "present time" is 1976. Thus the level equation for population should be written

$$POP.K = POP.J + (DT)(NBIRTH.JK)$$

This indicates that the population in 1976 (time K) equals the population in 1975 (time J) plus the number of net births between 1975 and 1976.

Taken together, then, the full model of the world population should be written:

$$POP.K = POP.J + (DT)(NBIRTH.JK)$$
$$NBIRTH.KL = POP.K*GF$$

The detailed steps involved in simulating world population can be carried out most easily by constructing a table similar to the table used in the Xanadu example. Table 13.2 depicts a table for computing world population.

Year 1975. When the simulation begins, the present time (time K) is 1975. The initial value of the world population (4 billion) is entered as the value of population in 1975, and then the net birth rate for the interval 1975 to 1976 can be calculated according to the equation:

$$NBIRTH.KL = POP.K*GF$$

Table 13.2 Table for computing world population

Time (years)	Change in population (people)	Population (people)	Net births (people/year)
1975	-----	4.00	0.08
1976	0.08	4.08	0.08
1977	0.08	4.16	0.08
1978	0.08	4.24	0.08
1979	0.08	4.32	0.09
1980	0.09	4.41	0.09
1981	0.09	4.50	0.09
1982	0.09	4.59	0.09
1983	0.09	4.68	0.09
1984	0.09	4.77	0.10
1985	0.10	4.87	0.10
1986			
1987			
1988			
1989			
1990			
1991			
1992			
1993			
1994			
1995			
1996			
1997			
1998			
1999			
2000			

The resulting net birth rate (0.08 billion persons per year) is then entered in the table, producing the following results.

Time	Change in population	Population	Net Births
1975	-----	4.00	0.08

Year 1976. At this point, the calculations for 1975 are complete, and the calendar is advanced one year. Thus the present time (time K) is now 1976. The year 1975 has become time J, and 1977 is time L. The population in 1976 can now be calculated, using the level equation:

$$POP.K = POP.J + (DT)(NBIRTH.JK)$$

The value used for net birth is, of course, the value for the period 1975 to 1976, which is the value calculated as the final computation of the 1975 simulated year. To calculate the value of population in 1976, the product (DT)(NBIRTH.JK) should be entered in the table under the column *Change in population,* and then the product can be added to the population in 1975.

Once the population for 1976 is calculated, net births for the period 1976 to 1977 can be computed, using the rate equation:

$$NBIRTH.KL = POP.K*GF$$

This produces the following results.

Time	Change in population	Population	Net Births
1975	-----	4.00	0.08
1976	0.08	4.08	0.08

These steps can then be continued for as many iterations as are desired.

Exercise 8. Computing World Population

Following the procedure outlined in Example IV, compute the world population through the year 2000. Graph the results.

ENDNOTE

1. This accounts for the occasional awkwardness in reading a causal-loop diagram, discussed in Chapter 3.

CHAPTER 14

SIMULATION USING DYNAMO

The computer can eliminate much of the tedium involved in performing simulations by hand, but using the computer requires a language in which simulation instructions can be given to the computer. This chapter introduces a simulation language called DYNAMO.[1] DYNAMO is an acronym for DYNAmic MOdels. Like the computer language BASIC, DYNAMO is used to direct the computer in the computations it should perform. Unlike BASIC, however, DYNAMO is not a general-purpose language. It is a special-purpose language to aid in building computer models. Other languages, like BASIC and FOR-TRAN, can be used to perform the simulations described in Parts V, VI, and VII. But, because DYNAMO is designed especially for simulation, it eases the task of building and running models.

The following example introduces some of the central elements of DYNA-MO.

EXAMPLE I: YEAST—A SIMPLE FEEDBACK SYSTEM

The growth of yeast in a sugar solution and the budding process by which yeast cells reproduce were discussed in Chapter 5, Exercise 7. Focusing only on the reproduction of yeast plants and ignoring factors affecting the death of cells, the reproduction process forms a one-loop feedback system depicted in the causal-loop diagram in Figure 14.1.

Exercise 1: Flow Diagram of Yeast Growth

Draw a flow diagram based on the causal-loop diagram in Figure 14.1. Begin by identifying the level and the rate.

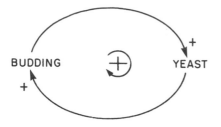

Figure 14.1 Causal-loop diagram of the growth of yeast

EQUATIONS FOR YEAST GROWTH

Working from a flow diagram, how can the equations be developed? The level equation, of course, has the standard form. (The name YEAST stands for the number of yeast cells, and the name BUDDNG stands for the budding rate, in cells per hour.)

$$\text{YEAST.K} = \text{YEAST.J} + (\text{DT})(\text{BUDDNG.JK}) \qquad \text{CELLS}$$

Now, what about the rate equation? As suggested by the flow diagram, the more cells there are, the more budding will take place. In fact, the situation is quite similar to the world population model discussed in Chapter 13. Recall that, in the world population model, the number of net births per year was formulated as a constant fraction of the world population:

$$\text{NBIRTHS.KL} = \text{POP.K} * \text{GF}$$

Thus it is plausible to formulate the yeast budding rate in a similar way, by making the assumption that a constant fraction of the yeast cells produce buds each hour. This constant fraction might be called the budding fraction BUDFR. For example, a budding fraction BUDFR of 0.1 would mean that 10 percent of the yeast cells in the population produce new yeast cells each hour.

$$\text{BUDDNG.KL} = \text{YEAST.K} * \text{BUDFR} \qquad \text{CELLS PER HOUR}$$

The budding fraction BUDFR plays exactly the same role in the yeast model as the growth fraction GF played in the world population model. Furthermore, BUDFR and GF have similar units. Recall that the growth fraction GF has units (persons/year)/person, which reduces to (1/year). Similarly, the budding fraction BUDFR has units (cells/hour)/cell, which reduces to (1/hour).

Exercise 2: Computing the Growth of Yeast

Using a table similar to Table 14.1, compute the growth of yeast through 20 hours. Assume that the budding fraction BUDFR is 0.1, and assume the initial number of yeast cells equals 10.

Table 14.1 Table for computing growth of yeast

Time (hours)	Change in yeast (cells)	Yeast (cells)	Budding (cells/hour)
0	—	10	
1			
2			
3			
4			
5			
6			
7			
8			
9			
10			
11			
12			
13			
14			
15			
16			
17			
18			
19			
20			

BUDDNG.KL = (YEAST.K)(BUDFR)
BUDFR = 0.10 1/hours
YEAST.K = YEAST.J + (DT)(BUDDNG.JK)
YEAST = 10 initially

Exercise 3: Graphing the Growth of Yeast

Now graph the yeast population calculated in Exercise 2.

WRITING THE MODEL IN DYNAMO

Below is a listing of the yeast model in the DYNAMO simulation language.

```
*       YEAST GROWTH
L       YEAST.K=YEAST.J+(DT)(BUDDNG.JK)
N       YEAST=10
NOTE            YEAST CELLS (CELLS)
R       BUDDNG.KL=(YEAST.K)(BUDFR)
NOTE            BUDDING (CELLS/HOUR)
C       BUDFR=0.10
NOTE            BUDDING FRACTION (1/HOUR)
PLOT    YEAST=Y/BUDDNG=B
PRINT   YEAST,BUDDNG
SPEC    DT=1/LENGTH=30/PLTPER=1/PRTPER=5
RUN
```

Each of the lines above is a DYNAMO statement. (See Figure 14.2.)

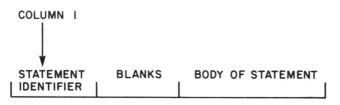

Figure 14.2 Format of DYNAMO statements

A DYNAMO statement begins with a letter or word to identify the type of statement. This letter or word is called the "statement identifier." One or more blank spaces follows the "statement identifier." DYNAMO uses blanks to separate different parts of a statement. DYNAMO accepts one blank or several blanks wherever a blank space is appropriate. Following the first set of blanks is the body of the statement. Often the body is an equation, but it may direct DYNAMO to perform an output or some other function. Blanks should *not* be inserted in the middle of equations or in the body of other types of statements, except NOTE and * statements. There may be up to 72 characters on a line. DYNAMO ignores any texts placed in column 73 and beyond. A line can be continued by putting an X in column 1 of the next line, leaving a space, and continuing the statement on that line. DYNAMO statements may occur in any order, since DYNAMO (unlike BASIC or FORTRAN) will automatically order equations for proper computation. Hence, a variable can be referred to in an equation before its defining equation is given. The last page of this chapter has a table of DYNAMO statement types.

Reading down the model listing, the statements are defined here and in the next several sections:

*   YEAST GROWTH

This statement, called a star (or asterisk) statement, named after its "statement identifier," provides the heading which DYNAMO prints at the top of each page of output at the terminal. In this case, DYNAMO will print "YEAST GROWTH" at the top of each page. Usually, a star statement begins each model and helps to identify it.

L   YEAST.K = YEAST.J + (DT)(BUDDNG.JK)

This statement defines the level YEAST. The L in column 1 indicates that this is a level variable. The rest of the statement is an equation that tells DYNAMO how to compute the value for the variable YEAST. In DYNAMO, variables or constant names can consist of one to six letters or numbers, with the first character being a letter. Equations use operators like $-$ and $+$, as described in Table 14.2.

Table 14.2 Arithmetic operations in DYNAMO

+	ADDITION
−	SUBTRACTION
*	MULTIPLICATION
/	DIVISION

Note: Multiplications can also be indicated by back-to-back parentheses. For example: (YEAST.K)(BUDFR)

In general, the style of writing equations is much like that of FORTRAN or other computer languages used for computing numerical formulas. However, DYNAMO also permits back-to-back parentheses to indicate multiplication. Thus, in the above line, (DT)(BUDDNG.JK) means that the variables DT and BUDDNG are multiplied together.

Another feature particular to DYNAMO is the use of "timescripts." Timescripts are the postscripts like ".K" and ".JK" in the above statement. They indicate the time relation among different variables as described in Chapter 13.

Exercise 4: Dynamo Statement Form

Some of the following statements are correct DYNAMO statements. Some are not. State whether each is correctly written. If not, rewrite the statement changing only enough to make it correct. (Warning: Some may have more than one error.)

a. LYEAST.K = YEAST.J + (DT)(BUDDING.JK)

b. L   YEAST.K = YEAST.J + (DT)(BUDDNG.JK)

c. L   YEAST.K = YEAST.J + DT*(BUDDNG.JK)

d. L   TEMP.K = TEMP.J + (DT)(−DECLNE.JK)

e. L   YEAST.K = YEAST.J + DT(BUDDNG.JK)

f. L   YEAST.K = YEAST.L + DT(BUDDNG.KL)

g. L   YEAST.L = YEAST.K + DT*BUDDNG.KL

h. LYEAST.K = YEAST.J + DT(BUDDNG.KL)

i. L   TEMP.K = TEMP.J + DT*(−DECLNE.JK)

j. L   POP.J = POP.K + (DT)(BIRTHS.JK)

k. L   INV.K = INV.J + DT*PROD.JK

MORE DYNAMO STATEMENT TYPES

N   YEAST = 10

This statement, an initial value statement, provides the starting value for the variable YEAST. Every level must have an initial value equation to provide the starting value for that level. This requirement is like the hand simulations' requirement of a starting value for each level computed in the tables.

NOTE   YEAST CELLS (CELLS)

NOTE statements provide a way of placing comments in a model. The form of the statement is NOTE followed by one or more blanks, and then any desired comment. A common use of NOTE statements, and one that should be used generously, is to provide a definition of each variable in the model. This definition should consist of the full variable name (e.g., YEAST CELLS) followed by the units of measure of the variable in parentheses. NOTE statements are also used later in the listing to comment on BUDDNG and BUDFR.

R   BUDDNG.KL = (YEAST.K)(BUDFR)

This statement is a rate statement defining the variable BUDDNG (which stands for "budding"). The R in column 1 indicates that the variable BUDDNG is a rate. As pointed out in Chapter 13, a statement that defines a rate equation has the form VARIABLE.KL =   .

C   BUDFR = 0.10

This statement is a constant statement defining the term BUDFR ("budding fraction"). The C in column 1 indicates that BUDFR is a constant.

Exercise 5: Dynamo Statement Types

Each of the following symbols is a valid DYNAMO statement identifier. State and illustrate for each its use in a DYNAMO program. That is, describe the kind of statement that would follow the given identifier and give an example different from the ones in the sample program.

  a. NOTE
  b. N
  c. *
  d. L
  e. C
  f. R

PLOT, PRINT, AND SPEC STATEMENTS

The following statement types are different from the preceding in that they are not part of the model itself, corresponding to parts of the flow diagram, but are detailed instructions to the DYNAMO compiler, informing it just how to run the model and for how long, as well as what results should be displayed. More details on this kind of "how to run" DYNAMO statements are given in the following sections.

    PLOT   YEAST = Y/BUDDNG = B

As indicated by the word PLOT starting in column 1, this statement indicates that a plot of the variables YEAST and BUDDNG is desired. DYNAMO plots are produced on the terminal after each simulation has been performed. The PLOT statement and the resulting plot are further described later.

    PRINT   YEAST,BUDDNG

This statement causes numerical values for the variables YEAST and BUDDNG to be printed out on the terminal in columns. The PRINT statement and the tabular output produced are further described later.

    SPEC   DT = 1/LENGTH = 30/PLTPER = 1/PRTPER = 5

The SPEC statement SPECifies values needed to perform the simulation and to produce output. DT, as explained in Chapter 13, is "difference in time" or "delta time," the time interval in level equations. LENGTH is the length of the simulation. The expression "LENGTH = 30" tells DYNAMO that the simulation, which will start at TIME = 0, should end at TIME = 30. PLTPER, or the PLoT PERiod, specifies the interval between times when the values of designated variables are plotted. In this example, since PLTPER is one, designated variables will be plotted each simulated hour (the unit of time used in this model). PRTPER, or PRinT PERiod, specifies the interval between times when the values of designated variables are printed. In this example, PRTPER is 5, so those values will be printed every fifth hour.

Exercise 6: The SPEC Statement

Here are some SPEC statements to interpret. Suppose that proper PRINT and PLOT statements accompany each SPEC statement. Answer for each SPEC statement the following questions:

1. What kinds of output will be produced (table or graph or both)?

2. What will be the time between successive lines in the table and successive lines in the graph?

3. Over what time period will the table and/or graphs extend?

4. How many lines will there be in each plot? In each table?
   a. SPEC   DT = 0.5/LENGTH = 10/PLTPER = 1/PRTPER = 0
   b. SPEC   DT = 2/LENGTH = 40/PLTPER = 4/PRTPER = 6
   c. SPEC   DT = 0.1/LENGTH = 20/PLTPER = 0/PRTPER = 2

THE RUN STATEMENT

      RUN

The RUN statement indicates the end of the model and is the signal to the computer to run the simulation as specified.

OUTPUT STATEMENTS

DYNAMO can produce both graphs and tables of the values of variables.

PRINT   Tabular output is specified by a PRINT statement, which has the form in this model of:

      PRINT   YEAST,BUDDNG

where any model variable can be named in place of YEAST and BUDDNG. Up to fourteen variables can be listed on a PRINT statement. If more are specified, the extra variables are treated as if they appeared on another PRINT statement. Table 14.3 depicts the tabular output from simulating the YEAST model.

The first row of the output contains the names of the variables being printed. The next row lists the scales, in terms of powers of ten. The actual value of a variable is the printed value times ten to the power listed in these

Table 14.3 Tabular output from yeast model

TIME	YEAST	BUDDNG
E 00	E 00	E 00
0.000	10.00	1.000
5.000	16.11	1.611
10.000	25.94	2.594
15.000	41.77	4.177
20.000	67.28	6.728
25.000	108.35	10.835
30.000	174.49	17.449

scales. (Note that for the three variables listed here, $10^0 = 1$.) Subsequent rows list the scaled variable values, printed to five significant digits. The values of time always appear in the left-most column. The interval of time between print lines is given by the parameter PRTPER specified on the SPEC statement.

PLOT   Graphs are specified with a PLOT statement, which has the form in this model of:

PLOT   YEAST = Y/BUDDNG = B

Any model variables can be named in place of YEAST and BUDDNG. Any letter or number can be used in place of the Y or the B as a plot character. Up to ten variables can be mentioned on a PLOT statement. The PLOT statement provides considerable flexibility:

1. DYNAMO can do the scaling for any or all of the plots.

2. Two or more variables can be forced to have the same scaling.

3. Either the upper scale or lower scale or both can be set by the user.

Figure 14.3 depicts the typical graph as it is produced at the terminal or line printer. At the top is a list of the variables plotted, each followed by an equal sign and its corresponding plotting character. Next are the plotting scales, each line consisting of the five scale values followed by the plot characters at the extreme right, indicating which scales apply to which variables. The main body of the plot follows, divided across the page into four sections.

DYNAMO positions the plotting characters on each line of output according to the values of the variables and their corresponding scales. Successive lines represent the values at later points in time. The value of time and a row of dashes are printed on every tenth line.

An unfortunate aspect of the plot output shown in Figure 14.3 is that the curves for YEAST and BUDDNG lie on top of each other. That is indicated on the right side of the plot by the YB (meaning that B lies in the same position as Y for that line). It might be clearer to see the two plot lines separated. When

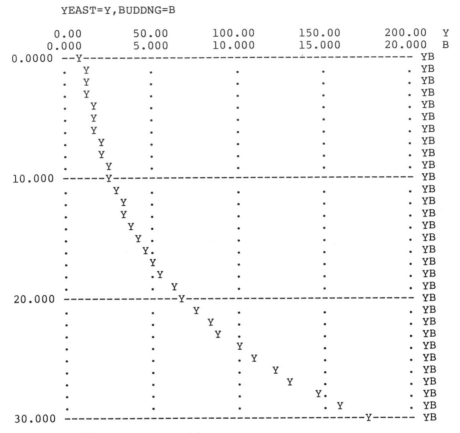

Figure 14.3 Plot from yeast model

this difficulty arises, DYNAMO may be given the scales for plotting each variable. To do so, the original model line

   PLOT   YEAST = Y/BUDDNG = B

may be changed to

   PLOT   YEAST = Y(0,180)/BUDDNG = B(0,25)

for example. This new PLOT statement tells DYNAMO to set the scale for YEAST from 0 to 180 (cells) and BUDDNG from 0 to 25 (cells/hour). Figure 14.4 gives the output with scales so set. None of the values for YEAST or BUDDNG have changed, just the PLOT scales. A summary of DYNAMO statement types is given at the end of this chapter.

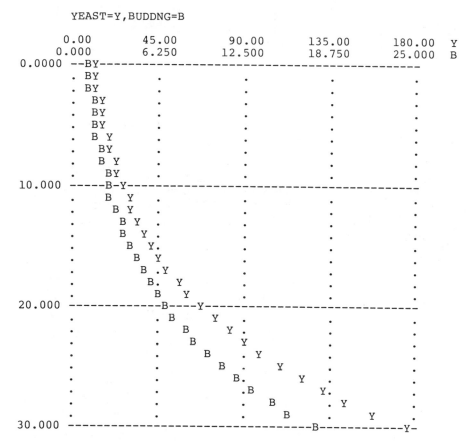

YEAST=Y,BUDDNG=B

```
       0.00          45.00          90.00         135.00         180.00   Y
       0.000          6.250         12.500         18.750         25.000   B
0.0000  --BY----------------------------------------------------------
        .  BY              .              .              .              .
        .  BY              .              .              .              .
        .    BY            .              .              .              .
        .    BY            .              .              .              .
        .    BY            .              .              .              .
        .    B  Y          .              .              .              .
        .     BY           .              .              .              .
        .    B  Y          .              .              .              .
        .     BY           .              .              .              .
10.000  -----B-Y----------------------------------------------------------
        .     B   Y        .              .              .              .
        .      B  Y        .              .              .              .
        .       B  Y       .              .              .              .
        .       B   Y  .              .              .              .
        .        B   Y.              .              .              .
        .         B   Y       .              .              .              .
        .          B  .Y       .              .              .              .
        .           B.   Y       .              .              .              .
        .           B     Y       .              .              .              .
20.000  -------------B----Y----------------------------------------------------
        .          .  B        Y  .              .              .              .
        .          .   B       Y .              .              .              .
        .          .    B        Y              .              .              .
        .          .     B    .  Y              .              .              .
        .          .       B  .      Y              .              .              .
        .          .        B.        Y  .              .              .
        .          .         .B             Y.              .              .
        .          .          .  B            Y.  .              .
        .          .          .     B           .  Y              .
30.000  -----------------------------------------------------B-----------Y-
```

Figure 14.4 Plot from yeast model with set scales

Exercise 7: Plotting

Suppose you wish to produce a plot of the output of a DYNAMO model. The
model is of the population of trees on a tree plantation and has levels for trees
of three ages named YOUNG, MIDDLE, and AGING. First, choose an answer to each question and then state how you will inform the computer of your
choice.

a. What variables to plot?

b. What symbols for each variable?

c. How much time between one plot line and the next?

d. For how long a time should the plot be made? (Note that your answer to
(d) will determine how long the model will run, and so will determine the
time span of any output printed in tabular as well as graphical form.)

e. Write your instructions to the computer in the form of two lines:
    SPEC
    PLOT

RERUNS

After the first set of output is produced, DYNAMO asks for rerun changes by typing at the terminal the message:

ENTER   RERUN   CHANGES

At this point, the user may change constants in the model. For example, to produce a simulation with the budding fraction BUDFR equal to 0.25 instead of 0.1, the user would enter the following statements at the terminal after the rerun request is made by DYNAMO:

C   BUDFR = 0.25
RUN

When the RUN statement is entered again, DYNAMO performs the requested rerun simulation and produces the indicated output. Since, as model parameters such as BUDFR are changed, any particular fixed plot scales will not necessarily be appropriate, it is best to do a first run with autoscaling (as shown in Figure 14.3), then change the PLOT line to fixed scales in a rerun. To end DYNAMO use, type QUIT when the rerun request appears.

   For further details on any aspect of DYNAMO operation or statements, see *DYNAMO User's Manual* (Pugh, 1976), or see the DYNAMO manual that accompanies your copy of the DYNAMO language.

Exercise 8: Running the Yeast Model

a. Input into the computer the YEAST GROWTH model as given following Exercise 3. Refer to the appropriate DYNAMO User's Guide for your particular computer for instruction on creating a model file.

b. Run the model. See if your outputs are the same as those shown in Table 14.3 and Figure 14.3.

c. Try changing the constant BUDFR = 0.1 to a new value in the rerun mode. How is the graph different and why?

EXAMPLE II: ADDING YEAST DEATHS TO THE MODEL

Figure 14.5 depicts a causal-loop diagram in which a second feedback loop involving yeast deaths has been added. As assumed in the second loop, the more yeast cells there are, the more cells will age and die.

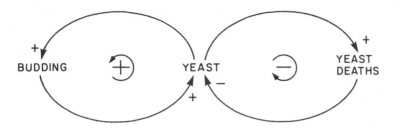

Figure 14.5 Causal-loop diagram with budding and yeast deaths

Exercise 9: Flow Diagram for Yeast Deaths

Draw a flow diagram based on the causal-loop diagram in Figure 14.5.

EQUATIONS FOR YEAST AND YEAST DEATHS

To simulate the system diagramed in Exercise 9, new equations for yeast deaths and yeast are needed. The level equation for yeast becomes:

$$YEAST.K = YEAST.J + (DT)(BUDDNG.JK - YDEATH.JK) \text{CELLS}$$

where YDEATH is yeast deaths. YDEATH appears in the level equation with a minus sign because it is an outflow. The net change in the yeast population is the budding rate minus the death rate, both multiplied by the time interval.

One way to formulate the rate equation for yeast deaths is to consider the average lifetime of yeast cells. Suppose yeast cells, on the average, live twenty hours. Under this assumption, one possible way of writing an equation for yeast deaths is:

$$YEAST.KL = YEAST.K/ALIFEY$$

where ALIFEY is the average lifetime of yeast cells (twenty hours).

One way to justify this formulation is the following. Assume there are 100 years cells evenly distributed by age. (See Figure 14.6.)

Thus 5 cells are 1 hour old, 5 are 2 hours old, and so forth. If each cell lives exactly 20 hours, then each hour, the cells that are exactly 20 will die. Since the number of cells that are exactly twenty hours old equals the total number of cells divided by twenty, the death rate (in cells per hour) can be expressed as the total number of cells divided by the average lifetime of cells.

Of course, in actual circumstances, cells do not all have the same lifetime, nor will cells be evenly distributed by age. Nevertheless, the formulation YDEATH.KL = YEAST.K/ALIFEY has the advantage of simplicity. In addition, under a variety of assumptions about the distribution of yeast lifetimes and ages, this formulation provides a reasonable approximation of the process of yeast deaths.[2]

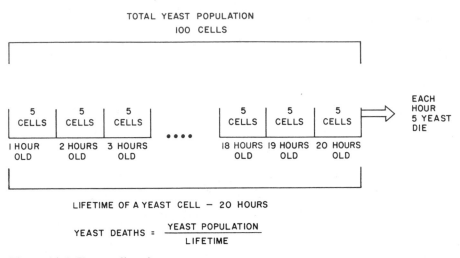

Figure 14.6   Yeast cells aging

Exercise 10: Computing Yeast Deaths by Hand

Using a table similar to Table 14.4, compute yeast population, budding, and yeast deaths. Graph the yeast population.

Exercise 11: Running the Model in DYNAMO

Incorporate yeast deaths into the DYNAMO model. You will need to modify the level equation for YEAST, add a rate equation for yeast deaths, and add a constant statement for the average lifetime of yeast. Modify the PRINT and PLOT statements so that yeast deaths is plotted and printed. Run the model. Compare the outputs with your results for Exercise 10 and contrast them with the outputs for Exercise 8(b) (or Figure 14.3).

Exercise 12: Computing Xanadu Population

From Chapter 13, Example III, input into the computer the model for Xanadu population growth. Run the model and compare your results to the answer for Chapter 13, Exercise 6.

Exercise 13: Computing World Population

From Chapter 13, Example IV, input to the computer the model for world population growth. Run the model and compare your results to the answer for Chapter 13, Exercise 8.

Table 14.4 Table for computing yeast population, budding, and deaths

Time (hours)	Change in yeast population	Yeast (cells)	BUDDNG (cells/ hour)	YDEATH (cells/ hour)
0				
1				
2				
3				
4				
5				
6				
7				
8				
9				
10				
11				
12				
13				
14				
15				
16				
17				
18				
19				
20				
21				
22				
23				

(continued)

Table 14.4 (continued)

Time (hours)	Change in yeast population	Yeast (cells)	BUDDING (cells/ hour)	YDEATH (cells/ hour)
24				
25				
26				
27				
28				
29				
30				

BUDDNG.KL = (YEAST.K)(BUDFR)   CELLS/HOUR
BUDFR = 0.1   1/HOURS
YDEATH.KL = YEAST.K/ALIFEY   CELLS/HOUR
ALIFEY = 20   HOURS
YEAST.K = YEAST.J + (DT)(BUDDNG.JK – YDEATH.JK)
YEAST = 10 initially

TYPICAL RATE EQUATIONS

The Xanadu, world population, and yeast examples have involved several different rate equation formulations. In the Xanadu case, the birth rate (in people per year) was simply a constant.

BIRTHS = 100

In the world population case, the net birth rate (in people per year) was written as a growth fraction GF multiplied by the size of the population POP.

NBIRTH.KL = POP.K*GF

The yeast budding example has precisely the same form as net births in the world population case.

BUDDNG.KL = YEAST.K*BUDFR

Finally, yeast deaths YDEATH was formulated as the number of yeast cells divided by the average lifetime of yeast ALIFEY.

YDEATH.KL = YEAST.K/ALIFEY

The rate formulations employed in these examples are fundamental: They appear in a large number of dynamic models, and they form the basis of other, more complex formulations. Given the widespread use of these formulations, it is worth calling attention to their general form. The Xanadu case offers an illustration of the simplest possibility—a rate that is constant.

RATE = CONSTANT

Given its simplicity, this is sometimes a good place to start trying to formulate a rate equation. Often rates *are* constant (or nearly so). For example, a private college might admit a constant number of new students per year; or a library might order a constant number of new books each year.

The world population and yeast budding cases illustrate a second possibility: a rate that equals a level multiplied by a growth fraction.

RATE = (LEVEL)(GROWTH FRACTION)

This formulation provides a precise representation of growth in a fair number of systems (for example, the interest earned on a bank balance). In addition, it offers a good approximation of the early stages of growth in many social and biological examples (such as the growth of cities or the growth of rabbits on a field).

Finally, the yeast deaths case illustrates a third possibility: a rate that equals a level divided by an average lifetime.

RATE = LEVEL/(AVERAGE LIFETIME)

This formulation offers a general representation of quantities that decay over time (analogous to the growth fraction formulation for quantities that grow). The average lifetime formulation provides a reasonable approximation to such diverse phenomena as the decay of housing, the loss of books from a library, or the dissipation of pollution from a lake.

Exercise 14: Forms of Rate Equations

Write DYNAMO equations for each of the following examples, using one of the three standard rate equation formulations discussed in the text.

a. The Hometown Dump currently contains 192,000 tons of nonbiodegradable garbage, and roughly 15,000 additional tons are dumped each year. Write equations for the level of garbage at the Hometown Dump over time.

b. The number of homes in Clean Air, Arizona, equipped with solar heating has been rising at 7 percent per year over the past few years, and 195 homes in the town now have solar heaters. Write equations for a model of the growth in the number of solar-heated homes in Clean Air, assuming the percentage increase continues.

c. The Metro City transportation department has just purchased a fleet of 100 new Sonic subway cars. Experience indicates that the average useful lifetime of cars similar to those purchased is 15 years. Write equations for a model of the number of Sonic subway cars remaining in use over time.

ENDNOTES

1. DYNAMO is available for the APPLE microcomputer and for most mini- and mainframe computers. The APPLE version is distributed by Addison-Wesley Publishing Company through computer stores. All other versions are distributed by Pugh-Roberts Associates, Inc., 5 Lee Street, Cambridge, MA 02139.
2. Alternative representations of processes involving an average lifetime are considered in Chapter 17.

DYNAMO statement types

*	page heading
L	Level
R	Rate
A	Auxiliary
N	iNitial value
C	Constant
T	Table
SPEC	SPECification of DT, LENGTH, PLTPER, and PRTPER
PLOT	PLOT statement
PRINT	PRINT statement
RUN	RUN statement (end of model)
NOTE	comment
X	continuation of previous line

CHAPTER 15

USING SIMULATION TO ANALYZE SIMPLE POSITIVE AND NEGATIVE LOOPS

The computer simulation techniques developed so far can be used to analyze the behavior of simple positive and negative loops. Because simple positive and negative loops form the building blocks of more complex models, it is important to understand the kinds of behavior they can generate in fairly rich detail. The examples that follow illustrate some of the most common positive and negative loop structures, and some useful ways of employing simulation to probe system behavior.

EXAMPLE I: YEAST BUDDING (POSITIVE LOOPS)

The simplest and most fundamental positive feedback loop consists of one level and one rate, and the rate is directly proportional to the level. An example, shown in Figure 15.1, is the model of yeast budding taken from Chapter 14. (The equations for this model are listed after Exercise 3 of Chapter 14.)

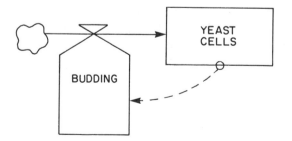

Figure 15.1 Flow diagram of yeast budding

In a simple positive feedback loop with one rate and one level, as the level increases, the rate increases as well, so the level grows at an increasing pace. If the rate is directly proportional to the level (as in the yeast example), the behavior generated is exponential growth.

A quantity that is growing exponentially will double in a fixed amount of time, no matter how long it has been growing or how large the quantity has become. For example, in the yeast budding loop, if the budding fraction BUDFR = 0.1, it can be shown that the doubling time for the number of yeast cells is roughly 7 hours. Thus, if the initial number of yeast cells is 10, the number of cells will reach 20 in about 7 hours, and it will reach 40 in another 7 hours.[1]

The following exercises provide an opportunity to explore the relationship between the growth fraction and the doubling time in simple positive loops.

Exercise 1: Simulation of a Positive Feedback Loop

Look up the equations for yeast growth from Chapter 14. From your first run of the model, describe the behavior of the level and rate.

Exercise 2: Doubling Time

a. Run the yeast model for forty simulated hours.

b. Measure the time needed for the number of yeast cells to double from the initial value of yeast.

c. Measure the time required for the number of yeast to double from its value at hour twenty.

Exercise 3: Effect of Budding Fraction

a. Run the yeast model with a budding fraction of 0.2.

b. Run the model with a budding fraction of 0.05. How do the results differ? Does the model still generate exponential growth? Is there a value of the budding fraction which will cause the model not to produce exponential growth?

c. If the budding fraction were negative, would the feedback loop still be positive?

Exercise 4: The Bank Account—Part I

Suppose you deposit $500 in a bank account earning 10 percent interest compounded annually.

a. Draw a causal-loop diagram and flow diagram for the bank account case. (Assume no money is withdrawn from the account.)

b. Write DYNAMO equations and simulate the bank account for a twenty-year period. (Set DT = 1 year).

c. How much money is in the account in year twenty? What is the doubling time for the account?

Exercise 5: The Bank Account—Part II

In Part I of the bank account problem, you deposited $500 in the bank and left it there to gather interest. Suppose the account earns interest exactly as before, but you must withdraw at the constant rate of $50 per year from the account.

a. Modify your flow diagram to include the withdrawal rate of $50 per year.

b. Modify your DYNAMO equations and run the model on the computer. How do the results differ from your results in Part I?

c. Suppose you begin with $600 in your account, rather than $500, and withdraw $50 per year. How do the results differ?

d. Suppose you begin with $400 in the account. How do the results differ?

STARTING A MODEL IN EQUILIBRIUM

In analyzing the behavior of a system, it is often helpful to begin by determining the system's equilibrium point. This can be done by trial and error, but it is often easier to determine the equilibrium point by examining the flow diagram and system equations. For example, in the preceding bank account case, equilibrium occurs when the money withdrawn each year exactly equals the amount of interest earned. If $50 is withdrawn per year, this means that equilibrium occurs when $50 interest is earned. If the rate is 10 percent, this corresponds to a bank account balance of $500.

Once you have determined the equilibrium point mathematically, it is easy to check your calculations by simulating the results on the computer. Just set the initial values of the system levels to their equilibrium points. If your calculations are correct, the model should remain in equilibrium.

Exercise 6: Calculating Equilibrium Values

a. Suppose you withdraw $60 per year from a savings account earning 10 percent interest. What is the equilibrium balance?

b. Suppose you withdraw $50 per year from an account earning 8 percent interest. What is the equilibrium balance?

EXAMINING A SYSTEM'S RESPONSE TO DISTURBANCES—PART I

Once you have started a system in equilibrium, it is often useful to see how the system responds to exogenous (i.e., outside the system) disturbances. For example, suppose you place $500 in a bank account earning 10 percent interest, and withdraw $50 per year. Then, however, starting five years from now, it becomes necessary to withdraw $75 per year. How will the system respond?

One way to test the response of the system is to use a special DYNAMO function called the STEP function to simulate the sudden $25 increase in the withdrawal rate beginning in five years. A reasonable set of equations for the model includes the following:

```
*     BANK ACCOUNT
L     BAL.K = BAL.J + (DT)(INT.JK − WDRW.JK) DOLLARS
N     BAL = 500
R     INT.KL = (0.10)(BAL,K) DOLLARS/YEAR
R     WDRW.KL = 50 + STEP(25,5) DOLLARS/YEAR
```

The expression "STEP(25,5)" instructs the computer to increase the withdrawal rate by 25 dollars in year 5. Thus a graph of the withdrawal rate would look like Figure 15.2.

The STEP function can be used whenever it is necessary to simulate a sudden step change in a system rate. The general DYNAMO form of the STEP function is:

STEP(HEIGHT,STTIME)

where HEIGHT is the height of the STEP, and STTIME is the abbreviation for step time, the time when the step occurs. Any numerical values can be used for HEIGHT and STTIME.

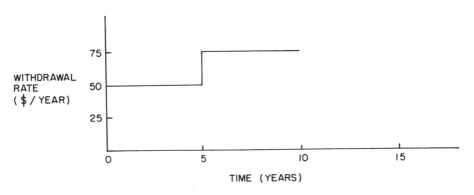

Figure 15.2 Step function graphed

Exercise 7: Using the STEP Function

a. Use the STEP function to test the response of the bank account model to an increase in the withdrawal rate from $50 to $75 at TIME = five years. What behavior does the system generate?

b. Use the STEP function to test the response to the model to a decrease in the withdrawal rate from $50 to $30 at TIME = three years. What behavior does the system generate?

EXAMPLE II: YEAST DEATHS (NEGATIVE LOOP)

The simplest and most fundamental negative loop contains one rate and one level. An example, shown in Figure 15.3, is the yeast deaths case taken from Chapter 14. If, in a simple negative loop, the rate is directly proportional to the level, the loop will generate exponential decay.

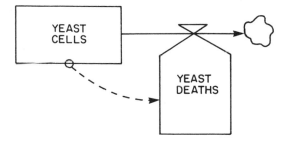

Figure 15.3 Yeast deaths

A quantity that is decaying exponentially will move half-way to its equilibrium value in a fixed amount of time, no matter how far from equilibrium it begins. For example, if the average lifetime of yeast is 20 hours, it can be shown that the halving time is 14 hours. Thus if the initial number of yeast cells is 10, the number of cells will fall to 5 in 14 hours; and it will fall to 2.5 in another 14 hours.[2]

The exercises that follow provide an opportunity to explore the relationship between the average lifetime and the halving time for simple negative loops.

Exercise 8: Simulation of a Negative Loop

Simulate the yeast deaths loop, using the equations developed in Chapter 14. (Assume an initial value of 10 yeast cells, and do not include yeast budding.) Examine the behavior of the level and the rate. How do they differ from the behavior of the yeast budding loop? What is the equilibrium point for the number of yeast cells?

Exercise 9: Halving Time

How long does it take the number of yeast cells to get half-way from its initial value to its equilibrium value? How long does it take it to get from half-way to one-quarter of the way?

Exercise 10: Effect of the Yeast Lifetime

Run the model with an average lifetime of yeast equal to 10 hours. Run the model again with the average lifetime of yeast equal to 40 hours. Does the behavior in each case still represent exponential decay? How do the results differ?

Exercise 11: Yeast Model with Budding and Deaths

Run the yeast model, including both yeast budding and yeast deaths. Does the model exhibit exponential growth or decay? If the model exhibits growth, try to find values of constants that will cause the model to show decay. If the model shows decay, try to find values of constants that will cause the model to show growth. Is there a set of constants that will cause the model to show growth and then decay? Why or why not?

Exercise 12: Central Library—Part I

Books in the Central Library in East Rapids are frequently stolen or lost, and often they just plain fall apart. In fact, the average lifetime of books in the library is just 10 years. The East Rapids City Council provides a library budget large enough for the purchase of 500 new books a year.

  a. Draw a causal-loop diagram and flow diagram for the Central Library case.

  b. Write equations for the model.

  c. Determine the equilibrium point for the number of books in the library, and start the model in equilibrium.

Exercise 13: Central Library—Part II

The City Council in East Rapids has just completed a lengthy analysis of its budget, and, as a result, the library budget is expected to fall sharply, starting in three years. With the planned budget cut, the library will be able to purchase only 300 books a year, rather than 500.

a. Use a STEP function to simulate the effects of the sharp reduction in the book purchase rate in three years.

b. What is the new equilibrium level of books? How long does it take the number of books to fall half-way to the new equilibrium level?

c. Suppose a wealthy donor provides an endowment that permits the library to purchase 700 books a year, beginning in six years. Use a STEP function to simulate the increase in the purchase rate. What is the new equilibrium? How long does it take the number of books to rise half-way to the new equilibrium?

d. How would the number of books in the Central Library be affected, if the average lifetime of books could be increased from 10 years to 20 years?

MORE COMPLEX RATE FORMULATIONS

The simple positive and negative loops considered so far all involve rate equations based on one of the following three forms:

RATE = CONSTANT
RATE = LEVEL*(GROWTH FRACTION)
RATE = LEVEL/(AVERAGE LIFETIME)

The following example introduces a rate formulation that is closely related to the average lifetime formulation, but is more complex.

EXAMPLE III: COFFEE COOLING (NEGATIVE LOOP)

This example involves performing a simple physical experiment and then modeling the system in the experiment. You will need the following equipment:

1. A cup of hot coffee or other liquid at a temperature well above room temperature;

2. A thermometer capable of measuring the temperature of the coffee (use one that will read at least as high as 100 degrees Celsius; a laboratory or candy thermometer should work);

3. A watch or stop watch;

4. A pencil and paper.

While the temperature is still hot, measure the coffee's temperature at regular intervals (every few minutes). Record the temperature and the time of the reading on a sheet of paper. Stop taking readings when the temperature is near room temperature.

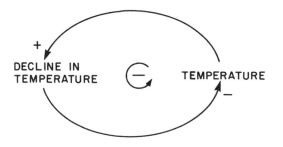

Exercise 14: Graphing Temperature

Plot the temperature versus time.

MODELING TEMPERATURE CHANGE

Development of a model of coffee cooling starts, of course, with a causal-loop diagram. Figure 15.4 depicts a possible diagram.

   The diagram says that the decline in temperature reduces the temperature, and the lower the temperature, the smaller the decline. (Is this hypothesis consistent with the data from your experiment?) Figure 15.5 depicts a flow diagram based on the causal-loop diagram in Figure 15.4.

Figure 15.4 Causal-loop diagram of temperature change

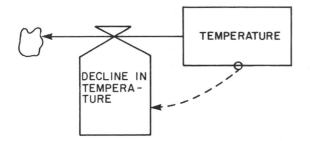

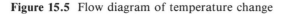

Figure 15.5 Flow diagram of temperature change

Using the name TEMP for temperature (in degrees Celsius) and the name DECLINE for the decline in temperature (in degrees Celsius per minute), the level equation for temperature can be written in the usual form:

TEMP.K = TEMP.J + (DT) (− DECLINE.JK)

Now, how should the rate equation for the decline in temperature DECLINE be formulated? Two assumptions are involved. First, according to the causal-loop diagram, the higher the temperature, the faster the decline. Second, common sense suggests that, when a cup of coffee reaches room temperature, it will not decline further. Therefore, perhaps the decline in temperature should be formulated as a function of the *difference* between the temperature of the coffee and room temperature. The further above room temperature a cup of coffee is, the faster its temperature will decline.

This suggests the following rate equation:

DECLNE.KL = (TEMP.K − ROOMTP)/T

where ROOMTP is room temperature, and T is a "cooling constant."

The "cooling constant" T determines how fast the adjustment of temperature occurs. (Thus T is measured in minutes.) The larger the value of T, the slower the decline in temperature. The value of T might depend on several things. One is the type of coffee container. For example, a glass generally will release heat more quickly than a ceramic cup. (You might redo the previous exercise, comparing several containers.) In addition, the larger the surface area of the container relative to the volume of liquid, the smaller T generally will be.

Exercise 15: Simulating the Coffee Cooling Case

Write DYNAMO equations and simulate the coffee cooling system. (Set DT = 1 minute.) Once you have simulated the values of temperature, plot them on the graph you used for Exercise 14. To carry out the simulation, you need to estimate T. There are at least two ways to do this. One is to start by simulating the system using a guess for T. If the simulated temperature drops faster than in the experiment, increase T and redo the simulation. Similarly, if the simulated temperature drops more slowly than the experimental data, decrease T. Keep trying values of T until the behavior of the model matches the behavior in the experiment.

Another way to estimate T is as follows. Let TEMP.J equal one observation in the experiment, let TEMP.K equal the next observation, and let DT equal the time between observations. Then T can be estimated using the formula:

$$T = (DT) (TEMP.J - ROOMTP)/(TEMP.J - TEMP.K)$$

For example, if a reading at one point is 50 degrees Celsius and two minutes later is 45 degrees Celsius, and if room temperature is 20 degrees Celsius, then T would be:

$$(2)(50 - 20)/(50 - 45) = 12$$

Averaging two or more estimates will give a more accurate value of T.

AUXILIARY VARIABLES

The coffee flow coding diagram in Figure 15.5 has one defect: much of the detail involved in the rate equation for the decline in coffee temperature is hidden in a singled dashed line connecting the level and the rate. One way to clarify the flow diagram is to define a new variable DIFF, which is the difference between the temperature of the coffee (TEMP) and the room temperature (ROOMTP): DIFF.K = TEMP.K - ROOMTP. This new variable is called an *auxiliary* variable because it aids in forming a rate. (This is similar to the role of auxiliary verbs in English, which aid in expressing an action verb.) Using the auxiliary variable DIFF, the equations for the coffee cooling model can be rewritten as follows:

```
L    TEMP.K = TEMP.J + (DT)( - DECLNE.JK)   DEGREES
N    TEMP = 50
R    DECLNE.KL = DIFF.K/T   DEGREES/MINUTE
C    T =   MINUTES
A    DIFF.K = TEMP.K - ROOMTP   DEGREES
C    ROOMTP = 20   DEGREES
```

Figure 15.6 indicates how DIFF can be added to the coffee cooling flow diagram.

Auxiliary variables are often useful in formulating complex rate equations. Auxiliaries arise when the formulation of a level's influence on a rate involves one or more intermediate calculations—similar to the calculation of DIFF in the coffee cooling case. This sort of intermediate calculation occurs quite frequently in complex models, as will be seen, and the use of auxiliaries

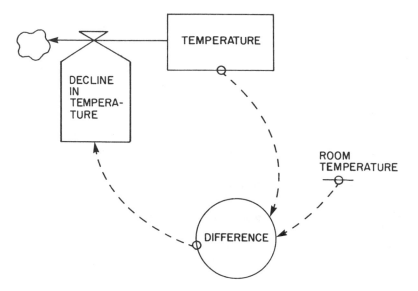

Figure 15.6 Flow diagram with auxiliary added

can clarify otherwise confusing formulations. A summary of flow diagrams symbols is given at the end of this chapter.

THE GOAL-GAP FORMULATION

The rate equation for the decline in temperature in the coffee-cooling model is an example of a quite general rate formulation:

$$\text{RATE} = (\text{LEVEL} - \text{GOAL})/(\text{ADJUSTMENT TIME})$$

In the coffee cooling example, the coffee temperature (the system level) can be viewed as drifting toward a "goal" (the room temperature). The "adjustment time" in this formulation plays a role analogous to the average lifetime in a simple decay formulation: It determines how rapidly the system level adjusts toward its goal. In fact, the simple decay formulation can be viewed as a special case of the goal formulation, in which the "goal" is zero.

$$\text{RATE} = \text{LEVEL}/(\text{ADJUSTMENT TIME})$$

The goal formulation is frequently useful in representing purposeful action. For example, suppose the manager of a department store wishes to maintain a certain fixed number of shoes in stock. The rate at which the manager orders new shoes might depend on the difference between the actual number of shoes in stock and the goal (i.e., the number the manager desires). Furthermore, if a gap exists between the number of shoes desired and the actual number in stock, the manager might not attempt to fill the gap all at once, but might prefer to close the gap gradually, over a period of time.

In this case, the shoe order rate might be written:

ORDERS.KL = (GOAL – SHOES.K)/ADJT

This indicates that the number of shoes ordered per month (ORDERS) depends on the gap between the desired number of shoes in stock (the GOAL) and the current level of shoes in stock (SHOES). Furthermore, the gap is not closed all at once, but instead is closed over a period of time, the adjustment time ADJT. (Note that, in this case, the order rate is formulated "GOAL minus LEVEL." In the coffee cooling example, the decline rate was formulated "LEVEL minus GOAL." The two formulations are equally useful. Which is chosen depends on the logic of the particular example.)

More generally, the goal-gap rate formulation is useful whenever an identifiable *goal* exists, alternatively seen as an objective, a target, a norm, or a desired condition. In comparison or in contrast to this goal is the actual *situation,* which is inevitably a level. The *difference* between the goal and the actual condition, the "goal-gap," is the motivator or driving force underlying corrective action. But the corrective action does not occur all at once; instead, it occurs over some adjustment time. (See Figure 15.7.)

The following exercise provides an opportunity to explore the goal-gap formulation.

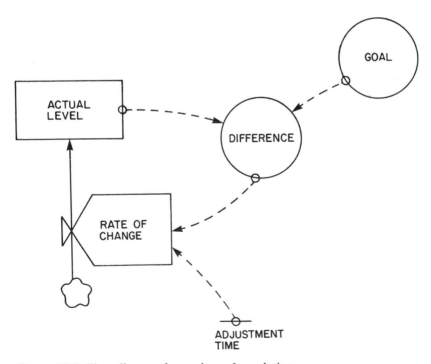

Figure 15.7 Flow diagram for goal-gap formulation

Exercise 16: Jobs and Migration—Part I

According to Example III in Chapter 3, the availability of job openings influences workers to migrate into the city; and as workers migrate into the city, they fill the available openings. This suggests the negative loop shown in Figure 15.8.

a. Formulate a flow diagram for the jobs and migration case, choosing a goal-gap structure. (*Hint:* It is easiest to view the population of workers in the city as the system level.)

b. Write equations for the model. (*Note:* You do not need to choose parameters or run the model on the computer. That will be treated in the last exercise of this chapter.)

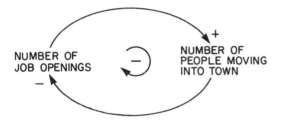

Figure 15.8 Effect of jobs on migration

EXAMINING A SYSTEM'S RESPONSE TO DISTURBANCES—PART II

Recall that earlier in the chapter, the response of the bank account system to an exogenous change in the withdrawal rate was examined, as well as the response of the library system to an exogenous change in the book acquisition rate. In a similar fashion, the response of the coffee cooling system to an exogenous change in room temperature can be analyzed.

For example, suppose a cup of tepid (20 degree Celsius) coffee that has been sitting on the kitchen table is suddenly placed in the refrigerator. How would it respond?

Once again, an easy way to examine the response of the system is to use a STEP function. In this case, the "variable" that needs to be stepped is the room temperature ROOMTP, which is currently a constant. In order to produce a step change in ROOMTP, it is necessary to redefine ROOMTP as an auxiliary variable and give it a subscript K. (ROOMTP must have a subscript, because it now will vary with time.) This produces the following equations:

L $TEMP.K = TEMP.J + (DT)(-DECLNE.JK)$
N $TEMP = 20$
R $DECLNE.KL = DIFF.K/T$
C $T =$

A   DIFF.K = TEMP.K − ROOMTP.K
A   ROOMTP.K = 20 + STEP(TCHG, 10)
C   TCHG = − 15

The equation for ROOMTP indicates that after 10 minutes, the room temperature drops 15 degrees, from 20 to 5.

Exercise 17: Stepping Room Temperature

a. Modify the coffee cooling-model to include the STEP change in ROOMTP. Run the model with step change from 20 degrees to 5 degrees as just mentioned.

b. Suppose the coffee was moved from the kitchen table to the sauna (35 degrees centrigrade). How would the system respond?

ADDITIONAL RATE FORMULATIONS

Sometimes, when trying to move from a causal-loop diagram to a flow diagram, it is difficult to decide which variables are levels, which are rates, and which are auxiliaries. In addition, once rates and levels have been identified, it is sometimes difficult to decide on an appropriate formulation for the rate equations. Frequently, none of the formulations discussed so far seem appropriate, and new formulations must be invented to fit the purpose. Often, formulating rate equations requires a certain amount of ingenuity—and a healthy willingness to try out alternative possibilities. The following example illustrates some of the problems involved in writing equations for a somewhat difficult model.

EXAMPLE IV: PUSHUPS AND PRACTICE (POSITIVE LOOP VERSION)

The causal-loop diagram in Figure 15.9 relates the number of pushups Jim can do, and the amount he practices. (See Exercise 1 in Chapter 4 for a discussion of this example.)

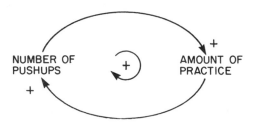

Figure 15.9 Pushups and practice cycle

Both the number of pushups Jim can do and the amount he practices seem at first glance to be levels; certainly neither appears to be a rate. This creates a problem, though, because levels cannot change unless there are rates to change them. Thus a rate must be "hidden" in the causal-loop diagram. One solution to the problem is the following. Assume that the number of pushups Jim can do is a level. Then there ought to be an "improvement rate" that causes the number of pushups to increase. (Jim's improvement rate would be the number of additional pushups he can do per month.) This produces the causal-loop diagram shown in Figure 15.10.

Now, what about "Amount of Practice"? One simplifying assumption would be that the amount Jim practices is a direct function of the number of pushups he can do; and the more he practices, the faster he improves. Under this simplification, "Amount of Practice" would be an auxiliary variable, and the complete flow diagram would look like Figure 15.11. Of course, more complicated assumptions about decisions affecting practicing would lead to quite different flow diagrams.

Now all that remains is to write the equations. According to the flow diagram, the number of pushups Jim can do influences the amount he practices. But what is the exact relationship between the two? Jim would have to be observed for some period to find out. Since this is a speculative model, a simple plausible relationship will be hypothesized. (In fact, one reason to build a simulation model is to examine the implications of plausible hypothesized relationships.)

One simple assumption is that the amount Jim practices is a linear function of the number of pushups he can do. For example, perhaps he practices one-half minute per day for each pushup he can do. This produces the following equation:

Amount of Practice(minutes) = 0.5(minutes/pushup)*Number of
     Pushups

Thus, for example, if Jim can do 30 pushups, he practices 15 minutes a day, under this assumption.

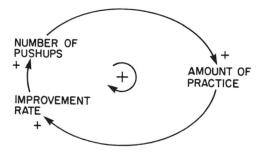

Figure 15.10 Pushups and practice—enlarged

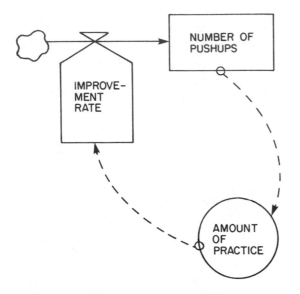

Figure 15.11 Flow diagram of pushups and practice

Now, the relationship between the amount Jim practices and his improvement rate must be formulated. It seems plausible to assume that Jim must practice at least a certain amount of time, simply to maintain his current level of pushups. Any practice above and beyond this maintenance amount would result in improved performance. It might be reasonable to assume that Jim must practice 10 minutes a day to maintain his performance; and for every minute he practices above 10, he improves at the rate of 0.2 pushups per month. This produces the following equation:

Improvement Rate (pushups/month) = (Amount of Practice – 10 minutes)*0.2 (pushups/month/minute)

This relationship implies that when Jim practices 10 minutes a day, he does not improve at all. If he practices 15 minutes a day, he improves at the rate of 1 pushup per month; and if he practices only 5 minutes a day, his performance drops by 1 pushup per month.

Exercise 18: A Model of Pushups and Practice

a. Write DYNAMO equations for the pushup and practice model. (Choose DT = 1 month.)

b. Run the model setting the initial number of pushups Jim can do to 30. Rerun the model, setting the initial number he can do to 20. Rerun the model again, setting the initial number of pushups to 10. How do the results differ? Why?

c. How does the model's equilibrium point depend on the amount of time Jim practices per pushup he can do? Run the model several times, choosing different values for this parameter, "Practice Time per Pushup."

d. How does the equilibrium point depend on the value 0.2 in the equation relating the amount Jim practices and his improvement rate? Try running the model with alternate values for this "Practice Effectiveness" parameter.

e. How does the equilibrium point depend on the value 10 in the equation relating the amount Jim practices and his improvement rate? Run the model with alternate values of the "Maintenance" parameter.

EXAMPLE V: PUSHUPS AND PRACTICE (NEGATIVE LOOP VERSION)

An alternative model can be formulated by assuming that Jim has a pushup goal. Assume that Jim would like to be able to do 50 pushups, and assume as well that the amount he practices is a direct function of how far he is from his goal. As before, assume that Jim's improvement rate depends on the amount he practices. This produces the causal-loop diagram shown in Figure 15.12.

Notice that this goal-gap formulation produces a negative feedback loop, which controls the amount of practice in an effort to achieve a goal of 50 pushups. In Example IV, on the other hand, the loop was positive, potentially resulting in unlimited growth in Jim's ability to do pushups.

The new flow diagram looks like Figure 15.13. (As before, still more complicated or different assumptions about decisions affecting practice would lead to different flow diagrams.)

Now all that remains is to write equations. According to the flow diagram, the difference between the number of pushups Jim would like to do (his goal) and the number he actually can do (the "actual") influences or activates the amount he practices. Jim would have to be observed for some period of

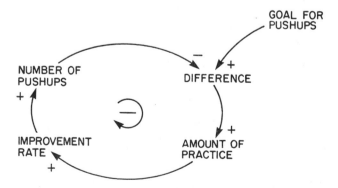

Figure 15.12 Pushups and practice, with goal

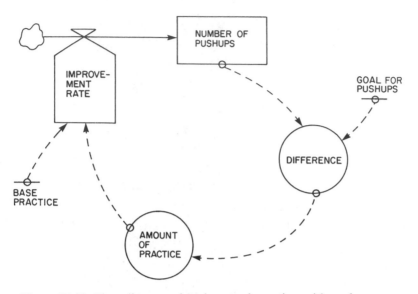

Figure 15.13 Flow diagram of pushups and practice, with goal

time to find out the exact relationship between the size of the "gap" and the amount he practices. But a plausible hypothesized relationship might be that Jim will practice one-half minute per day for each pushup he desires but cannot do:

Difference(pushups) = Goal for Pushups − Number of Pushups = 50 − Number of Pushups

Amount of Practice(minutes) = 0.5(minutes/pushup)*Difference (pushups)

Thus, for example, if Jim can do 20 pushups, while wanting to do 50, he practices (0.5)(50 − 20) = 15 minutes a day.

Now, all that remains is to specify the relationship between the amount Jim practices and the rate at which he improves. For simplicity, we might as well retain the assumption used in Example IV.

Improvement Rate(pushups/month) = (Amount of Practice − 10 minutes)*0.2(pushups/month/minute)

Exercise 19: Pushups and Practice Model, with Goal

a. Write DYNAMO equations for the pushup and practice model. (Choose DT = 1 month.)

b. Run the model setting the initial number of pushups Jim can do to 30. Rerun the model, setting the initial number he can do to 20. Rerun the model again, setting the initial number of pushups to 10. How do the results differ? Why?

c. How does the model's equilibrium point depend on the amount of time Jim practices per desired pushup he cannot do? Run the model several times, choosing different values for this parameter, Practice Time per Desired Pushup.

d. How does the equilibrium point depend on the value 0.2 in the equation relating the amount Jim practices and his improvement rate? Try running the model with alternate values for this parameter.

e. How does the equilibrium point depend on the value 10 in the equation relating the amount Jim practices with his improvement rate? Run the model with alternate values of this Practice Effectiveness parameter.

f. What changes in equation structures or parameters would enable Jim to reach his goal of 50 pushups? Run the model to demonstrate this.

CHOOSING A VALUE FOR DT

In the exercises and examples so far, an important technical issue has been treated lightly. The simulation technique used proceeds iteratively, stepping through time in intervals of length DT. In the library example, for instance, DT = one year; in the pushups example, DT = one month; in the yeast case, DT = one hour; and in the coffee cooling example, DT = one minute. Why were these values chosen to use in simulating the models? What would have happened had different values for DT been used?

The yeast example will be used to examine some of these issues more carefully. Table 15.1 shows four simulations of the yeast budding positive feedback loop, each using a different value of DT. (The four values are DT = 10, DT = 1, DT = 0.5, and DT = 0.1.) Table 15.2 shows four simulations of the yeast deaths negative feedback look, using the same four values of DT.

Table 15.1 Yeast budding simulations

Time (hours)	DT = 10	DT = 1	DT = .5	DT = .1
0	10.000	10.000	10.000	10.000
5		16.11	16.29	16.45
10	20.000	25.94	26.53	27.05
15		41.77	43.22	44.48
20	40.000	67.28	70.40	73.16
25		108.35	114.67	120.32
30	80.000	174.49	186.79	197.88

Table 15.2 Yeast deaths simulations

Time (hours)	DT = 10	DT = 1	DT = .5	DT = .1
0	10.000	10.000	10.000	10.000
5		7.738	7.763	7.783
10	5.000	5.987	6.027	6.058
15		4.633	4.679	4.715
20	2.500	3.585	3.632	3.670
25		2.774	2.820	2.856
30	1.250	2.146	2.189	2.223

As can be seen, in both the yeast budding and yeast deaths examples, the precise numerical results produced by the simulations differ depending on the values chosen for DT. For example, in the yeast budding case, the number of yeast cells grows most rapidly when DT = 0.1, and least rapidly when DT = 10. At first glance, this seems puzzling, since the budding fraction in all four cases is identical: BUDFR = 0.1. But a moment's reflection suggests an explanation: The situation is exactly analogous to compound interest. When DT = 1, for example, new cells are added to the yeast population exactly once per hour. When DT = .5 new cells are added every half hour, and thus new cells produced in one half-hour interval can themselves produce new yeast buds in the next half hour. (Hand-simulating the results for two or three hours, using DT = 1 and DT = 0.5, should illustrate this adequately.)

Which (if any) of these values for DT is correct? Certainly, yeast cells do not wait until the exact stroke of each hour to bud. For that matter, they surely do not bud exactly on the half hour, or at quarter-hour intervals. Presumably, individual yeast cells bud at various times throughout the hour.

For all practical purposes, it seems appropriate to assume that yeast cells bud more or less continuously. That is, at any moment, some yeast cells are in the process of budding. Similarly, at any moment, some are in the process of dying. But if this is true, how small a value of DT is needed to represent appropriately the yeast cell behavior?

It would be possible, of course, to simulate yeast budding using a DT of one minute, or even one second, if necessary. The main restriction is a practical one. The smaller the value selected for DT, the more computer time required to run the model (since more iterations are required). Thus choosing small values of DT increases the cost of running a model and lengthens the time spent sitting at the computer waiting for a model run to be completed.

In general, the proper approach in choosing DT is to select a value small enough to provide a reasonable approximation of the process being modeled, but not a value so small that it requires unnecessary computation time.

How small a value of DT is small enough? Consider the yeast deaths case first. Notice that when DT = 10, the number of yeast cells drops most quickly, falling to 1.25 cells in 30 hours. When DT = 1, the number of cells falls somewhat less rapidly, reaching about 2.1 in 30 hours. The results for DT = 0.5 and DT = 0.1 are very similar to the results for DT = 1. Thus in simulating the process of yeast deaths, it seems reasonable to choose DT = 1. Choosing a value of DT less than one would not produce a noticeably better approximation of the process of yeast deaths, and it would use up unnecessary computer time.

Now turn to the yeast budding loop. When DT = 10, the number of cells rises most slowly, reaching 80 cells in 30 hours. When DT = 1, the number of cells grows to about 174; when DT = 0.5, the number of cells reaches about 187; and when DT = 0.1, the number of cells rises to 197. The results for DT = 0.5 and DT = 0.1 are fairly similar to one another, although they still differ a bit. In simulating the process of yeast budding, a value of DT = 0.5 or possibly DT = 0.1 might be best. (The value used in Chapter 14, DT = 1, is perhaps a bit large.)

When formulating a model, it is important to give some attention to the choice of DT *before* using the model to draw any final conclusions about system behavior. The easiest way to choose an appropriate value of DT is to try various values, until one small enough is found, such that still smaller values do not produce noticeable changes in the simulated results. Once an appropriate choice of DT has been chosen, that value can, in general, continue to be used when employing the model to analyze system behavior and to test proposed policies. (Of course, experiments with DT should *not* be conducted when the model is in equilibrium, or nothing will happen!)

A few general rules of thumb can often provide a good starting point in selecting DT. Models that generate exponential growth require a DT that is much smaller than the doubling time involved. It is often a good first step to choose a DT from 1/5 to 1/10 the doubling time. (For example, the doubling time in the yeast budding case is roughly 7 hours. Thus a value of DT around one-half hour is a reasonably conservative choice.) Models that generate exponential growth over an extended period of time are especially sensitive to the choice of DT. (For example, the yeast population doubles roughly 4 times in a 30-hour simulation run, and thus, small errors mount up fairly quickly.)

Models involving negative loops require a DT smaller than the halving-times associated with the negative loops. A DT of roughly 1/3 or 1/4 the halving time is often a reasonable choice. (For example, the halving time in the yeast deaths case is roughly 15 hours. Thus any value of DT shorter than 4 or 5 hours is appropriate.)

One difficulty in trying to apply these rules of thumb is that, for complex models, it is often hard to determine, in advance of simulating the model, what the exact doubling times or halving times involved in various loops might be. In many cases, the rules of thumb provide only a rough general idea, and model experiments must be used to select an appropriate DT.[3]

Exercise 20: Experiments with DT

a. Run the coffee cooling model, trying various values of DT.

b. How does the choice of DT affect model behavior?

c. Which value of DT do you think is best?

Exercise 21: Jobs and Migration—Part II

Review Exercise 16 earlier in the chapter, and then do the following:

a. Use your judgment to select some plausible parameters for the model of jobs and migration you developed there. (You might want to choose parameters that reflect a hypothetical city.)

b. Write DYNAMO equations for your model.

c. Simulate the model, experimenting with various values of DT.

d. Determine the equilibrium level of population in the city.

e. Suppose the population in your city is in equilibrium, but in two years a major industrial company in the town closes, causing the number of jobs to fall by 10 percent. Use a STEP function to analyze the response of the system to the sudden decline in jobs.

ENDNOTES

1. Students with some background in calculus will recognize that the yeast budding model can be written in differential equation form as

$$\frac{dy}{dt} = by(t),$$

where $y(t)$ = the number of yeast cells at time t, and b = the budding fraction. The solution to the equation is $y(t) = y(0)e^{bt}$, where $y(0)$ = the initial number of yeast cells.

   This solution to the differential equation can be used to derive the doubling time for the number of yeast cells. The number of cells will double when $y(t) = 2y(0)$, and this takes place when $e^{bt} = 2$. Taking logarithms, the number of cells will double when $bt = ln2$. Since $ln2$ is roughly 0.7, and the budding fraction $b = 0.1$, the doubling time $t = (0.7/0.1) = 7$ hours.

2. The differential equation for the yeast deaths loop can be written $dy/dt = -y(t)/a$, where $y(t)$ = the number of yeast cells at time t, and a = the average lifetime of yeast. The solution to this equation is $y(t) = y(0)e^{-t/a}$, where $y(0)$ = the initial number of yeast cells.

   This solution can be used to derive the halving time for the number of yeast cells. The number of cells will fall to half its initial value when $y(t) = \frac{1}{2}y(0)$, and this takes place when $e^{-t/a} = \frac{1}{2}$. Thus taking logarithms, the number of cells will reach one-half its initial value when $-t/a = ln(\frac{1}{2})$. Since $ln(\frac{1}{2})$ is roughly equal to -0.7, and the average lifetime of yeast $a = 20$, the halving time $t = (20)(0.7) = 14$ hours.

3. In discussing the choice of DT, it has been assumed that the process being modeled takes place continuously in time. In the yeast case, for example, at any moment, some yeast cells are in the process of budding and others are in the process of dying. The coffee cooling case is a particularly good example of a continuous process. For an actual cup of coffee, a certain amount of cooling occurs every second, or every micro-second for that matter.

    In some cases, the assumption that the process being modeled is continuous may seem less valid. For instance, in the pushup case, Jim might practice exactly once each day, at 2:00 in the afternoon. Thus at first glance, it may seem that any improvement that takes place in the number of pushups Jim can do is not continuous, but occurs in once-a-day jumps. It is nevertheless possible that some of the improvement in Jim's ability to do pushups occurs following his practice session each day, while he is eating, sleeping, and so forth. Although the daily practice sessions are not continuous, the overall process of improvement can be viewed as at least approximately continuous.

    By and large, for most social and economic systems, it is reasonable to assume that the process being modeled is roughly continuous, at least to a first approximation. In cases where the processes being modeled are quite clearly not continuous, the methods of analysis described in this chapter are not strictly appropriate.

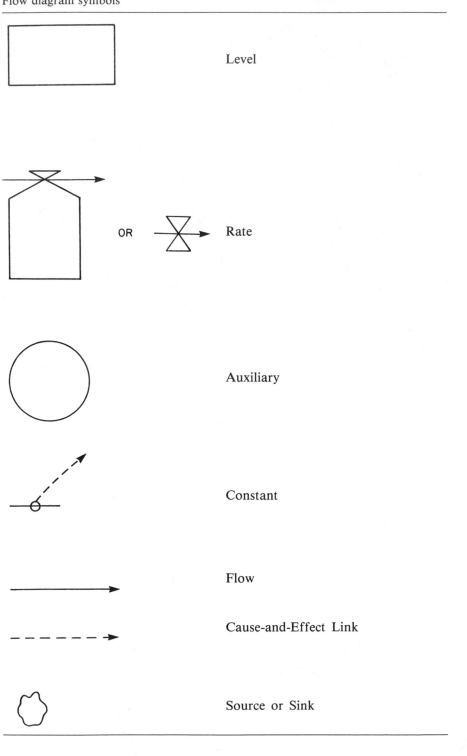

Level

OR   Rate

Auxiliary

Constant

Flow

Cause-and-Effect Link

Source or Sink

CHAPTER 16

REPRESENTING MORE COMPLEX CAUSAL RELATIONS

One of the central elements of the art of simulation modeling involves specifying causal relationships. Causal relations can take on a variety of forms. Linear causal relationships—those that can be written as linear equations and graphed as straight lines—are ordinarily the easiest to represent. For example, the relationship between the number of yeast cells and the budding rate discussed in Chapter 15 is a linear equation. (Recall that the equation for the budding rate is: BUDDNG.KL = YEAST.K*BUDFR.) When the number of yeast cells doubles, the number of buds formed per hour also doubles. The gap formulation in the coffee cooling model is also linear, although a bit more complex. The coffee cooling rate depends in a linear way on the coffee temperature and the room temperature.

Unfortunately, many interesting causal relationships are not linear; that is, they cannot be written as linear equations and graphed as straight lines. This chapter focuses on a model of urban growth to introduce some of the issues involved in formulating and analyzing nonlinear causal relationships.[1]

GROWTH OF A CITY

A city begins as a settlement near a river, a harbor, some natural resources, or other geographical features that make living there more attractive than other places. In addition to providing a place for people to live, settlements foster manufacturing, retailing, mining, or other economic activities. The initial concentration of industrial activity attracts further industry. The increased concentration of people attracts retail and service-oriented business. Suppliers to industry will settle near those industries that they service. In turn, the concentration of suppliers is an incentive to other companies in the same business to locate nearby. Hence, the presence of industry results in the growth of industry.

The growth of a city cannot continue forever. Economic or business activities need to be housed in some kind of structure. As more and more industrial buildings (used here to include all types of structures for business, whether a general store or a factory) are constructed, the best sites become occupied. Eventually, construction of buildings slows because, for one reason, finding favorable sites becomes increasingly difficult.

Of course the construction of buildings is not the only thing that affects the number of buildings in a city. Industrial buildings do not last forever. As they become obsolete and deteriorate, they are eventually demolished.

The following series of exercises and examples demonstrate the formulation of a model based on this short description.

Exercise 1: The Construction of Buildings

As stated, businesses tend to attract additional businesses. Therefore, industrial buildings to house businesses tend to foster the construction of additional buildings to house new business attracted to the city. Figure 16.1 depicts a causal-loop diagram of this process.

Construct a flow diagram based on the causal-loop diagram in Figure 16.1. Notice that there are two arrows pointing toward construction in the causal-loop diagram—industrial buildings and the construction fraction. This means that the construction rate depends on two variables: the current number of industrial buildings and the percent of the current industrial buildings being added each year (the construction fraction). Remember to show both these factors influencing the construction rate in your flow diagram.

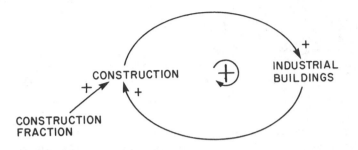

Figure 16.1 Causal-loop diagram of building construction

Exercise 2: Equations for Building Construction

Write a DYNAMO model based on the flow diagram developed for the preceding exercise. Assume the construction fraction is a constant 0.1, and set the initial number of buildings equal to 10. What kind of behavior does the model generate?

Exercise 3: Demolition

Figure 16.2 depicts a causal-loop diagram showing the feedback between demolition of buildings and the level of buildings. Construct a flow diagram based on this causal-loop diagram.

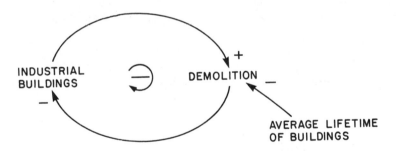

Figure 16.2 Causal-loop diagram of demolition

Exercise 4: Demolition Equations

  a. Write the equations necessary to add demolition to the model developed in Exercise 2. Assume the average lifetime for a building is fifty years. The formulation for demolition is analogous to the formulation for yeast deaths in Chapter 14. Run the model on the computer. What kind of behavior does it generate?

  b. Rerun the model with the construction fraction set equal to zero. What type of behavior does the model generate?

  c. What value must be chosen for the construction fraction to produce a construction rate exactly equal to the demolition rate?

A NONLINEAR REPRESENTATION OF LIMITED LAND

Growth of buildings does not continue forever. As noted in the description of the growing city, the land available for industrial buildings eventually becomes nearly filled. The difficulty of finding suitable sites slows construction. Figure 16.3 depicts a causal-loop diagram of the process, and Figure 16.4 depicts the associated flow diagram.

In Figure 16.4, the construction fraction is no longer a constant, but an auxiliary variable depending on the fraction of land occupied. The fraction of land occupied is a number ranging from zero to one, which indicates what fraction of the land allotted to industrial buildings is actually built on. If this fraction is zero, then there are no industrial buildings; if the fraction is one, then all the land available for industrial buildings is occupied.

A possible DYNAMO equation for the fraction of land occupied is:

A   FLANDO.K = (BUILD.K∗AVAREA)/LAND
NOTE   FRACTION OF LAND OCCUPIED (DIMENSIONLESS)

where FLANDO is the fraction of land occupied, BUILD is the number of industrial buildings, AVAREA is the average area per building, and LAND is the land available for industrial buildings.

It is now necessary to specify how the fraction of land occupied influences the construction fraction. At first this may seem like a difficult task, but beginning with a graph makes the task easier. Figure 16.5 depicts a graph of one possible nonlinear relationship between the construction fraction and the fraction of land occupied.

The relationship depicted in the graph is hypothetical: the graph was drawn without gathering any data about the precise numbers involved. But even without doing any research on land availability and urban growth, it is

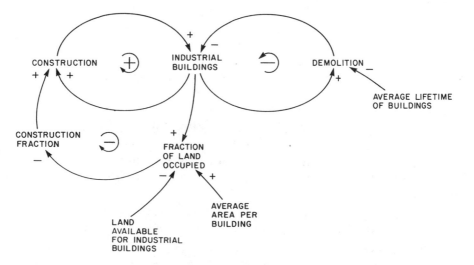

Figure 16.3 Causal-loop diagram of city growth with land limitation

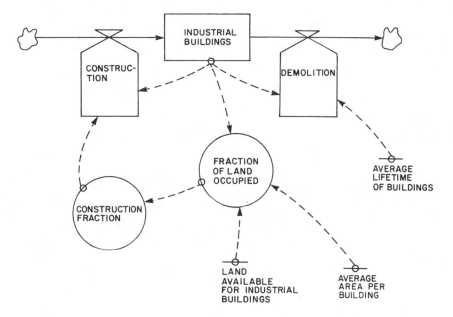

Figure 16.4 Flow diagram of city growth with land limitation

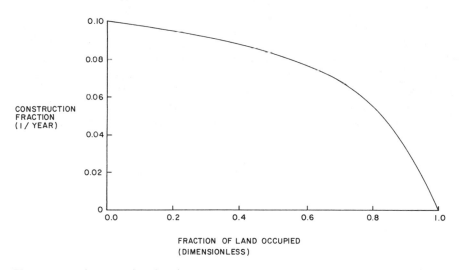

Figure 16.5 Construction fraction

possible to make some plausible assumptions about the nature of the relationship. One feature of the relationship is clear: As the fraction of land occupied approaches one, the construction fraction must decline to zero. When no land is left, nothing new can be built. Also, as Figure 16.5 indicates, as the land fills up (i.e., as the fraction of land occupied approaches one), construction slows because it is harder to find suitable sites. By making some assumptions about land availability and urban growth, it is possible to formulate a graph that represents a plausible causal relationship between two variables. The precise numbers on the graph are somewhat arbitrary, but the shape of the curve seems reasonable.

USING THE TABLE FUNCTION

Having specified a graph, you can now convert it into a DYNAMO equation using a special feature built into the DYNAMO program, called the TABLE function. A set of two equations is used to construct a graphical relation in DYNAMO.

The first equation needed here is an auxiliary equation:

A   CNSTF.K = TABLE(TCNSTF,FLANDO.K,0.0,1.0,0.2)
NOTE   CONSTRUCTION FRACTION (1/YEAR)

This equation says that the construction fraction CNSTF is to be found in a table of numbers, and that this table is called TCNSTF. Further, the value for the construction fraction CNSTF depends on the fraction of land occupied FLANDO. Both the construction fraction and the fraction of land occupied refer to a specific point in time, K. The value for the construction fraction at any given point in time depends upon the fraction of land occupied by industry at the same time.

To use the TABLE function, take a graph of the relationship and mark off values along the x-axis (horizontal axis) at equal intervals. Then, for each value of the independent variable marked on the x-axis, put a dot on the graph to mark the corresponding value of the dependent variable (on the y-axis) as shown in Figure 16.6. Thus pairs of values are identified, one value of the pair from the independent variable and one value from the dependent variable. Figure 16.7 shows a set of values obtained from the graph in Figure 16.6.

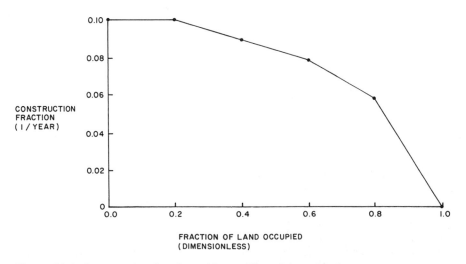

Figure 16.6 Construction fraction with specific points marked

Three numbers complete the first equation for the TABLE function, as illustrated in Figure 16.7. The first number refers to the minimum value on the graph of the independent variable. For FLANDO this value is 0.0. The second number describes the maximum value of the independent variable, which for FLANDO is 1.0. The last number specifies how large the intervals are between the values marked off on the x-axis of the graph. For the graph in Figure 16.6, the x-axis begins with 0.0 and is marked off in intervals of 0.2, up to a maximum of 1.0.

At this point, the actual values assumed by the dependent variable have not been specified in equation form. The independent variable FLANDO has been described, but more information is needed about the dependent variable CNSTF. The second equation fills in the needed information:

T   TCNSTF = 0.10/0.10/0.09/0.08/0.06/0.00
NOTE   TABLE FOR CONSTRUCTION FRACTION

This second equation, beginning with the letter T to signify a table statement, provides the table of values TCNSTF for use in determining CNSTF. One value is given for every point marked on the graph in Figure 16.6. As noted before, the set of two equations is equivalent to a pairing of values, as shown in Figure 16.7. For example, the value of 0.0 for the fraction of land occupied is

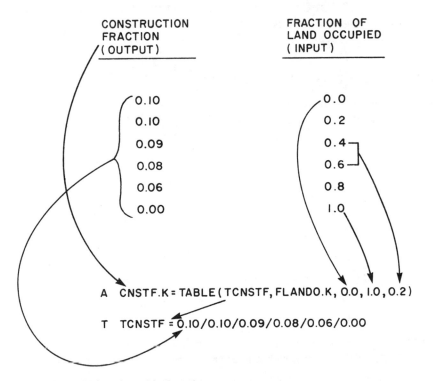

Figure 16.7 Using the table function

paired with the value 0.10 for the construction fraction. When the two equations describing the relationship are entered, DYNAMO is able to reconstruct the entire graph by using the following rules:

1. If the value of the independent variable is one of the points at the specified interval, then the TABLE function uses the corresponding value of the dependent variable taken from the table statement. For example, if the fraction of land occupied FLANDO were 0.8 then TABLE would return a value of 0.06 for CNSTF.

2. If the value of the independent variable is part way between the selected points, then TABLE does a linear interpolation. For example, if FLANDO were 0.9, then TABLE would return a value of 0.03. Since 0.9 is halfway between 0.8 and 1.0, TABLE calculates the value that is halfway between 0.06 and 0.0.

3. If the value of the independent variable is larger than the largest value specified in the TABLE function, then the function returns the last value on the TABLE statement. For example, if FLANDO were 1.2 (a meaningless number for this variable), then TABLE would return 0.0.

4. If the value of the independent variable is smaller than the smallest value specified in the TABLE function, then the function returns the first value in the table statement. For example, if FLANDO were -1 (again, a meaningless number in this case), then TABLE would return a value of 0.10

EXAMPLE I: EQUATIONS FROM A GRAPH

Figure 16.8 shows another possible relationship between the construction fraction and the fraction of land occupied. To convert this relationship into DYNAMO equations, write the auxiliary equation in the form:

> A   _____.K = TABLE(_____,_____.K,_____,_____,_____)

All the needed information for the auxiliary is available in the graph in Figure 16.8. Begin by filling in the name of the dependent variable. The dependent variable is located on the vertical or y-axis. In Figure 16.8, it is the construction fraction CNSTF.

> A   CNSTF.K = TABLE(_____,_____.K,_____,_____,_____)

The name of the table TCNSTF is also given in Figure 16.8. Fill in the next blank.

> A   CNSTF.K = TABLE(TCNSTF,_____.K_____,_____,_____)

Now fill in the information about the independent variable. Its name FLANDO is located on the horizontal axis.

> A   CNSTF.K = TABLE(TCNSTF,FLANDO.K,_____,_____,_____)

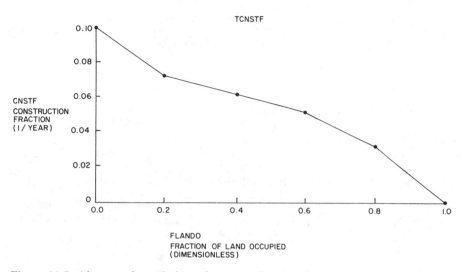

Figure 16.8 Alternate formulation of construction fraction

Then fill in the information for the values of the independent variables. The minimum value is 0.0, the maximum value is 1.0, and the interval is 0.2.

A   CNSTF.K = TABLE(TCNSTF,FLANDO.K,0.0,1.0,0.2)

Now for the table. The format of the table equation is:

        Name   Values

T   _____ = _____/_____/_____/ . . . /_____/_____

First, determine how many values are needed on the right side of this equation. One easy way to tell is to count the number of points on the graph, being careful to remember any points directly on the axes. There are six data points plotted. A look at the x-axis should reveal exactly the same number of marks there, and it does (at values 0.0, 0.2., 0.4, 0.6, 0.8, and 1.0). There must be space for six y-values on the right side of the table equation, separated by slashes.

T   _____ = _____/_____/_____/_____/_____/_____

Next, fill in the table name, making sure it is the same name that appeared in the auxiliary equation:

T   TCNSTF = _____/_____/_____/_____/_____/_____

Then fill in the blanks for the values of the dependent variable CNSTF. Begin with the leftmost value:

T   TCNSTF = 0.10/_____/_____/_____/_____/_____.

294   **Introduction to Simulation**

When all the values are filled in, the statement looks like this:

T   TCNSTF = .010/0.07/0.06/0.05/0.03/0.00

Exercise 5: Table Functions

Figure 16.9 depicts a possible linear relation between the construction fraction and the fraction of land occupied. Convert this relation into DYNAMO statements.

Exercise 6: Simulating the Effects of Land Limits

Included here is a listing of the DYNAMO model depicted in the flow diagram in Figure 16.4. Enter and run the model on the computer. The behavior of the model should be similar to the behavior shown in Figure 16.10.

```
*            URBGM - URBAN GROWTH MODEL - INDUSTRIAL SECTOR
NOTE
NOTE    ------------------------------------------------------------------
NOTE
NOTE    NAME      - URBAN GROWTH MODEL
NOTE                 INDUSTRIAL BUILDING SECTOR
NOTE
NOTE    THIS MODEL EXHIBITS THE GROWTH IN THE
NOTE    NUMBER OF INDUSTRIAL BUILDINGS IN AN URBAN AREA.
NOTE    INCLUDED IS THE LIMITATION OF GROWTH FROM AVAILABILITY
NOTE    OF LAND
NOTE
NOTE    ------------------------------------------------------------------
NOTE             INDUSTRIAL BUILDING SECTOR
NOTE    ------------------------------------------------------------------
NOTE
L        BUILD.K=BUILD.J+(DT)(CONST.JK-DEMO.JK)
N        BUILD=10
NOTE             INDUSTRIAL BUILDINGS (BUILDINGS)
R        DEMO.KL=BUILD.K/ALTB
NOTE             DEMOLITION OF IND. BUILDINGS (BUILDINGS/YEAR)
C        ALTB=50
NOTE             AVERAGE LIFETIME OF BUILDINGS (YEARS)
R        CONST.KL=(BUILD.K)(CNSTF.K)
NOTE             CONSTRUCTION (BUILDINGS/YEAR)
A        CNSTF.K=TABLE(TCNSTF,FLANDO.K,0,1,0.2)
NOTE             CONSTRUCTION FRACTION (1/YEAR)
T        TCNSTF=0.10/0.10/0.09/0.08/0.06/0.00
NOTE             TABLE FOR CONSTRUCTION FRACTION
A        FLANDO.K=(BUILD.K*AVAREA)/LAND
NOTE             FRACTION OF LAND OCCUPIED (DIMENSIONLESS)
C        AVAREA=1
NOTE             AVERAGE AREA PER BUILDING (ACRES/BUILDING)
C        LAND=1000
NOTE             LAND AVAILABLE FOR BUILDINGS (ACRES)
NOTE
NOTE    ------------------------------------------------------------------
NOTE             CONTROL STATEMENTS
NOTE    ------------------------------------------------------------------
NOTE
PLOT     BUILD=B/CONST=C,DEMO=D
SPEC     DT=2/PLTPER=10/LENGTH=100
RUN
```

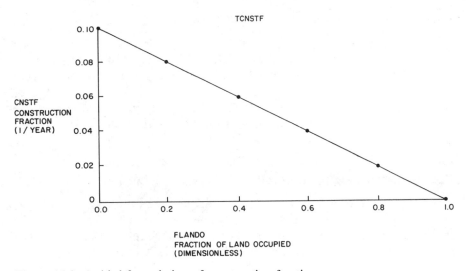

Figure 16.9 A third formulation of construction fraction

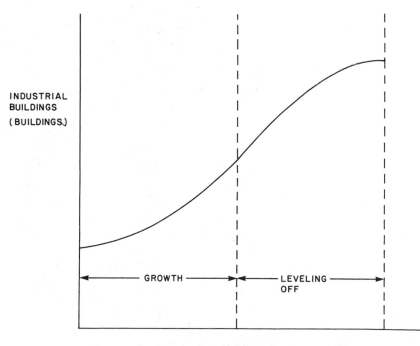

Figure 16.10 Behavior of urban growth model

LOOP DOMINANCE AND S-SHAPED GROWTH

The curve for buildings, Figure 16.10, shows a type of growth called sigmoidal (or S-shaped) growth. Sigmoidal growth is a combination of two kinds of behavior. The first part of the curve shows exponential growth, characteristic of a positive feedback loop. The second part of the curve shows a leveling off characteristic of a negative feedback loop. Thus the behavior of the model results from a shift in loop dominance, from the positive loop to the negative loop.

Exercise 7: Checking for Dominance of the Positive Loop

From the causal-loop diagram in Figure 16.3 and the flow diagram in Figure 16.4, you can see that the model has only one positive loop—the loop relating industrial buildings and construction. To examine the role of this loop in generating S-shaped growth, change the table TCNSTF to the following:

T   TCNSTF $= 0/0/0/0/0/0$

This change cancels out the positive feedback loop in the model, since an increase in industrial buildings no longer causes an increase in construction. What kind of behavior results from this change?

Modifying a TABLE function to "cut open" a positive or negative loop is often a helpful strategy in trying to analyze a model's behavior. In this case, the role of the construction loop is fairly easy to determine—since the construction loop is the only positive loop in the model. But models with multiple loops can sometimes be quite different to analyze, and "cutting open" loops can be a helpful strategy.

Exercise 8: Checking the Dominance of a Negative Loop

As seen from the causal-loop diagram of Figure 16.3, the Urban Growth Model has two negative feedback loops. One loop relates industrial buildings to demolition. The second loop relates industrial buildings, the fraction of land occupied, the construction fraction, and construction. Which loop accounts for the number of industrial buildings moving toward equilibrium in the latter period of the simulation? To find out, cancel out each loop, one at a time, and see the effect. Start by eliminating the effect of the feedback loop involving demolition. Run the model with the average lifetime of buildings set to some very large number like 1E12 (which means 1×10^{12}). This change reduces demolition to an insignificant number, rendering the relationship between buildings and demolition, for all practical purposes, inoperative. Does the model still move toward equilibrium? Now run the model with the original value of ALTB and the following values for the table TCNSTF:

T   TCNSTF $= 0.10/0.10/0.10/0.10/0.10/0.10$

Does the model still move toward equilibrium? Which negative feedback loop dominates the behavior during the second part of the simulation?

Exercise 9: Why Does the Dominance Shift?

A deeper understanding of the urban model can be gained by looking more carefully at the relationship between the number of buildings in the city and the construction rate. This analysis is easiest to carry out when the relationship between the fraction of land occupied FLANDO and the construction fraction CNSTF is assumed to be linear.

  a. Examine the linear version of the TABLE function relating FLANDO and CNSTF, described in Exercise 5. The following algebraic expression is equivalent to the table:

  $$CNSTF.K = 0.1*(1 - FLANDO.K)$$

  Recall that the fraction of land occupied FLANDO is a function of the number of buildings BUILD, the average area used per building AVAREA, and the total land area of the city LAND. Rewrite the algebraic equation for CNSTF in terms of BUILD, AVAREA, and LAND.

  b. Recall that the equation for the construction rate is CONST.KL = BUILD.K*CNSTF.K. Rewrite this equation, substituting for CNSTF the algebraic expression you obtained in part (a).

  c. Is the equation you obtained in part (b) linear or nonlinear?

  d. Use the equation you obtained in part (b) to calculate the construction rate when the number of buildings in the city equals 10. When the number of buildings equals 100? When the number equals 500? When the number equals 900? When the number equals 1000?

  e. Examine your results from part (d). At what level of buildings does the construction rate reach its maximum value? Why? Which loop is dominant before the construction rate reaches its maximum value? Which is dominant after it reaches its maximum value?

  f. Simulate the model, using the linear version of the construction fraction; and then compare the results using several nonlinear versions with different shapes. How are the results affected?

MULTIPLIERS AND NORMALS

The relationship between the fraction of land occupied and the construction rate can be expressed in a somewhat different form that provides additional insight into the model structure. Recall that as the fraction of land occupied FLANDO increases from zero to one, the construction fraction CNSTF falls from 10 percent per year to zero. And as the construction fraction falls, the construction rate falls correspondingly. Another way to express this relationship in equation form is to write the construction rates as the product of three terms: the number of buildings, a "construction normal" CNSTN, and a "construction multiplier" CNSTM.

```
R      CONST.KL = BUILD.K*CNSTN*CNSTM.K
NOTE      CONSTRUCTION RATE (BUILDINGS/YEAR)
C      CNSTN = 0.1
NOTE      CONSTRUCTION NORMAL (PERCENT/YEAR)
A      CNSTM.K = TABLE(CNSTMT,FLANDO.K,0,1,0.2)
T      CNSTMT = 1/1/.9/.8/.6/0
NOTE      CONSTRUCTION MULTIPLIER (DIMENSIONLESS)
```

The basic idea underlying this formulation is that the growth of buildings in an urban area would proceed at a "normal" or natural rate of 10 percent per year were land widely available. As land is used up, growth proceeds at a rate below this normal value, and this is represented in the model by the "construction multiplier."

When formulations like this are used, it is common to call the constant involved a "normal value." For example, in this case, the constant term is called the "construction normal" CNSTN. Of course, this does not imply that cities normally grow at 10 percent per year—only that they would, if land, among other things, were not a constraint. This is a somewhat peculiar use of the term "normal," but the terminology is widespread in system dynamics modeling.

One advantage of the "normal and multiplier" formulation is that it often makes it easier to decide upon the shape and numerical values for TABLE functions. For instance, in the urban model, it is clear that the maximum value in the construction multiplier table must be one, and this value must occur when the fraction of land occupied FLANDO = 0. Similarly, the minimum value of the construction multiplier must be zero, and this must occur when FLANDO = 1.

Exercise 10: Testing the Construction Normal

a. Rewrite the urban model using the normal and multiplier formulation.

b. Run the model. The behavior should be identical to the behavior you obtained in Exercise 6.

c. Rerun the model, setting the construction normal CNSTN = 0.2. How do the results differ? Rerun the model, setting CNSTN = 0.05. How do the results differ?

d. What determines the equilibrium value of the number of buildings?

Exercise 11: Comparing Gap and Multiplier Formulations for Urban Growth

It is possible to formulate a somewhat different urban growth model, in which the construction rate depends on the gap between the number of buildings which can be built on the city's land area (the "goal") and the number which currently exist (the "actual").

a. Draw a causal-loop diagram and flow diagram for an urban growth model based on the "gap" formulation.

b. Write DYNAMO equations, chosing values for the parameters to represent a city similar to the one discussed previously.

c. Run the model. How does the behavior differ from the behavior of the model using the multiplier formulation? Does the model employing the gap formulation exhibit S-shaped growth? Why or why not?

d. Which version of the model do you think provides a better representation of the process of urban growth?

Exercise 12: Jobs and Migration—Part III

Recall that in Chapter 15, Exercise 16 and 22, you developed a model of the relationship between jobs and urban migration, using a formulation in which the migration rate into the city depends on the "gap" between the number of jobs in the city and the number of workers. It is also possible to represent the relationship between jobs and migration using a "multiplier and normal" formulation.

Consider the causal-loop representation of urban migration shown in Figure 16.11. According to this view, migration into the city depends on the job ratio, or the ratio of the number of jobs to the number of workers. When the number of jobs exactly equals the number of workers (i.e., when the job ratio equals one), in-migration equals its normal value (in percent per year). When the job ratio is greater than one (i.e., when there are more jobs than workers), in-migration rises above its normal value; and when the job ratio is less than one (i.e., when there are fewer jobs than workers), in-migration falls below its normal value. (For simplicity, the causal-loop diagram indicates that out-migration does not depend on the job ratio. You might want to add this complication to the model, once you have formulated the in-migration loops.)

a. Draw a flow diagram corresponding to the causal-loop diagram.

b. Write DYNAMO equations, choosing whatever numerical values you think are plausible for the constants and TABLE function. (*Hint:* The form of the equations is quite similar to the urban construction model.)

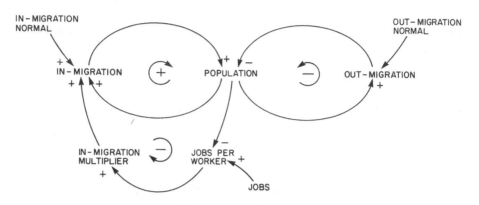

Figure 16.11 Causal-loop diagram of urban migration

    c. Determine the equilibrium value for population, and start the model in equilibrium.

    d. Examine the response of the model to a 10-percent step decline in jobs. Does the behavior of the model differ noticeably from the behavior of the model you formulated in Chapter 15?

    e. Examine the response of the model to a 100-percent step increase in jobs. A 200-percent step increase. How large a step increase is necessary to produce an S-shaped response? Why?

    f. How would you add the effect of jobs on out-migration?

A more complete model of urban growth is presented in Chapter 20.

ENDNOTE

    1. Models containing only linear causal relations are sometimes called "linear systems." Of course, linear systems can quite easily generate *behavior* that is nonlinear. For example, exponential growth and exponential decay can both be generated by linear systems.

CHAPTER 17

INTRODUCTION TO DELAYS*

Delays occur frequently in social and economic systems. When a business organization orders supplies, the supplies usually arrive only after a delay. When a pollutant is dumped into a river, it takes time to dissipate. When the price of gasoline rises, consumers take time to adjust by driving less or by purchasing more fuel-efficient cars. And, of course, when you mail a letter, the letter will be delivered only after a delay.

Delays are conveniently divided into two types: delays resulting from the time involved in processing physical materials and delays resulting from the time involved in perceiving and acting upon information. As these two types of delays—called *material delays* and *information delays*—abound in social and economic systems, some of their properties are investigated in this chapter.

EXAMPLE I: THE MARTAN CHEMICAL COMPANY

Recall from Example I in Chapter 7 that the Martan Chemical Company, which manufactures the pesticide Nobug, dumps a quantity of Nobug into the Sparkill River once a week. During the course of the week, the pollutant is absorbed by the river's natural clean-up processes. A causal-loop and flow diagram of the Martan case are shown in Figure 17.1.

The equations for the model are similar to the equations for the yeast deaths system, discussed in Chapters 13 through 15.

L	NOBUG.K = NOBUG.J + (DT)(DUMP.JK − ABSORB.JK)
N	NOBUG = NOBUGN
C	NOBUGN = 0
R	ABSORB.KL = NOBUG.K/NAT
C	NAT = 2

*Students wishing immediately to try modeling a more complete problem situation might do Chapters 18 and 19 before this chapter. However, the contents of this chapter are critical for the models contained in Chapter 20 and beyond.

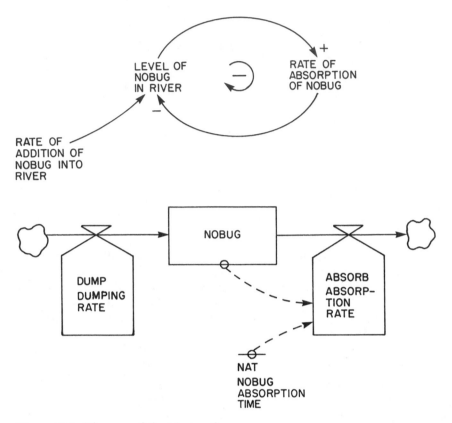

Figure 17.1 Diagram of the Martan Case

These equations indicate that the level of NOBUG in the Sparkill River is influenced by the dumping rate DUMP and the absorption rate ABSORB; and the absorption rate ABSORB is equal to the level of NOBUG divided by the Nobug Absorption Time NAT (2 days).

The only equation that remains to be specified is the dumping rate DUMP. According to the description in Chapter 7, the Martan Company releases Nobug into the river in once-a-week batches, producing a Nobug concentration in the river of about 420 parts per million (ppm). Assuming that the river contains 1 million gallons of water, this amounts to a dumping rate of 420 gallons each week.

For modeling purposes it is easier to begin by assuming that Martan dumps Nobug into the river continuously, at an overall rate of 420 gallons per week. This amounts to a continuous daily dumping rate of 420/7 = 60 gallons per day.

Exercise 1: Preliminary Nobug Model

a. Write DYNAMO equations for the Nobug case, adding the DUMP equation and other needed specifications. Run the model, setting the initial level of NOBUG = 0, and choosing DT = 0.25 days. What behavior does the model generate?

b. Rerun the model, setting the Nobug Absorption Time NAT = 4. How do the results differ? Rerun the model, setting NAT = 1. How do these results differ?

USING THE PULSE FUNCTION TO REPRESENT THE DUMPING RATE

The model developed so far is somewhat inadequate, because it assumes that Nobug is continuously released into the Sparkill River at a rate of 60 gallons per day. The DYNAMO PULSE function permits modifying the model to represent the dumping of Nobug in once a week batches. The following equation indicates that 420 gallons of Nobug are dumped into the river on day 1 of the simulation, and 420 gallons are dumped again at regular intervals of 7 days.

$$R \quad DUMP.KL = (1/DT)*PULSE(420,1,7)$$

Figure 17.2 shows the dumping rate over the first 10 days of the simulation. On day one, the dumping rate rises to 1680 gallons per day for the duration of a time interval of one DT (i.e., during one quarter-day). It then falls to zero, and remains there until day 8, when it again rises to 1680 for a time interval of one DT. It will rise again on day 15, and then once again fall to zero.

One aspect of the equation for the dumping rate may seem puzzling at first glance. Why is the factor (1/DT) included in the formulation? On the surface this may seem odd, since it produces a dumping rate of 1680 gallons per day rather than 420. To understand the use of (1/DT) in the dumping equation, it is necessary to take a closer look at the level equation for NOBUG.

$$L \quad NOBUG.K = NOBUG.J + (DT)(DUMP.JK - ABSORB.JK)$$

Note that in the level equation for NOBUG, the dumping rate DUMP is multiplied by DT to produce the amount of NOBUG added to the river during one time interval DT. Thus during the first time interval of one day, the amount dumped is equal to DT*1680 = 0.25*1680 = 420 gallons. In general, the amount dumped during any time interval equals

$$(DT)*(1/DT)*PULSE(420,1,7) = PULSE(420,1,7)$$

The factor (1/DT) is necessary in the PULSE rate equation to cancel the factor DT included in the level equation. If the dumping rate that takes place during the first quarter of day one were maintained *for the entire day*, 1680 gallons of NOBUG would be dumped into the river, but the rate is not main-

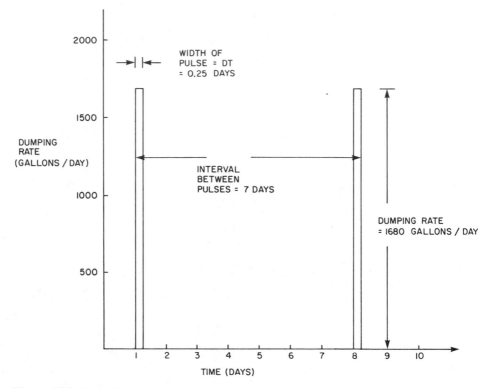

Figure 17.2 Dumping rate

tained for the entire day. It continues for just one time period DT and then it falls to zero. Thus, the *total amount dumped* is 420 gallons—exactly what the formulation is supposed to produce.

The PULSE function can also be used to examine the response of the system to the dumping of just one batch of NOBUG. It is possible to do this by making the following change in the equation for DUMP.

R   DUMP.KL = (1/DT)*PULSE(420,1,1000)

This equation indicates that the dumping rate rises from zero to 420/DT or 1680 on day one and every 1000 days thereafter (rather than every 7 days). Thus if the model is run for a period shorter than 1000 days, only one pulse in the dumping rate will occur.

Figure 17.3 shows a simulation of the Nobug system, with one batch of Nobug released into the river on day one. As can be seen, the level of NOBUG in the river rises sharply to 420 gallons on day one, and then drifts slowly toward zero.

The general form of the PULSE function is:

PULSE(AMOUNT,FIRST,INTERVAL)

```
          NOBUG=N,ABSORB=A,DUMP=D

          0.00        150.00       300.00       450.00       600.00  NA
          0.0         600.0       1200.0       1800.0       2400.0   D
0.0000   N-----------------------------------------------------------  NAD
         N                .            .            .             .  NAD
         N                .            .            .             .  NAD
         N                .            .            .             .  NAD
         N                .            .          D  .            .  NA
         D              .    A          .          N  .             .
         D              .  A            .      N        .             .
         D             .A               .  N            .             .
         D            A.              N.                .             .
         D           A  .          N       .            .             .
2.5000   D----------A-------N---------------------------------------
         D          A   .  N           .            .             .
         D         A    .N            .            .             .
         D        A     N             .            .             .
         D       A    N  .            .            .             .
         D     A     N   .            .            .             .
         D     A    N    .            .            .             .
         D    A   N       .            .            .             .
         D    A  N        .            .            .             .
         D    A N         .            .            .             .
5.0000   D-A--N-----------------------------------------------------
         D A N            .            .            .             .
         D AN             .            .            .             .
         D AN             .            .            .             .
         DA N             .            .            .             .
         DAN              .            .            .             .
         DAN              .            .            .             .
         DAN              .            .            .             .
         DAN              .            .            .             .
         DN               .            .            .             .  NA
7.5000   DN--------------------------------------------------------  NA
         DN               .            .            .             .  NA
         AN               .            .            .             .  AD
         AN               .            .            .             .  AD
         AN               .            .            .             .  AD
         AN               .            .            .             .  AD
         AN               .            .            .             .  AD
         N                .            .            .             .  NAD
         N                .            .            .             .  NAD
         N                .            .            .             .  NAD
10.000   N-----------------------------------------------------------  NAD
```

Figure 17.3 Release of one batch of NOBUG

where AMOUNT indicates the amount to be inputted in the pulse; FIRST indicates the time at which the first pulse occurs; and INTERVAL indicates the time interval between pulses.

Exercise 2: The Halving Time for NOBUG

Revise your DYNAMO equations to include a PULSE function for the dumping rate. Choose a time between pulses of 1000 days, in order to examine the effects of just one pulse.

a. What is the halving time for the amount of NOBUG in the Sparkill River?

b. Experiment with various values of NAT. How does the choice of NAT influence the halving time?

c. Using your results for part (b), select a value of NAT to produce a halving time that corresponds to the data shown in Chapter 7, Figure 7.1.

Exercise 3: Simulating Repeated Batches

a. Select NAT equal to the value you determined in Exercise 2, part (c). Use the PULSE function to simulate the effect of dumping 420 gallons of Nobug into the river at 7-day intervals. Compare your results with the data shown in Chapter 7, Figure 7.1.

b. Reread Example II in Chapter 7, "Martan Chemical—Part II". According to that example, Martan Chemical changed the chemical composition of Nobug, resulting in an increase in the Nobug absorption time NAT. Change the value of NAT in your model to reflect the change in the chemical composition of Nobug. How does the system respond? Compare your results with the data shown in Figure 7.3, Chapter 7.

NOBUG AND MATERIAL DELAYS

In the Nobug case, Nobug is dumped into the river in once-a-week batches. The absorption of Nobug does not occur immediately, however. Instead, it occurs over a period of time. In fact, the absorption of Nobug by the river can be viewed as a *time delay* process. The absorption rate is the river's delayed response to the dumping rate.

The Nobug model structure is called a *first-order material delay,* because it describes the flow of a material substance into and out of a single level. Figure 17.4 shows a general flow diagram for a first-order material delay,

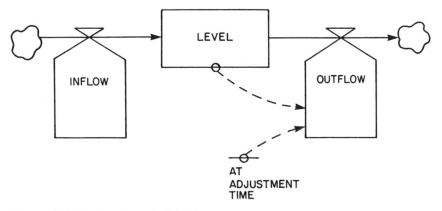

Figure 17.4 First-order material delay

along with the corresponding equations. The general form of a first order material delay is exactly analogous to the Nobug structure. An inflow rate accumulates in a level; and an outflow rate is equal to the amount in the level, divided by a time constant for adjustment of the level.

Because first-order material delays are widely used in system dynamics models, they are often given a special flow diagram notation. Figure 17.5 shows the usual flow diagram symbol for a first order material delay, and Figure 17.6 shows how the symbol can be used to represent the Nobug case.

When a first-order material delay is used in a model, there are two ways to write the equations. One approach is simply to write out individual equations, exactly as was done in the Nobug case. But because first-order material delays are frequently used, a special DYNAMO function is available that can be used to replace the set of individual equations. The following equation can be used to indicate that the OUTFLW rate is a first-order delayed response to the INFLOW rate, with an adjustment time AT.

R   OUTFLW.KL = DELAY1(INFLOW.JK,AT)

The DELAY1 function is simply a shorthand notation. When the model is run, DYNAMO will substitute the full level and rate formulation for a first-order delay, whenever the DELAY1 function appears in the model.

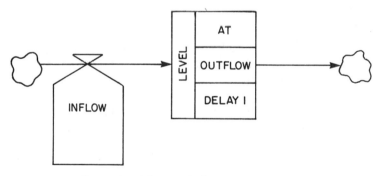

Figure 17.5 First-order delay symbol

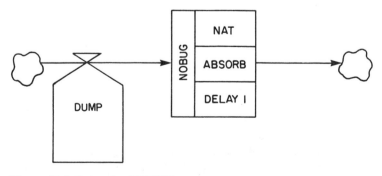

Figure 17.6 Delay for NOBUG

The DELAY1 function can be used quite easily to represent the Nobug case. The full set of equations for the Nobug case becomes:

R   ABSORB.KL = DELAY1(DUMP.JK,NAT)
C   NAT = 2
R   DUMP.KL = (1/DT)*PULSE(420,1,7)

This indicates that the absorption rate is a delayed response to the dumping rate, with an absorption time NAT = 2.[1] Note that in this formulation the NOBUG level equation is not needed, and is also not available for use elsewhere, such as output printing or plotting.

Exercise 4: Nobug and the DELAY1 Function

a. Modify your equations for the Nobug model, using the DYNAMO DELAY1 function to replace the level of NOBUG and the Nobug absorption rate.

b. Run the model. It should behave exactly as it did in the previous Exercises 2 and 3.

c. Experiment with various values of NAT. How does the model respond?

Exercise 5: The Mail Delay

The Nifty Department Store sends out bills to its charge-card customers once a month, and the credit department has learned that, on the average, it takes about three days for the bills to arrive in the mail.

a. Draw a causal-loop diagram, flow diagram, and equations for the Nifty Department Store case. (*Hint:* See Figure 7.5 in Chapter 7.) Assume that NIFTY has 1000 charge customers.

b. Run the model and examine the results.

c. How many bills take more than six days to arrive?

EXAMPLE II: THE GOAL-GAP FORMULATION, DELAYS, AND CYCLES IN APARTMENTS

Seemingly simple models can often generate surprising behavior when a delay is introduced. One interesting example is the construction of apartment buildings in a large city. Suppose builders construct apartments in response to the gap between the total number of apartments desired by people in the city and the total number of apartments available. A causal-loop diagram and flow diagram for this system are shown in Figure 17.7. The system described by the flow diagram is exactly analogous to the coffee cooling system in Chapter 15.

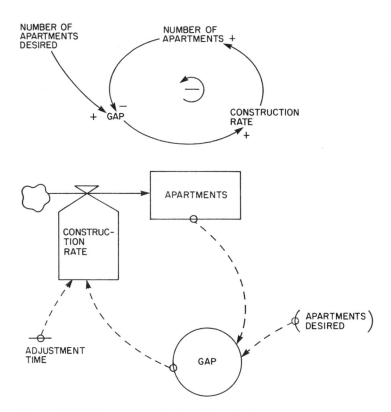

Figure 17.7 Diagrams expressing a housing gap

Exercise 6: Apartments—Part I

a. Write DYNAMO equations for the apartment model shown in Figure 17.7. Assume that the desired number of apartments is 10,000 and the time required to respond to the gap is one year. (Do not include an explicit construction delay in your formulation—it will be added in the next exercise.)

b. Run the model and examine the behavior. Determine the equilibrium value for the number of apartments.

c. Start the model in equilibrium, and use a STEP function to test the response of the system to an increase in the number of apartments desired from 10,000 to 15,000.

The apartment model developed so far has ignored an important delay: it takes time to construct apartments. Once an apartment builder makes the decision to build a new apartment house, it may take roughly four years to purchase appropriate land, obtain building permits, complete architectural drawings, and build the apartments.

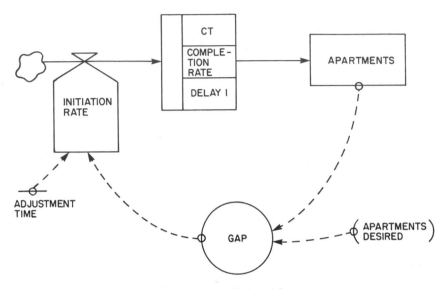

Figure 17.8 Apartment model with completion delay

This suggests that a delay should be added to the model. One way to do this is shown in Figure 17.8. The diagram indicates that the apartment completion rate is a delayed response to the apartment initiation rate. A delay exists between the times apartments are initiated and completed.

Exercise 7: Apartments—Part II

a. Modify your model of apartment construction to include a first-order material delay in the construction completion rate.

b. Start the model in equilibrium.

c. Examine the response of the system to a STEP increase in desired apartments from 10,000 to 15,000. What behavior does the system generate?

d. What do you think accounts for this behavior?

e. Experiment with the model, choosing various values for the time required to construct apartments. What happens when this value is made longer than 4 years? What happens when it is made shorter than 4 years?

f. At times, the apartment construction rate may become negative. What does this mean? Is it realistic? Why or why not? Predict how the system would behave for a period twice as long as your run. Run the model to check your prediction. If your prediction does not fit the run, try to formulate a new prediction for the long-term behavior.

INFORMATION DELAYS

Delays are frequently involved in transmitting and acting on information. One common information delay occurs in the labor market. It takes time for information about the availability of jobs to reach people who are seeking them. Similarly, information delays occur in organizational management. For example, the information is delayed from the time sales occur in a department store until the manager learns of the sales and uses the information to reorder depleted stock.

Information delays also occur in individual and organizational decision-making for another, more subtle reason. Often, individuals and organizations make decisions on the basis of information that has been *averaged*, and averaging implicitly involves delays. For example, data on U.S. unemployment are often presented on an annual basis, which is generally an average of the percent of the work-force that has been unemployed over the last twelve months. Thus a rise in current monthly unemployment must persist for several months before it has a large impact on the average annual unemployment figure.

Decisions based on averaged information are widespread. For example, upon filling up the tank with gasoline, the driver notices that her car has not traveled as far on a tank-full as it usually does. She is unlikely to immediately conclude that something is the matter with the car. Instead, she probably waits to see what happens on the next few fill-ups before sending the car to the shop. Thus the car-owner has informally computed an average. Similarly, the owner of a baseball team does not fire the manager when he loses the first game. The owner usually waits to see what the balance of wins and losses looks like over the longer run.

While it is certain that delays and information averaging are frequently involved in individual and organizational decision-making, it is less clear how to represent these processes in a model. After all, some business organizations use complex information averaging formulas in making decisions, while individuals often simply weigh whatever information is available in an informal, intuitive manner. One simple formulation that might be used to represent the delays involved in perceiving and acting on information is the moving average, which is considered in the following example.

EXAMPLE III: MARINA'S BAKE SHOP

Marina's Bake Shop bakes and sells sourdough French bread Monday through Friday. Each morning Marina has to decide how many loaves to bake, and she relies on the average sales over the past five days in making her decision. She collects the daily sales from the previous five days, adds them together, and divides by five.

Sales data for Marina's Bake Shop for a twenty-five-day period are shown in Table 17.1. Since it takes five days of data to compute an average, the first

Table 17.1 Marina's
sales data and moving
average sales

Day	Moving average sales	Sales
1	—	94
2	—	116
3	—	87
4	—	104
5	—	107
6	101.6	90
7	100.8	102
8		96
9		108
10		121
11		123
12		130
13		135
14		113
15		117
16		117
17		128
18		97
19		109
20		123
21		116
22		117
23		121
24		128
25		122

morning on which average sales can be computed is day six. The average on day six is just the sum of the first five days' sales, divided by five. The average for day seven is the sum of the sales for days two through six, divided by five; and so on.

Another way to state the formula for the moving average is:

$$AVERAGE\ SALES(today) = AVERAGE\ SALES(yesterday)$$
$$+ SALES(yesterday)*(1/5)$$
$$- SALES(six\ days\ ago)*(1/5)$$

This simply means that to compute the average sales, take the value of the average sales computed yesterday, subtract the portion of the average contributed by the sales six days ago, and add on the portion of the average contributed by the most recent day.

Exercise 8: Computing the Moving Average

Calculate the moving average for Marina's sales for days 8–25, based on the data in Table 17.1.

The moving average is fairly simple, conceptually, and easy to calculate by hand. Unfortunately, however, to represent it precisely in a simulation model requires that data be maintained on each past event included in an average. The longer the period of averaging, the more data to be retained. Furthermore, the moving average has the additional defect that all data over the time period of the average are weighed equally. However, when individuals and organizations make decisions, they tend often to rely more heavily on recent information, and less heavily on older data.

A second type of average, called the exponential average, resolves this latter problem by explicitly weighing recent information more heavily. Suppose Marina, on the morning of day two, wants to compute average sales using an exponential average. To compute the average, she has only one piece of information: sales on day one. Thus Marina sets the average on day two equal to the sales on day one:

AVERAGE SALES(day two) = SALES(day one)

If Marina wants to compute the average sales on the morning of day three, she has one more piece of information. She now has the average sales she computed on day two, along with the new sales figure for day two. To combine these two pieces of information to produce a five-day exponential average, Marina should use the following formula:

AVERAGE SALES(day three) = (4/5)∗AVERAGE SALES(day two)
                         + (1/5)∗SALES(day two)

The new sales figure should be weighted by a factor of one-fifth, and the previous average should be weighted by a factor of four-fifths.

Similarly, to calculate the average sales on the morning of day four, Marina should use the formula:

AVERAGE SALES(day four) = (4/5)∗AVERAGE SALES(day three)
                        + (1/5)SALES(day three)

Once again, the new sales figure is weighted by a factor of one-fifth, and the previous average is weighted by a factor of four-fifths.

The formula for a five-day exponential average of sales in Marina's Bake Shop is:

AVERAGE SALES(today) = (4/5)∗AVERAGE SALES(yesterday)
                     + (1/5)∗SALES(yesterday)

Hence, in computing today's average, yesterday's average is given a weight of four-fifths, and yesterday's new sales figure is given a weight of one-fifth. Implicitly, this procedure weights recent data more heavily than old data. By the twenty-first day, for example, the twentieth day's sales figure is given a weight of one-fifth; and all nineteen previous days' sales are given a total weight of only four-fifths.[2] Table 17.2 shows the exponential averages calculated for the first several days.

Exercise 9: Computing the Exponential Average

a. Calculate the exponential average for Marina's sales for days five through twenty-five, based on the data in Table 17.2.

b. Compare the exponential average and the moving average you obtained for each day. In what ways are they similar? How do they differ?

Table 17.2 Marina's exponentially averaged sales

Day	Exponential average sales	Sales
1	—	94
2	94.0	116
3	98.4	87
4	96.1	104
5		107
6		90
7		102
8		96
9		108
10		121
11		123
12		130
13		135
14		113
15		117
16		117
17		128
18		97
19		109
20		123
21		116
22		117
23		121
24		128
25		122

AVERAGING AS A FIRST-ORDER DELAY

The formula used to calculate the exponential average of the sales in Marina's Bake Shop can be rewritten in a form that clarifies the structure involved. The equation:

AVERAGE(today) = (4/5)*AVERAGE(yesterday) + (1/5)SALES(yesterday)

can be rewritten:

AVERAGE(today) = AVERAGE(yesterday) − (1/5)*AVERAGE(yesterday)

+ (1/5)SALES(yesterday)

or:

AVERAGE(today) = AVERAGE(yesterday) + (1/5)SALES(yesterday)
− (1/5)*AVERAGE(yesterday)

This indicates that the average computed today is simply the average computed yesterday, plus one-fifth of yesterday's sales figure, minus one-fifth of the average computed yesterday.

The structure can be clarified further by noticing that the expression

(1/5)*SALES(yesterday) − (1/5)*AVERAGE(yesterday)

is simply one-fifth of the gap between yesterday's sales and yesterday's average. Thus the formula for today's average might be written:

AVERAGE(today) = AVERAGE(yesterday) + (1/5)*GAP(yesterday)

where GAP(yesterday) represents the gap between yesterday's sales and yesterday's average.

The formulation should now look quite familiar: it is simply a goal-gap structure. This can be seen more easily by writing the equations in DYNAMO:

```
L      AVG.K = AVG.J + (DT)(ADJ.JK)
NOTE     AVERAGE SALES (loaves)
R     ADJ.KL = GAP.K/5
NOTE     ADJUSTMENT RATE (loaves/day)
A      GAP.K = SALES.K − AVG.K
NOTE     GAP (loaves)
```

Average sales AVG is a level, and the adjustment rate tends to move the average toward the daily sales figure. But the average does not adjust immediately to daily sales. Instead, the gap is closed over a period of time.

Figure 17.9 shows a flow diagram for the exponential averaging process, corresponding to the DYNAMO equations. The flow diagram is identical in

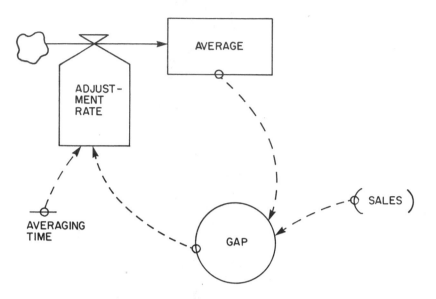

Figure 17.9 Flow diagram for exponential averaging

structure to the coffee cooling model discussed in Chapter 15. Coffee cools to room temperature over a period of time T. Similarly, the average sales in Marina's Bake Shop adjusts toward the daily sales over a period of time AT = 5. Thus the 5-day exponential average is simply a goal-gap formulation with an adjustment time of 5 days.

One striking difference exists between the coffee cooling example and that for Marina's Bake Shop. The coffee cooling case represents the adjustment of the coffee temperature to a constant outside temperature. Marina's average sales is adjusting to a fluctuating sales rate, but the structures are identical.

To demonstrate the similarity of the averaging structure and the coffee cooling structure, it is helpful to examine how the averaging process responds to a simple step-change in the sales rate. Table 17.3 shows sales figures for Evelyn's Bake Shop. As can be seen, sales in Evelyn's shop are much more regular than sales in Marina's Bake Shop. In fact, the only change in sales occurs on day ten, when sales jump from 100 to 120 loaves per day. Table 17.3 also shows computations for the five-days exponential average for days 1–25; and Figure 17.10 shows a plot of both actual sales and average sales. As expected, the exponential average responds to a step change exactly as a simple goal-gap structure.

Table 17.3 Evelyn's bake shop sales and average

Day	Exponential average sales	Sales
1	100	100
2	100	100
3	100	100
4	100	100
5	100	100
6	100	100
7	100	100
8	100	100
9	100	100
10	100	120
11	104	120
12	107	120
13	109.8	120
14	111.8	120
15	113.5	120
16	114.8	120
17	115.8	120
18	116.6	120
19	117.3	120
20	117.9	120
21	118.3	120
22	118.6	120
23	118.9	120
24	119.1	120
25	119.3	120

Exercise 10: Equations for the Exponential Average

a. Write DYNAMO equations to compute the average sales in Evelyn's Bake Shop, using a five-day exponential average.

b. Start the model in equilibrium.

c. Use a STEP function to represent the rise in sales from 100 to 120 on day 10.

d. How long does it take for average sales to rise half-way from 100 to 120?

e. Rerun the model, using as averaging time of 10 days rather than 5. How do the results differ?

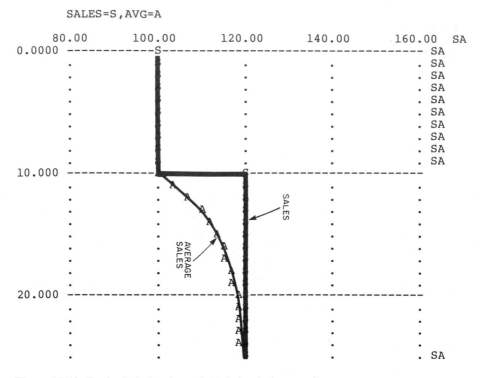

Figure 17.10 Evelyn's bake shop plotted simulation results

THE SMOOTH FUNCTION

The exponential average can be used whenever it is necessary to represent information averaging in a model. Figure 17.9 also shows the general structure involved in exponential averaging, with the input being averaged substituting for SALES in that diagram. The average to be computed is a level, which adjusts toward an input value over an averaging time.

Like first-order material delays, information averages are widely used in system dynamics modeling, and it is convenient to have a special flow diagram symbol for them. Figure 17.11 shows the symbol that is used to represent the exponential averaging process. Because this averaging process "smooths" the disturbances in the input, the function is often called a smoothing equation.

Whenever an exponential average is used in a model, it can be computed explicitly, using the level and rate equations previously described. In addition, DYNAMO includes a special function called SMOOTH, which can be used to

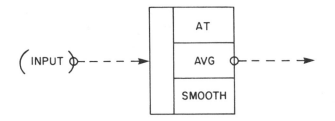

Figure 17.11 Flow diagram symbol for exponential averaging or smoothing

calculate the exponential average directly. The SMOOTH function takes the following form:

A   AVG.K = SMOOTH(INPUT.K,AT)

where AVG represents the average to be computed, INPUT is the variable (rate, level, or auxiliary) to be averaged, and AT is the adjustment time to be used.[3]

Exercise 11: The SMOOTH Function

a. Use the SMOOTH function to represent the averaging process in Evelyn's Bake Shop.

b. Run the model and compare your results with your model in the previous exercise. The results should be identical.

c. Rerun the model, setting the adjustment time AT = 10. How do the results differ?

NEGATIVE LOOPS, INFORMATION DELAYS, AND CYCLES

Like material delays, information delays can often cause surprising behavior in seemingly simple structures. For example, consider the negative loop shown in Figure 17.12, describing the relationship between jobs and urban migration. (This loop was discussed in more detail in Exercises 16 and 22, Chapter 15.)

According to the loop, the availability of job openings influences people to migrate into the city; and as people migrate into the city, they fill the available openings. The simplest level and rate formulation for this structure involves just one level and one rate, shown in Figure 17.13. It should be evident that this simple structure will generate goal-seeking behavior: the number of people in the city will adjust smoothly to the number of jobs.

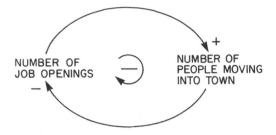

Figure 17.12 Relationship between jobs and migration

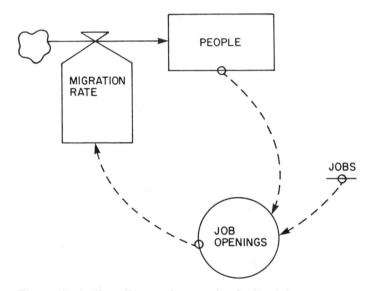

Figure 17.13 Flow diagram for negative feedback loop

While this formulation provides a rough description of the relationship between jobs and migration in a city, it ignores one important aspect of the problem. Undoubtedly, it takes time for people to learn about new job openings, and it takes even more time for them to relocate. Furthermore, people probably respond to the average number of job openings in a city, not to short-run increases or decreases. Thus the structure shown in Figure 17.14 is probably a more sensible representation of the relationship between jobs and migration.

The introduction of an information delay can cause the behavior of the model to change substantially. If people respond instantaneously to information about job openings, the adjustment process is smooth. Migration declines until any gap between the number of jobs and the number of workers is closed. But, if it takes time for people to respond to information about jobs, the model can generate cycles, as the following exercise indicates.

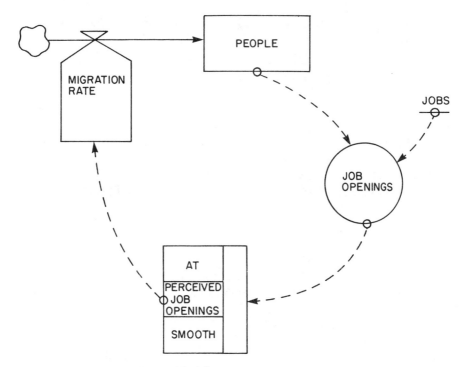

Figure 17.14 Negative loop with delay

Exercise 12: Information Delays, Migration, and Jobs

a. Review your model of migration and jobs developed in Exercise 21, Chapter 15. (It should correspond to the flow diagram structure shown in Figure 17.13.) Run the model and note the behavior it generates.

b. Revise the model to include a first-order information delay, as indicated in Figure 17.14. Choose an averaging time that you believe might represent the time required for people to respond to information about jobs.

c. Run the model and examine the results. (Be sure to examine the value of DT you have chosen. Make certain it is no more than one-third the value of the averaging time in your delay. See the discussion of DT at the end of Chapter 15.)

d. Rerun the model, choosing various values for the information averaging time. How do the results differ?

e. What do you believe causes the model to behave as it does?

EXAMPLE IV: STEVE'S ICE CREAM PARLOR—SEEMINGLY POSITIVE LOOPS THAT CONTAIN HIDDEN INFORMATION DELAYS

On occasion, a loop appearing in a causal-loop diagram may seem not to contain any level variables. Consider, for example, the loop shown in Figure 17.15, which represents the relationship between advertising, revenues, and sales in Steve's Ice Cream Parlor. According to the diagram, the amount Steve spends on advertising influences the number of ice cream cones sold; the number sold influences revenues; and the amount Steve earns in revenues influences the amount he spends on advertising.

At first glance, all the variables around the loop seem to be auxiliaries. This would produce the flow diagram shown in Figure 17.16. Equations for the model might be written:

```
A    AD.K = FRSA*REV.K
NOTE        ADVERTISING (DOLLARS/MONTH)
C    FRSA = 0.1
NOTE        FRACTION OF REVENUE SPENT ON ADVER-
NOTE        TISING (DIMENSIONLESS)
A    SALES.K = SALESN + ADEFF*AD.K
NOTE        SALES (CONES/MONTH)
C    SALESN = 1000
NOTE        SALES, NORMAL VALUE (CONES/MONTH)
C    ADEFF = 5
NOTE        ADVERTISING EFFECTIVENESS (ADDITIONAL
NOTE        CONES PURCHASED PER DOLLAR
NOTE        SPENT ON ADVERTISING)
A    REV.K = SALES.K*PRICE
NOTE        REVENUE (DOLLARS/MONTH)
C    PRICE = 1
NOTE        PRICE (DOLLARS/CONE)
```

The equations indicate that the amount spent on advertising each month is equal to one-tenth of the sales revenue earned per month. Furthermore, the

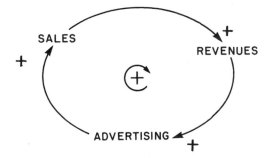

Figure 17.15 Causal relationships for Steve's Ice Cream Parlor

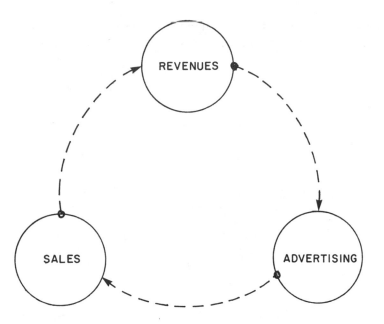

Figure 17.16 Flow diagram for Steve's Ice Cream Parlor

number of ice cream cones sold is a function of the amount Steve spends on advertising. For example, if Steve spends $100 on advertising, the number of cones sold = 1000 + 5*100 = 1500. Finally, the revenue earned is equal to the number of cones sold, multiplied by the price of a cone.

While this information seems sensible, it suffers from an important difficulty, each variable has an instantaneous effect on the next. According to the equations, advertising has an instantaneous effect on sales; sales has an instantaneous effect on revenue; and revenue has an instantaneous effect on the amount spent on advertising.

Because all of the relationships in the loop are instantaneous, it is not possible to draw any conclusions about behavior over time. Technically, the model is written as a set of simultaneous equations. It is possible to analyze the equations algebraically, to determine whether there are values of advertising, sales, and revenue that are consistent with the entire set of equations. But, since none of the equations include an explicit formulation for a level and a rate of change, the equations cannot be used to generate behavior over time.[4]

Loops such as this one, which seem at first glance not to contain any level variables, often involve hidden information delays. For example, in the advertising loop, two information delays might be involved. First, Steve undoubtedly does not respond immediately to increases in sales by increasing the amount he spends on advertising. Instead, he probably uses average sales over a month or two to determine how much should be spent. And, as previous sections have indicated, averaging implicitly introduces a delay. Second,

Steve's advertising probably does not have an immediate effect on the number of ice cream cones purchased. It probably takes consumers some time to notice and to react with purchases.

The following equations might be added to the model to represent the fact that Steve uses a monthly average of the number of ice cream cones sold to determine the amount he spends on advertising:

A   AVGREV.K = SMOOTH(REV.K,RAT)
NOTE   AVERAGE   REVENUE   (DOLLARS/MONTH)
C   RAT = 1
NOTE   REVENUE   AVERAGING   TIME   (MONTHS)
A   AD.K = FRSA*AVGREV.K
NOTE   ADVERTISING   (DOLLARS/MONTH)

The equations indicate that average revenue AVGREV is a first-order exponential average, with an averaging time or time constant of one month; and the amount spent on advertising is one-tenth of the average revenue each month.

As usual, it is necessary to provide an initial value to begin the simulation. In this case, however, since the only level in the loop is contained in the SMOOTH function, it is necessary to choose an initial value for the SMOOTH, which, in this case, is REVN = 2000.

N   AVGREV = REVN
C   REVN = 2000

Figure 17.17 is a modified flow diagram for the advertising and sales model, showing the first-order information delay involved in determining the average revenue. The flow diagram indicates that the model now contains two loops: one positive and one negative. As the following exercise demonstrates, the advertising and sales model can generate either exponential decay or exponential growth, depending on the values of the parameters chosen. Thus a structure that initially seemed to be a simple positive loop is actually more complex. It contains a hidden negative loop, and this has a critical influence on the model's behavior.

Exercise 13: Steve's Ice Cream Parlor

a. Write DYNAMO equations for the version of the advertising and sales model that does *not* include an information delay.

b. Try to run the model on the computer. What error message does DYNAMO generate?

c. Add the formulation for average revenue to your model. (As mentioned in the text, it is necessary to choose an initial value for the SMOOTH

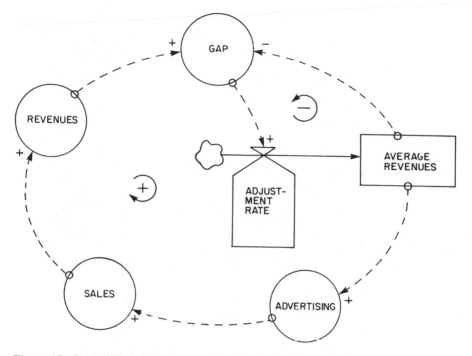

Figure 17.17 Modified flow diagram for Steve's Ice Cream Parlor

function. Choose an initial value that will place the system in equilibrium.) Test the response of the model to a STEP increase in normal sales,
from 1000 cones per month to 2000 cones per month.

d. Repeat part (c), choosing a revenue averaging time RAT = 2 months.
How do the results differ? Rerun the model, setting RAT = 0.5 months.
How do the results differ?

e. Choose a value of zero for normal sales, and choose an initial value for
the input to the SMOOTH function to start the system in equilibrium.
How does the system respond to a STEP increase in normal sales, from
zero to 1000 cones per month? (Compare your results with those in part
(c).)

f. Repeat part (e), setting the fraction of revenue spent on advertising
FRSA = 0.05. How do the results differ? Set FRSA = 0.2. How do the
results differ? Rerun the model, setting FRSA = 0.3. Which loop
dominates if FRSA is less than 0.2? Which loop dominates if FRSA is
larger than 0.2. Which loop dominates if FRSA is *exactly* 0.2?

EXAMPLE V: TREE HARVESTING—HIGHER-ORDER DELAYS

The material and information delays considered so far have all been first-order delays: that is, they have involved only one level. Although first-order delays provide a useful representation of the delays involved in many social and economic systems, delay processes are frequently encountered that do not seem to resemble the behavior of a first-order delay. Recall, for example, the tree harvesting case in Chapter 7 (Example III). According to the example, Lester Splintz planted 10,000 saplings in 1930, and the particular species of trees he chose was supposed to reach harvesting size after an average of twenty years. Figure 17.18 (copied from Figure 7.6) indicates the harvest rate Lester Splintz obtained.

It seems somewhat plausible to represent the tree harvesting process as a first-order delay, as shown in the following causal-loop diagram and flow diagram (Figure 17.19), and equations (Figure 17.20). The behavior generated by this model is shown in Figure 17.21. Unfortunately, however, the behavior obtained does not resemble the behavior shown in Figure 17.18.

As the discussion in Chapter 7 indicated, the problem lies in the causal-loop representation. The initial causal-loop diagram fails to take into account the fact that trees come in different ages and sizes; for example, saplings (trees that are 0 to 1 inch in diameter); small trees (1 to 3 inches in diameter); medium-sized trees (3 to 6 inches in diameter); and harvestable trees (6-plus inches in diameter). Ordinarily, trees are not harvested until they reach

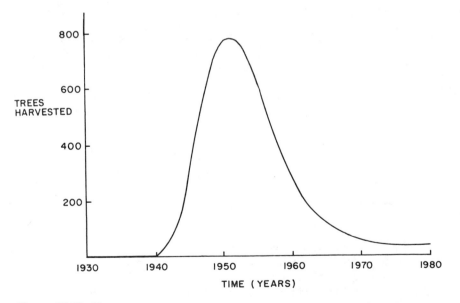

Figure 17.18 Harvest rate

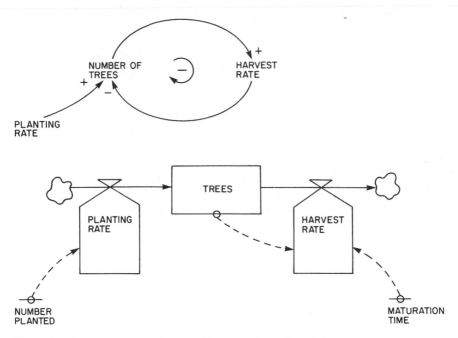

Figure 17.19 Causal-loop and flow diagrams for Splintz's farm

harvestable size. This distinction is incorporated in the causal-loop diagram in Figure 17.22 (copied from Figure 7.9).

The following flow diagram and equations, Figures 17.23 and 17.24, are based on the causal-loop diagram in Figure 17.22. This new model contains four levels, each with an adjustment time of 5 years (4*5 = 20). The behavior generated by the model is shown in Figure 17.25.

The harvest rate results (Figure 17.25) are quite similar to the data shown in Figure 17.18. The number of trees harvested remains at zero for the first 10 years. It then begins to rise, reaching a peak shortly after 1950. It then falls to zero by about 1975.

Exercise 14: Harvesting Model

  a. Enter the DYNAMO equations for the model as shown in Figure 17.24.

  b. Run the model and examine the behavior. It should be identical to the behavior in Figure 17.25.

  c. Rerun the model, using a tree growth time of 40 years. Rerun the model, choosing a growth time of 10 years. How do the results differ?

```
*           TREE HARVESTING MODEL
NOTE
NOTE
L           TREES.K=TREES.J+(DT)(PLANT.JK-HRVSTR.JK)
N           TREES=TREESN
NOTE             TREES (TREES)
C           TREESN=0
NOTE             TREES, INITIAL (TREES)
R           PLANT.KL=(1/DT)*PULSE(NPLANT,PTIME,TBP)
NOTE             PLANTING RATE (TREES/YEAR)
C           NPLANT=10000
NOTE             NUMBER PLANTED DURING EACH PULSE (TREES)
C           PTIME=1930
NOTE             PLANTING TIME (YEARS)
C           TBP=1000
NOTE             TIME BETWEEN PULSES (YEARS)
R           HRVSTR.KL=TREES.K/MT
NOTE             HARVESTING RATE (TREES/YEAR)
C           MT=20
NOTE             MATURATION TIME (YEARS)
NOTE
NOTE             SIMULATION SPECIFICATIONS
NOTE
SPEC        DT=0.5/PLTPER=1/LENGTH=1980
N           TIME=1930
PLOT        TREES=T(0,10000)/HRVSTR=H(0,800)
RUN
```

Figure 17.20 Equations for Splintz's Farm

    d. Modify your model to include only two levels: young trees and harvest-able trees. Set the adjustment time for each level equal to half the total growth time (20 years). Run the model. How do the results differ from your results in part (b)?

    e. Modify your model again, this time choosing three levels rather than two. How do the results differ?

    f. Modify your model, using 5 levels rather than three. What happens?

    g. What do you think would happen if you chose 10 levels? 20 levels? 100 levels?

DYNAMO FUNCTIONS FOR THIRD-ORDER DELAYS

Many social and economic processes resemble the tree-growing process pre-viously discussed, in that they can be represented by a sequence of smaller delays. For example, while the apartment construction delay discussed earlier in this chapter was represented as a first-order delay, it is probably more accurate to consider it a higher-order delay, representing a sequence of smaller

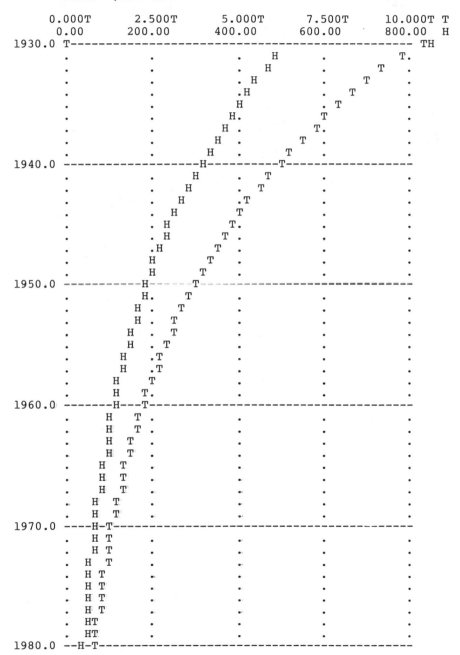

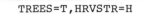

Figure 17.21 Plot for Splintz's farm

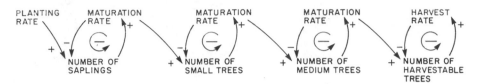

Figure 17.22 Causal-loop diagram showing levels of tree growth

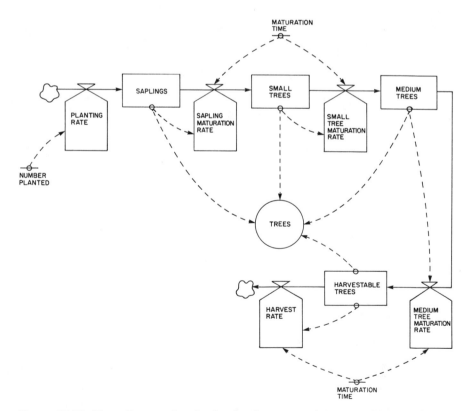

Figure 17.23 Flow diagram showing levels of tree growth

```
*          TREE HARVESTING MODEL
NOTE
NOTE
L          SAPLNG.K=SAPLNG.J+(DT)(PLANT.JK-SAPMR.JK)
N          SAPLNG=SPLNGN
NOTE               SAPLINGS (TREES)
C          SPLNGN=0
NOTE               SAPLINGS, INITIAL (TREES)
L          SMALL.K=SMALL.J+(DT)(SAPMR.JK-SMAMR.JK)
N          SMALL=SMALLN
NOTE               SMALL TREES (TREES)
C          SMALLN=0
NOTE               SMALL TREES, INITIAL (TREES)
L          MEDIUM.K=MEDIUM.J+(DT)(SMAMR.JK-MEDMR.JK)
N          MEDIUM=MDIUMN
NOTE               MEDIUM TREES (TREES)
C          MDIUMN=0
NOTE               MEDIUM TREES, INITIAL (TREES)
L          HRVST.K=HRVST.J+(DT)(MEDMR.JK-HRVSTR.JK)
N          HRVST=HRVSTN
NOTE               HARVESTABLE TREES (TREES)
C          HRVSTN=0
NOTE               HARVESTABLE TREES, INITIAL (TREES)
A          TREES.K=SAPLNG.K+SMALL.K+MEDIUM.K+HRVST.K
NOTE               TOTAL TREES (TREES)
R          PLANT.KL=(1/DT)*PULSE(NPLANT,PTIME,TBP)
NOTE               PLANTING RATE (TREES/YEAR)
C          NPLANT=10000
NOTE               NUMBER PLANTED DURING EACH PULSE (TREES)
C          PTIME=1930
NOTE               PLANTING TIME (YEARS)
C          TBP=1000
NOTE               TIME BETWEEN PULSES (YEARS)
R          SAPMR.KL=SAPLNG.K/MT
NOTE               SAPLING MATURATION RATE (TREES/YEAR)
R          SMAMR.KL=SMALL.K/MT
NOTE               SMALL TREE MATURATION RATE (TREES/YEAR)
R          MEDMR.KL=MEDIUM.K/MT
NOTE               MEDIUM TREE MATURATION RATE (TREES/YEAR)
R          HRVSTR.KL=HRVST.K/MT
NOTE               HARVEST RATE (TREES/YEAR)
N          MT=TMT/4
NOTE               MATURATION TIME (YEARS)
C          TMT=20
NOTE               TOTAL MATURATION TIME (YEARS)
NOTE
NOTE               SIMULATION SPECIFICATIONS
NOTE
SPEC       DT=0.5/PLTPER=1/LENGTH=1980
N          TIME=1930
PLOT       TREES=T(0,10000)/HRVSTR=H(0,800)
RUN
```

Figure 17.24 Model of tree growth

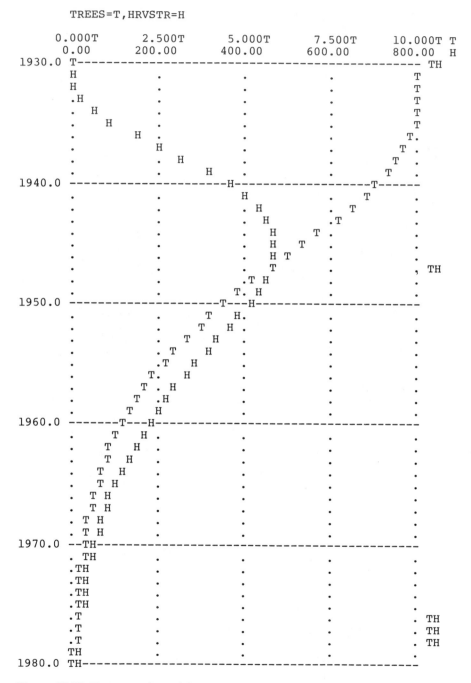

Figure 17.25 Tree growth model

delays (the delays involved in getting permits, hiring an architect, negotiating loans, and so forth). Similarly, the mail delay discussed in Exercise 5 might more realistically be considered a higher-order delay, since it undoubtedly involves a sequence of smaller delays (the delays involved in sorting the mail, getting it from post-office to post-office, and delivering it).

When modeling a delay process, one immediate question arises: How many levels should be included in the delay? The answer, of course, depends on the exact process being modeled. However, to a rough approximation, it is often sufficient to choose either a first-order delay or a third-order delay (that is, a delay with only one level or a delay with three).

As the preceding exercise indicates, for example, the behavior of a third-order delay is reasonably similar to the behavior of a fourth-order delay. Thus either might be a reasonable choice in modeling the tree-harvesting case.

DYNAMO includes special functions to represent third-order material and information delays, analogous to the DELAY1 and SMOOTH first-order delays. The DELAY3 function can be used to represent third-order material delays, and the DLINF3 function can be used to represent third-order information delays. The DELAY3 and DLINF3 functions are simply shorthand notations, exactly identical to the full set of equations for third-order delays. For example, the following equations might be used to represent the tree-harvesting case:

    R   HARVST.KL = DELAY3(PLANT.JK,GT)
    C   GT = 20

This indicates that the harvest rate HARVST is a third-order material delay of the planting rate PLANT, with a total growth time GT = 20 (years). The following equations might be used to indicate that perceived job openings PJO is a third-order information delay of actual job openings JO, with an adjustment time of two years.

    A   PJO.K = DLINF3(JO.K,AT)
    C   AT = 2

Exercise 15: Using the DELAY3 and DLINF3 Functions

a. Use the third-order material delay function (DELAY3) to represent the tree-harvesting case.

b. Reread Exercise 2 in Chapter 7 (Tree Harvesting—Part II). Use a third-order delay function to represent Warren's tree harvesting process. (*Hint:* Use a STEP function to represent the planting rate.)

c. Modify your model of the Warren Splintz case, using a first-order delay rather than a third-order delay. How do the results differ? Revise the model, using a fourth-order delay. (One way to do this is to combine a third-order and first-order delay. What delay times should you choose for each?)

d. Modify your apartment construction model (Exercise 7), using a third-order material delay, rather than a first-order delay. How do the results differ?

e. Modify your model of job-migration cycles (Exercise 12), using a third-order information delay (DLINF3) rather than a first-order delay. How do the results differ?

ENDNOTES

1. When the DELAY1 function is used, DYNAMO automatically calculates an initial value for the level equation in the delay, using the formula: LEVELN = INFLOW*AT. This produces an initial outflow rate exactly equal to the inflow rate. For example, in the Nobug case, DYNAMO selects an initial value of 0, since the initial value of the dumping rate is 0. It is possible to initialize the level at whatever value is desired, by using an initial value equation for the inflow rate. For example, the following equations could be used to set the initial value of the level of NOBUG equal to 100:

    R   ABSORB.KL = DELAY1(DUMP.JK,NAT)
    C   NAT = 2
    N   DUMP = 100/NAT
    R   DUMP.KL = (1/DT)*PULSE(420,1,7)

    DYNAMO will calculate the initial value of the level equation
    LEVELN = DUMP*NAT = (100/NAT)*NAT = 100.

2. For an infinite sequence of observations, it is not hard to show that in a five-day exponential average the most recent day's sales receives a weight of 1/5; the second most recent, a weight of 4/25; the third, 16/125; and in general, the nth most recent previous observation receives a weight of

$$\frac{4^{n-1}}{5^n}$$

3. As previously explained, all averages really are levels. A peculiarity of DYNAMO processing requires that the average be treated as an auxiliary equation (A) when the SMOOTH function is used. Using the proper L (for Level) with the SMOOTH equation will generate a strange DYNAMO error message. In order to avoid problems, the model-builder should also provide an initial condition (N equation) for AVG, as would be done with any other level equation.

4. DYNAMO checks to make sure there is a level equation in all loops. If one or more loops occur without levels, DYNAMO generates an error message similar to the following:

    "SIMULTANEOUS EQUATIONS IN THE AUX EQUATIONS FOR AD"
    which indicates that the auxiliary variable "AD" is part of a simultaneous equation loop.

    If the equations of the advertising model are analyzed as a set of simultaneous equations, the solution is:

    SALES = 2000 cones/month; REV = 2000 dollars/month; and AD = 200 dollars/month.

PART VI

FORMULATING AND ANALYZING SIMULATION MODELS*

OBJECTIVES

This part draws together the concepts developed in Parts I through V. When this part is completed, the student should be able to use simulation to examine topics in such areas as sociology, biology, economics, and ecology. In completing this part the student will be:

1. Formulating a problem definition from a brief verbal description;

2. Developing a causal-loop diagram;

3. Drawing a flow diagram;

4. Writing equations for a model;

5. Running the model on a computer;

6. Testing the model and using the model to analyze the problem that has been defined.

---

*All information needed about DYNAMO is available either in Part V or VI of this text or the appropriate DYNAMO user's guide.

CHAPTER 18

MODELING THE ECOLOGY OF THE KAIBAB PLATEAU*

INTRODUCTION: THE PROBLEM

The Kaibab Plateau is a large, flat area of land located on the northern rim of the Grand Canyon. The plateau has an elevation of about 6,000 feet and it is bounded on all sides by steep slopes and escarpments. It has an area of about 800,000 acres, and it is the natural home of rabbits, deer, mountain lions, wolves, coyotes, and bobcats. The plateau was the fall meeting place for Navahoes, Piutes, and other Southwestern Indian tribes who came to the Kaibab to hunt deer, an important article of commerce. Early white settlers in the area called the plateau "Buckskin Mountain," but it has become more widely known by the Indian name "Kaibab," which means "mountain lying down."

In 1907, President Theodore Roosevelt created the Grand Canyon National Game Preserve, which included the Kaibab Plateau. Deer hunting was prohibited. At the same time, a bounty was established to encourage the hunting of mountain lions and other natural deer predators. During the period from 1906 to 1931, nearly 800 mountain lions were trapped or shot. As a result of the extermination of the Kaibab mountain lions and other natural enemies of the deer, the deer population began to grow quite rapidly. The deer herd increased from about 4,000 in 1906 to nearly 100,000 in 1924.

As the deer population grew, Forest Service officials and other observers began to warn that the deer would exhaust the food supply on the plateau. One observer wrote: "Never before have I seen such deplorable conditions. . . . But one conclusion could be reached, that from 30,000 to 40,000 deer

---

*The model discussed in this chapter is based, in part, on the work of Professor Donella Meadows of Dartmouth College and Michael Goodman of Pugh-Roberts Associates, Inc., for which the authors express deep appreciation.

   Historical information about the Kaibab Plateau is, in part, based on Edward J. Kormondy, *Concepts of Ecology* (Englewood Cliffs, N.J.: Prentice-Hall, 1976).

were on the verge of starvation." Indeed, over the winters of 1924 and 1925, nearly 60 percent of the deer population on the plateau died.

The deer population on the Kaibab continued to decline over the next fifteen years, and it finally stabilized at about 10,000 in 1939. (See Figure 18.1.)

Imagine you were an official of the National Forest Service in 1930, and you were interested in the fate of the deer population on the Kaibab Plateau. To examine some alternative approaches to the problem, you have decided to build a simulation model. Your main concern is the growth and rapid decline of the deer population observed over the period from 1900 to 1930, and the future course of the population from 1930 to 1950. Thus the time frame for the model you will build is the fifty years from 1900 to 1950, and the problematic behavior (reference mode) is the behavior shown in Figure 18.1. Once you have developed an adequate model, you can use it to examine some alternative ways of controlling the size of the deer population on the plateau.

AN INITIAL MODEL

In building a model, it is often helpful to begin with a simple model and then expand it in several steps. Start with a small model focusing on a few of the factors that influenced the size of the deer population on the Kaibab Plateau. Later, the model can be enlarged to include additional elements of the plateau ecology, finally obtaining a model that generates the behavior shown in Figure 18.1.

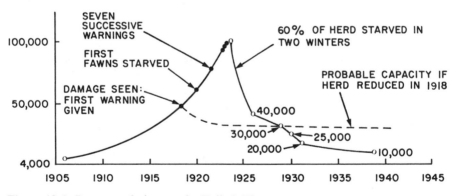

Figure 18.1 Deer population on the Kaibab Plateau

Exercise 1: Deer Births

a. The size of the deer population on the plateau is increased by deer births. Draw a causal-loop diagram representing the deer population and the deer birth rate. Is the loop positive or negative?

b. The next step in formulating the model is to translate the causal-loop diagram developed so far into a flow diagram that explicitly identifies which variables are levels and which are rates. Draw a flow diagram showing the deer population DEER and deer births DBRTHS.

c. Write equations for the model, to produce a simulation of the deer population over the period from 1900 to 1950. To begin write the level equation for the deer population DEER. What is the initial value for DEER? (*Hint:* See Figure 18.1.)

d. How should the rate equation for deer births be formulated? The number of deer births DBRTHS per year equals the size of the deer population DEER multiplied by the number of births per year per deer. Assume that each female deer gives birth to one deer per year. Since one-half the deer population is female, this corresponds to a fractional birth rate for deer FBRD = 0.5 deer born per deer per year. Write the rate equation for deer births.

e. To simulate the model we have developed, we need to add the simulation specifications. Write the appropriate specifications, setting the solution interval DT = 0.1 year, the LENGTH = 1950, and the PLTPER = 1 year. Plot the deer population DEER and deer births DBRTHS, choosing a plot scale of (0,1E6), i.e., a lower scale of 0 and an upper scale of 1,000,000. It is also necessary to write an initial value equation to set TIME = 1900. (An initial value for TIME is necessary only because we want the simulation run to begin in the year 1900. If you do not specify an initial value for TIME, DYNAMO will automatically set an initial value of TIME = 0.) The initial value equation can be placed anywhere in the model listing, and it should take the following form:

N   TIME = 1900

f. Run the model and examine the output. Label the model run Exercise 1(f). What behavior mode does the deer population exhibit?

g. Rerun the model, setting the fractional birth rate for deer FBRD = 0.3. Label the run Exercise 1(g). How do the results differ from Exercise 1(f)?

Exercise 2: Deer Deaths

a. Now add deer deaths to the model developed so far. Draw a causal-loop diagram representing the deer population and the deer death rate. Is the loop positive or negative?

b. Copy the flow diagram from Exercise 1(b), and add deer deaths DDTHS

c. What is the equation for deer deaths? Assume the average lifetime of deer AVGLD = 5 years. This means that about one-fifth of the deer population dies each year of disease or other natural causes (other than predators). Write the rate equation for deer deaths DDTHS.

d. Add the equation for deer deaths DDTHS to the model developed in Exercise 1(f), and run the new model. (Be sure to add DDTHS to the PLOT statement). Label the run Exercise 2(d). How does the behavior of the new model differ from Exercise 1(f)?

e. Now rerun the model with the fractional birth rate for deer FBRD = 0.2. Label the run Exercise 2(e). How do the results differ from Exercise 2(d)?

Exercise 3: Deer Killed by Mountain Lions

The rate at which deer are killed by mountain lions is a bit more complicated than the natural birth and death rates. How many deer might be killed by mountain lions on the plateau in a year? The answer, of course, depends on the number of mountain lions on the plateau. The more mountain lions there are, the more deer will be killed. In addition, the number of deer killed by mountain lions in a year also depends on the number of deer there are per acre on the plateau. The fewer deer there are per acre, the more difficult it is for mountain lions to track them down.

a. Draw a causal-loop diagram representing the number of deer killed by mountain lions. (Assume that the number of mountain lions on the plateau is constant.) Is the loop positive or negative?

b. Copy the flow diagram from Exercise 2(b) and add the deer killed by mountain lions DKILLD. Use the label DKPL for "Deer Killed Per mountain Lion" and DDEN for "Deer DENsity" per acre.

Now what are the equations for the deer killed by mountain lions? Remember that the number of deer killed by lions depends on both the number of mountain lions and the number of deer each mountain lion can kill. And the number of deer killed per mountain lion DKPL in turn depends on the deer density DDEN, measured in deer per acre, on the plateau.

In 1900 there were 4000 deer on the plateau, which measured about 800,000 acres. Thus the deer density DDEN in 1900 was 4000/800,000, or .005 deer per acre. Assume that at a deer density DDEN = .005 deer per acre, each mountain lion can kill 3 deer per year. As the density of deer on the plateau increases, it becomes easier for the lions to find and kill more deer. When the deer density is DDEN = .025, or five times the density in 1900, many deer are unable to find cover, and as a result each lion can kill 30 deer per year. Eventually, at a deer density DDEN = .05 deer per acre (ten times the density in

1900), the number of deer each lion can kill reaches its maximum value of 60 deer per year. Figure 18.2 is a table function showing the relationship between the deer density DDEN and the deer killed per mountain lion DKPL.

c. Using the table-function values shown in Figure 18.2, write the equations for DKPL.

d. The number of deer killed by mountain lions per year DKILLD is equal to the number of mountain lions LIONS multiplied by the number of deer killed per mountain lion per year DKPL. Write the equation for deer killed DKILLD. (Assume for the moment that there were 400 mountain lions on the plateau over the period from 1900 to 1950. Later, we will include a representation of the changes in the mountain lion population due to the establishment of bounty killing.)

e. Add the equations for DKILLD to the model developed previously in Exercise 2(d) and run the new model. (Be sure to add DKILLD to the PLOT statement.) Choose a PLOT scale of (0,40000) for DEER, DBRTHS, DDTHS, and DKILLD. Label the run Exercise 3(e). What behavior mode does the deer population exhibit?

f. Now rerun the model, setting the number of mountain lions LIONS = 600. Label the Run Exercise 3(f). How do the results differ?

g. Rerun the model again, this time setting the number of mountain lions LIONS = 200. Label the Run Exercise 3(g). How do the results differ?

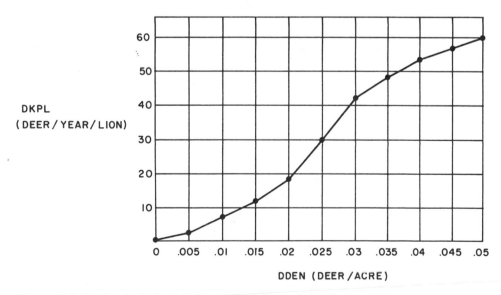

Figure 18.2 Table of relationship between DDEN and DKPL

h. (Optional) To understand the model behavior in Exercises (e)–(g), it is helpful to compute the number of deer births, deer deaths, and deer killed for various values of the deer and lion populations. This will help you understand when the deer population will grow, when it will decline, and when it will remain constant. The easiest way to do this is to draw some graphs, which we will do in Exercises (h) through (k). In Figure 18.3, draw a graph showing the relationship between deer births DBRTHS and the deer population DEER. (*Hint:* The relationship is a straight line.) On the same graph, draw the relationship between deer deaths DDTHS and the deer population DEER.

i. (Optional) The difference between the two lines in Exercise 3(h) indicates the number of *net deer births.* It represents the rate at which the deer population would grow (in deer/year) if no deer were killed by lions. On a similar graph draw the relationship between net deer births and the deer population.

j. (Optional) Suppose the number of mountain lions on the plateau LIONS = 200. On a graph similar to Figure 18.3, draw a graph showing

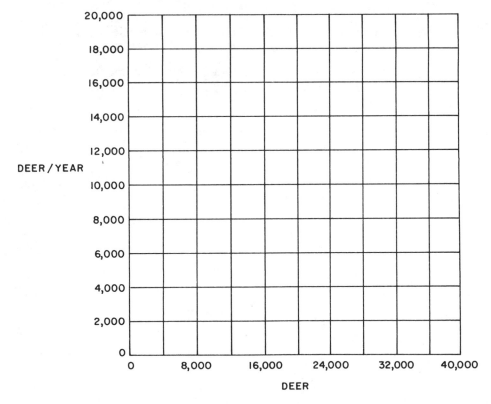

Figure 18.3 Relationship between DBRTHS, DDTHS, and DEER

the relationship between the number of deer killed by mountain lions DKILLD and the deer population DEER. (*Hint:* Examine the equation for deer killed by mountain lions: DKILLD.KL = LIONS∗DKPL.K. Remember that DKPL is a table function of deer density DDEN. Also, remember that DDEN.K = DEER.K/AREA, and AREA = 800,000 acres. We can figure out the value for deer killed by mountain lions DKILLD when the deer population DEER = 4000. When DEER = 4000, deer density DDEN = 4000/800,000 = .005. According to the table function for DKPL shown in Figure 18.2, DKPL has a value of 3 when DDEN has a value of .005. Thus when DEER = 4000, DKILLD = 200∗3 = 600 deer per year. To complete the graph, figure out the values for DKILLD when DEER equals 8000, 12,000, and so forth, and plot them.)

k. (Optional) Copy the graph relating net deer births and the deer population from Exercise 3(i) onto the graph relating deer killed by mountain lions and the deer population from Exercise 3(j). At what deer population do the two lines intersect? What happens at the point of intersection? (*Hint:* Examine the simulation run in Exercise 3(g).) What happens on the left of the point of intersection? What happens on the right?

l. (Optional) On a separate sheet of paper, redo Exercises 3(h)–3(k), assuming the number of mountain lions on the plateau LIONS = 600. How do the results differ from the results in Exercise 3(k)? (*Hint:* Examine the simulation run in Exercise 3(f).)

m. (Optional) Rerun the model with LIONS = 400 and the initial deer population DEERN = 6000. Why does the deer population decline toward 4000? Label this run Exercise 3(m).

n. (Optional) Rerun the model with LIONS = 600 and the average lifetime of deer AVGLD = 20. What happens? Why? Label this run Exercise 3(n).

A SECOND MODEL

In the initial version of the model worked out in the preceding sections, the number of mountain lions LIONS on the Kaibab Plateau was considered constant. Now examine the factors that influence the size of the mountain lion population. Three factors are particularly important: mountain lion births, mountain lions deaths, and mountain lions killed by hunters.

Exercise 4: Formulating the Second Model

a. The mountain lion births and deaths can be formulated exactly like the birth and death rates for deer. Draw a causal-loop diagram and flow diagram and write equations for mountain lion births and deaths. Assume the fractional birth rate for mountain lions FBRL = 0.2 mountain lion born per lion per year. (This means that each female mountain lion gives birth to 0.4 lion per year. How did we come to this conclusion?) Also, assume that the average lifetime of mountain lions AVGLL=10 years.

b. Now consider the rate of mountain lions killed by hunters per year LKILLD. One way to represent hunting in the model is to assume that hunters can kill a constant fraction of the mountain lion population each year. Suppose in 1900 the fraction of mountain lions killed FLK = 0.1. Draw a causal-loop diagram and flow diagram and write equations for mountain lion hunting.

Examine the effects of the large increase in mountain-lion hunting that took place in 1906 as a result of the bounty imposed by the Forest Service. One way to do this is to use a STEP function to represent an increase in the fraction of mountain lions killed FLK in 1906.

A	FLK.K=STEP(FLKB,BBTIME)−STEP(FLKB,BETIME)
X	+FLKN
C	FLKB=0.2
C	BBTIME=1906
C	BETIME=2000
C	FLKN=0.1

FLK	Fraction of Lions Killed (1/year)
FLKB	Fraction of Lions Killed due to Bounty (1/year)
BBTIME	Bounty Beginning Time (years)
BETIME	Bounty Ending Time (years)
FLKN	Fraction of Lions Killed, Initial Value (1/year)

According to this formulation, the per-year fraction of mountain lions killed due to the bounty FLKB = 0.2, and the bounty is imposed in the year BBTIME = 1906.

The history of the Kaibab Plateau described in the Introduction of this chapter indicated that the bounty remained in force indefinitely, once it was imposed. It may be interesting, however, to examine what might have happened had the bounty been stopped some time after 1906. This can be tested in the model by choosing a value for the bounty ending time BETIME. In the preceding equations, the bounty ending time BETIME is set equal to 2000, which occurs after the end of the simulation run (1950). With BETIME = 2000 (or any value larger than 1950), the bounty remains in force over the entire simulation period. Later, the effects of ending the bounty earlier can be examined.

According to the previous formulation, the initial annual fraction of mountain lions killed FLKN = 0.1. After 1906, the total annual fraction of mountain lions killed FLK = FLKB + FLKN = 0.2 + 0.1 = 0.3.

c. (Optional) How many mountain lions were born in 1900? How many died? How many were killed by hunters? What does this indicate about the mountain lion population?

d. (Optional) How many mountain lions were born in 1906? How many died? How many were killed by hunters? What does this indicate about the mountain lion population?

e. (Optional) What do you think will happen to the deer population beginning in 1906?

Exercise 5: Running the Second Model

Now add the equations developed in the previous exercise to the model worked out in Exercise 3(e). (*Note:* Recall that in the prior model, the mountain lion population LIONS was a constant. In the current model, it is a level. Make sure to change "LIONS" to "LIONS.K" in the equation for deer killed by mountain lions DKILLD.) Also, add the mountain lion population LIONS to the PLOT equation, choosing a scale of (0,2000), and change the scale of DEER to (0,1E6).

a. Run the model and examine the output. Label the run Exercise 5(a). What behavior mode does the deer population exhibit after 1906?

b. Now, rerun the model with the bounty beginning time BBTIME set equal to 1920 rather than 1906. How does the behavior in this run compare with the behavior in the initial run? Label the run Exercise 5(b).

c. (Optional) Rerun the model setting the average lifetime for deer AVGLD = 20 years and the mountain lion population LIONS = 600. Label this run Exercise 5(c). How does the behavior differ from the initial run? Why? (See Exercise 3(n).)

A THIRD MODEL

The model discussed so far shows exponential growth in the deer population after 1906. The history of the Kaibab Plateau shown in Figure 18.1, however, is rather different. There is a period of exponential growth from about 1906 to 1923. Then, beginning in 1924 the deer population reaches a peak and starts to decline. This discrepancy between the behavior of the model and the data shown in Figure 18.1 indicates that something is missing from the model.

According to the discussion at the beginning of this chapter, the main reason for the decline in the deer population from 1924 to 1925 was starvation; the deer population had exhausted the vegetation supply on the plateau. This section adds a simple representation of the deer food supply to the model.

Exercise 6: Formulating the Third Model

One way to do this is to assume a constant food supply FOOD on the plateau. As the deer population increases, the food available per deer FAPD declines, and this in turn reduces the average lifetime of deer on the plateau.

a. Draw a causal-loop diagram and flow diagram representing this process. Is the loop positive or negative?

Call the normal amount of vegetation growing on 4 acres one food unit. Thus in 1900 there were 800,000/4 = 200,000 units of vegetation on the plateau. The 200,000 units of vegetation supported 4000 deer, and so each deer was able to draw on 200,000/4000 = 50 food units.

Suppose that a Kaibab deer requires about 1 food unit in order to survive and live 5 years. As the amount of food available per deer falls below 1 unit, the average lifetime falls as well. Assume that at 0.6 food units per deer, the average lifetime is four years; and at 0.4 food units per deer the average lifetime is two years.

b. Draw a table function showing the relationship between food available per deer FAPD and the average lifetime of deer AVGLD.

c. Now write the equations for average lifetime of deer AVGLD and deer deaths DDTHS.

d. (Optional) Without actually running the model, what growth pattern in the deer population will the model described in this Exercise generate? Why?

e. (Optional) What fraction of the deer population dies each year when the average lifetime of deer AVGLD = 5? When AVGLD = 4? When AVGLD = 3? When AVGLD = 2? When AVGLD = 1? Recall that the fractional birth rate for deer FBRD = 0.5. For what value of AVGLD will the deer birth rate and deer death rate be equal?

f. (Optional) What value of food available per deer FAPD will result in the value of AVGLD determined in Exercise 6(e)? What size of the deer population is required to produce this value for FAPD? What does this imply?

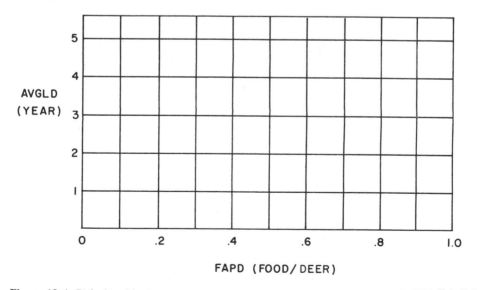

Figure 18.4 Relationship between FAPD and AVGLD

Exercise 7: Running the Third Model

a. Add the equations developed in the previous exercise to the Second Model. (See Exercise 5(a).) Add AVGLD to the PLOT equation, choosing a scale of (0,5). (Also, be sure to change "AVGLD" to "AVGLD.K" in the equation for deer deaths DDTHS.)

b. Now run the model and examine the output. Label the run Exercise 7(b). What behavior mode does the deer population exhibit? Why?

c. Run the model again, setting FOOD = 300,000. Label the run Exercise 7(c). How does the behavior differ from the run in Exercise 7(b)? Why?

d. (Optional) Rerun the model with FOOD = 100,000. Label this run Exercise 7(d). How does the behavior differ from the behavior obtained when FOOD = 200,000? What determines the equilibrium level of DEER? (*Note:* See Exercises 6(e) and 6(f).)

A FOURTH MODEL

The model discussed in the last section generates S-shaped growth in the deer population beginning in 1906. The discrepancy between the model behavior and the data in Figure 18.1 indicates that something is still missing from the model. In Exercise 6 and 7 a constant food supply on the plateau of 200,000 food units was assumed. Now examine the food supply in more detail.

Exercise 8: Formulating the Fourth Model

a. The food supply on the Kaibab can be represented as a level influenced by two principal rates: the food consumed by deer and the rate of food growth. Let's first consider food consumption. The amount of food consumed by deer each year is determined by both the number of deer and the amount of food each consumes. The amount of food consumed per deer, in turn, is influenced by the food available per deer. Draw a causal-loop diagram.

Assume that the maximum a Kaibab deer can eat, when food is freely available, is 1 food unit per year. When the food available per deer FAPD falls below 2 units, grazing becomes more difficult, and the amount each deer consumes falls slowly. When the food available per deer reaches 1 unit, each deer consumes 0.8 units. As the food available per deer falls below 1 unit, the amount each deer consumes falls more rapidly.

b. Draw a table function showing the relationship between food available per deer FAPD and food consumed per deer FCPD. (See Figure 18.5.)

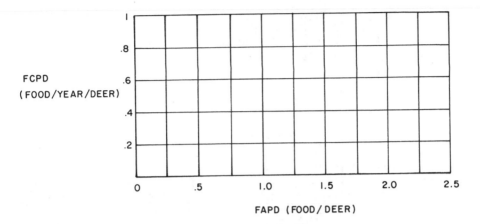

Figure 18.5 Relationship between FAPD and FCPD

   c. Now draw a flow diagram and write equations for the food supply FOOD
       and food consumed FCON.

      Now consider the rate of food growth FGRTH. Assume that under nor-
mal rainfall, soil, and other conditions on the Kaibab, the plateau can support
a maximum of 200,000 units of vegetation. This might be called the plateau's
food capacity FCAP. The amount of food growth each year depends on the
differences between the level of vegetation on the plateau and the maximum
capacity FCAP. When the level of vegetation equals FCAP, no net growth
occurs. When the amount of vegetation on the plateau falls below FCAP, it requires a period of time for the food level to return to full capacity.

   d. Assume that the amount of time required for the vegetation to return to
       full capacity increases as the amount of vegetation on the plateau de-
       clines. Draw a causal-loop diagram.

      Call the time required for the vegetation on the plateau to return to full
capacity the "Time for Food Regeneration" or TFR. How might TFR depend
on the level of vegetation on the plateau? Assume that when the food level on
the plateau is zero, the regeneration time TFR is 40 years. When the food level
is one-half the capacity of the plateau, the regeneration time is 5 years, and
when the food level is near capacity, the regeneration time is 1 year.

   e. Define the food ratio FRATIO as the level of food divided by the pla-
       teau's food capacity, and draw a table function showing the relationship
       between FRATIO and TFR. (See Figure 18.6.)

   f. Now draw a flow diagram for FOOD and food growth FGRTH. Write
       equations. (*Note:* The rate of food growth equals the difference between

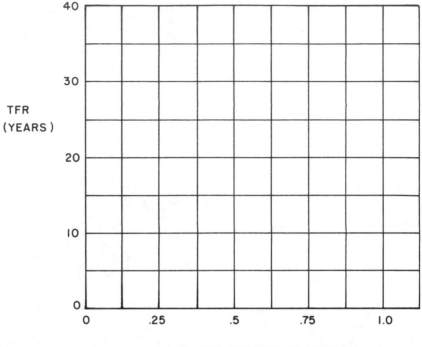

Figure 18.6 Relationship between FRATIO and TFR

food capacity FCAP and the food level FOOD, divided by the time for food regeneration TFR.) Assume an initial value for the level of food FOODN = 196,000. To see why this is an appropriate choice, see Exercise (g) below.

g. (Optional) How much food is consumed by deer in 1900? How much food is generated? Suppose we had chosen 200,000 for the initial food supply. How much food would have been generated in 1900?

Exercise 9: Running the Fourth Model

a. Now add the equations developed in Exercise 8 to the third model (Exercise 7(b).) Add the food supply FOOD to the PLOT equation, choosing a scale of (0,2E5). Change the scale for DEER to (0,1.2E5). Run the model and examine the output. Label the run Exercise 9(a). What behavior does it exhibit?

b. How well do you think the behavior matches the data shown in Figure 18.1? (*Note:* There are several aspects of the behavior generated by the model that can be compared with the data shown in Figure 18.1. For example, what is the maximum value reached by the deer population in the simulation run, and what is the maximum value according to the data in Figure 18.1? In what year does the turning point in the deer population occur in the simulation run, and in what year does it occur according to the data? What is the level of deer in 1940 in the simulation run, and what is the level according to the data?)

c. Another important question to ask about the model is how sensitive the behavior generated is to errors in assumptions. For example, consider the food capacity on the plateau. The model we worked out in Exercise 8 assumes a food capacity FCAP = 200,000 and an initial food supply FOODN = 196,000. Suppose that we are uncertain whether these numbers are correct, and we think the actual value for the food capacity might be FCAP = 150,000 and the initial food supply FOODN = 146,000. Run the model with these new values, and label the run Exercise 9(c). What behavior mode does the deer population exhibit?

d. How does the behavior differ from the behavior in Exercise 9(a)? Does this difference seem important? How might you decide which of the values for FCAP is correct?

e. (Optional) How much confidence do you have that the model provides a reasonable explanation for the growth and rapid decline of the deer population on the plateau? Are there additional factors that you think should be included in the model? Are there assumptions you think should be changed?

Exercise 10: Using the Model to Examine Policies

Recall that in the introduction, you were asked to imagine that you were a Forest Service officer in 1930, interested in the fate of the deer population on the Kaibab Plateau. Now that you have completed a working model, you can use it to test the effects of alternative policies.

The main problem on the plateau in 1930, of course, was starvation: The deer population had become too large to be supported by the available food supply. One approach to solving this problem might be to permit hunters to kill a portion of the deer herd each year; to reduce the size of the herd and to balance new deer births.

One way to implement this policy might be to license hunters to kill a certain number of deer each year. Test the effects of this policy by adding a "Deer Killed by Hunters" rate to the model developed in Exercise 9(a). The following equations permit initiating deer hunting in 1930.

R	DKHUNT.KL = STEP(DHPK.K,HTIME)
C	HTIME = 1930
A	DHPK.K = ?

DKHUNT   Deer Killed by Hunters (deer/year)
HTIME   Hunting Beginning Time (years)
DHPK   Deer Hunters are Permitted to Kill (deer/year)

a. How many deer should the Forest Service permit hunters to kill each year? Should the number be a constant? Should it depend on the size of the deer population? Try various formulations for the number of deer that hunters are permitted to kill DHPK and decide which formulation is best.

b. Do you think licensing hunters to kill deer is a good policy? Why or why not? What other policies do you think should be tested using the model?

c. You can also use the model to examine some "what if" questions. For example, what might have happened if the bounty on mountain lions, which was imposed in 1906, had been ended before the mountain lion population was completely exterminated? For example, suppose the bounty had been ended in 1910. What might have happened? This can be tested by setting the bounty ending time BETIME = 1910. Run the model with the new value for BETIME and label the run Exercise 10(c). What behavior mode does the deer population exhibit?

d. How does the behavior generated with BETIME = 1910 differ from the behavior in Exercise 9(a)? Why?

e. Notice that in both Exercises 9(a) and 10(d) the deer population eventually settles into equilibrium. But the equilibrium conditions in the two simulation runs are quite different. In Exercise 9(a) a food shortage has reduced the average lifetime of deer to two years. In Exercise 10(d) the average lifetime of deer remain at its initial value of five years, but some deer are killed by mountain lions. Which simulation run do you think generates more desirable conditions? Why?

f. (Optional) Rerun the model with the bounty ending time BETIME = 1909. Label this run Exercise 10(f)(1). Rerun it with BETIME = 1911. Label this run Exercise 10(f)(2). How do the results differ? Why?

g. (Optional) If you were a Forest Service officer in 1930, what policies would you recommend? Why?

CHAPTER 19

DYNAMIC CHARACTERISTICS OF FLU EPIDEMICS

Almost everyone has had the flu, a common respiratory infection. Influenza is caused by viruses that invade and multiply in the cells lining the upper and lower respiratory passages. Usually, when a person contracts the flu it lasts about five days.

When a person becomes infected with the flu, viruses liberated from damaged cells eventually contaminate the saliva. Transfer of infection from one person to another is caused by the emission of droplets of infected saliva into the air. These droplets evaporate quickly, and it is during evaporation that the disease is most likely to be transferred. Contact between susceptible and infectious individuals is necessary for the disease to spread.

As part of the process of recovering from the flu, antibodies are produced in the blood stream. Once a person has had the flu, he or she is immune from another attack of the same virus for a period of a year or two. Immunity can also be produced by vaccination.

Influenza epidemics occur periodically. In part, epidemics recur because the viruses responsible for the flu never fully die out, even after an epidemic has passed. In addition, new influenza viruses are often created by the genetic mutation of normally harmless organisms.

Whether to immunize or not to immunize against newly emerging strains of the flu has recently taken on importance in political as well as medical circles (see the following news synopsis). Because immunization programs are expensive and may have unintended side-effects, policy makers are interested in strategies other than mass immunizations for the control of epidemics.

Strategies such as the selected closing of schools in local districts and "spot immunizations" for certain vulnerable populations (such as persons over sixty-five years of age in institutions) have been suggested as alternatives to mass immunization strategies, such as the one adopted by the Ford administration in 1976.

Imagine you are a health systems analyst for the U.S. Department of Health and Human Services (HHS) and you have been charged with constructing a system dynamics simulation model of a flu epidemic. To begin with, only concern yourself with the dynamics of the spread of the flu through a single, homogeneous population. Do not attempt to group the population by age or by geographic units. If this initial model can be constructed, passing the initial feasibility test, a larger, more detailed model may be constructed.

Your first task is to write a draft report discussing your initial problem definition. Your report should include at least the following:

1. A description of the perspective or point of view of the preliminary model;

2. A description of the model's time frame;

3. A description of the dynamic problematic behavior to be studied, including the major variables and the reference mode;

4. A description of the possible policies that may be tested in the proposed model;

5. A description (preliminary) of the model structure underlying the behavior of interest. Comment on questions of system boundary, the basic causal relationships underlying the spread of an epidemic, and how some of these relationships might be coded into equations. What difficulties might you encounter in trying to estimate parameters for the proposed model? You may find it convenient to sketch several of the major levels and rates inherent in the spread of an epidemic.

Your next task is to write equations for the model, and run the model on the computer. Finally, your third task is to use the model to examine several possible strategies to control flu epidemics. (This may require expanding your initial model a bit.) Write a report to the Secretary of HHS, outlining your conclusions.

NEWS SYNOPSIS: FLU-SHOT CLAIMS

The General Accounting Office reports that damage claims against the federal government from its short-lived swine flu programs could very well top $1 billion.

When Gerald Ford was President, he regarded the swine flu program as a political and effective move.

Several researchers thought otherwise. But Ford decided the government had nothing to lose but money and that money was more expendable than lives.

He therefore let the government assume ultimate liability for injury and death arising from the vaccination program. The General Accounting Office report reveals that insurance companies may earn an $8.65 million profit due to the way the insurance program was structured. Damage claims against the swine flu vaccinations have already topped $500 million. The vaccinations were intended originally to avert a possible epidemic; but after paralysis developed in several persons who took the injections, the program resulted in an epidemic of damage claims.

There are some politicians who believe that if an epidemic of swine flu had broken out in the U.S. and had been halted by the vaccination program, Gerald Ford would have been elected President of the U.S. as a farsighted, cautious, and humane politician.

CHAPTER 20

URBAN GROWTH*

INTRODUCTION

In the past few decades, the number of people living in the cities has increased tremendously compared to the number living in the countryside. It is estimated that currently over 96 percent of the U.S. population reside in cities. Many cities grew out of small settlements that became attractive locations for business and industry, due to their endowments of natural resources or due to their proximity to market towns. Construction of transportation infrastructures aided the urbanization process by providing links between established market towns and the naturally endowed areas.

The growth of small settlements into large cities has generally not been smooth. Many settlements experienced a sudden growth boom, often short lived, followed by a period of stagnation and decline. For example, a number of coastal towns in New England showed tremendous growth over the early part of this century. In fact, the region became an industrial center as businesses moved in and people followed to take up the jobs these businesses created. The industrial boom in New England transformed many sleepy settlements along the coast into cities humming with activity. But many of these cities are now losing population and are having difficulty attracting new businesses, while older businesses are slowly moving out. (See Examples IV through VI of Chapter 6.)

This chapter provides a basis for modeling the growth behavior of a typical small town, which experiences a boom when the construction of a highway nearby makes it an attractive location for business and industry. Recall that a simple urban growth model was presented in Chapter 16. It may be help-

*The authors deeply appreciate the assistance of Khalid Saeed in the writing of this chapter.

ful to read that chapter again. Then carefully read the following sections on model purpose and problem definition, paying attention to both the qualitative and quantitative information provided.

MODEL PURPOSE

You have been appointed an advisor to the major of Boomtown, once a growing industrial and commercial center but now experiencing a decline. Few new businesses are being established in Boomtown, and the old businesses are slowly moving out. Jobs are becoming more and more scarce, and an increasing number of people are leaving town for good.

During the recent campaign for Mayor, Mr. Bentubo promised to guide the city out of stagnation. So far, however, he has made little progress, and if the current declining trends continue, Mayor Bentubo will have serious problems getting reelected for a second term. Bentubo has seen the failures of several apparently promising development programs that had been introduced in other cities. He is skeptical of the schemes being suggested by the City Planner's Office. He has hired you to find out what is really causing the decline before he considers any new programs.

THE PROBLEM

Old files and documents at the City Archives Department have revealed a reasonable amount of information about Boomtown's history. About fifty years ago, Boomtown was a small vegetable farming village of about 200 inhabitants. About one-third of the village population was of working age. Five farm-owning families employed most of the people. Some workers were self-employed and served as handymen and construction people whenever farm buildings and homes needed to be built or repaired. About 90 percent of the village area was being used as farmland, the farm buildings and housing occupying only a small fraction of total available land.

In those days, life rarely changed in Boomtown except for the occasional events of births, deaths, someone leaving town, or someone returning. Three or four babies were born almost every year, while one or two senior citizens passed away. Almost a dozen youth left town each year to find jobs elsewhere or to attend college. One out of six who left never returned. It appears that the net emigration from the town was balanced by the net increase in population from births and deaths. Thus the town population remained constant for a long time.

In 1935, a major highway was constructed near Boomtown, connecting Boomtown to Zetroit, a nearby industrial city. The large manufacturing companies in Zetroit required large supplies of metal parts to use in making their products. In the past, these metal parts were imported from relatively far-off cities at a high cost. But the new direct transportation link made Boomtown an ideal location to produce parts for Zetroit's industry. New businesses started pouring into Boomtown.

As new jobs were created by the incoming businesses, fewer and fewer people left town in search of jobs. In fact, a steady stream of people from adjoining areas started arriving when they heard about the job opportunities in Boomtown. These immigrants often came with their families, and this further increased the town's population. In less than ten years, the population and the number of businesses and houses in Boomtown grew several-fold.

Credit for Boomtown's growth was taken by Mr. Marshal, who began his first term as the Mayor of Boomtown in 1946. Marshal once owned the largest vegetable farm in the community, but had diversified into the banking and liquor businesses after the boom started. As also true of other original landowning families of Boomtown, Marshal had been quite active in politics, and considered himself a founding father of Boomtown. The boom provided an excellent political opportunity for Marshal, and he had no difficulty getting reelected for three consecutive terms.

Toward the end of Marshal's third term in office, Boomtown started showing signs of stagnation. The construction industry was the first to experience decline. A large number of construction workers became jobless. Some of these workers could be absorbed by other businesses, but the other businesses were growing at a much slower rate than they had been just a few years earlier. As a result, the number of people leaving town for jobs elsewhere started rising, while the number of new immigrants began to fall. Mr. Marshal ran for his fourth term, but failed to get reelected. He was replaced by Mr. Thompson.

Mr. Thompson had the same vegetable farming background as Mr. Marshal, and had been a contestant in the mayoral race in the past two elections. During his successful campaign, Mr. Thompson had identified himself as a concerned citizen with positive plans to save the city from decline. Mr. Thompson undertook to step up housing construction, with the hope that increased housing availability would dissuade people from leaving town. Unfortunately, Thompson's plans did little to stop the town's declining trends. Nevertheless, he was able to hold office for two consecutive terms until he was defeated in 1966.

Thompson was followed as Mayor by Mr. Kelley, who served from 1966 to 1978. During Mayor Kelley's three terms, Boomtown continued its slow decline in population and industry, in spite of a special new jobs program instituted by the Mayor. Then, in the 1978 election, voters turned to a somewhat obscure contestant, Mr. Bentubo, who was an early immigrant to Boomtown and impressed people during his campaign as a person with good ideas and realistic attitudes.

Bentubo has no preconceived notions about what to do. He first wants to find out what has been going wrong in the past before considering any development programs, and that is why he has hired you.

Exercise 1: Developing a Model of Boomtown

One way to analyze the problems in Boomtown is to formulate a system dynamics model. The following are some suggestions about how to proceed:

a. Reread the section entitled "The Problem," paying particular attention to the behavior of businesses, population, and housing in Boomtown from 1930 to 1980. Then draw a graph showing the *reference mode* for the model you will build. What variables do you think should be included in the reference mode?

b. Use the information to propose an hypothesis that might explain the behavior of Boomtown from 1930 to 1980.

c. To formulate a model of Boomtown, it will probably be easiest to proceed in steps. One good place to begin is to concentrate on the growth of businesses. You can then add population and housing once the initial model is working.

Using the urban growth model introduced in Chapter 16 as a guide, construct a model of the growth of businesses in Boomtown. Choose initial values to reflect the conditions in Boomtown in 1930, before the development of the highway nearby. Assume there were 5 business structures in Boomtown in 1930. (*Note:* Exercises 1 and 2, Chapter 16, may help here.)

Adjust the parameters for business construction and business demolition so that the business construction and demolition rates exactly balance each other in 1930, then simulate the model. It should remain in equilibrium. (See Exercises 3 and 4, Chapter 16, for help.)

Now use a STEP function to increase the business construction fraction in 1935 to represent the effects of the new highway built near Boomtown. Simulate your model, picking a LENGTH long enough to permit business structures to reach a new equilibrium level. The number of business structures should exhibit S-shaped growth.

In order to add population to the model you have developed so far, it will be helpful to include a representation of the number of jobs available in Boomtown. Assume that each business in Boomtown employs 10 workers. Add a representation of jobs to the model. (*Hint:* It is easier to represent the number of jobs as an auxiliary variable.) Run the model and discuss the results. How does the model-generated behavior differ from the reference mode?

d. Now add population to the model. Include a representation of births and deaths, as well as migration into and out of Boomtown. According to the data, one of the main influences on migration into and out of Boomtown is the availability of jobs. How should this be represented in the model?

You may also wish to consider the influence of worker availability on the construction of businesses. Assume that the construction rate increases when workers are readily available and decreases when there is a shortage.

Once you have worked out diagrams and equations to add population to the model, adjust the parameters and initial conditions so that business structures and population are in equilibrium in 1930. Then step up the business construction fraction in 1935 to represent the new highway. Run the model, picking a LENGTH long enough to permit population and business to come to a new equilibrium. They should exhibit S-shaped growth. Discuss your results.

e. Finally, add houses to the model. Develop a representation for both housing construction and housing demolition. (*Hint:* Use a representation similar to the formulation for business construction and demolition. A STEP function can be used to increase the housing construction fraction in 1935, to represent the effects of the highway built near Boomtown.)

Consider the influence of housing availability on the housing construction rate. Assume that the construction rate increases when there is a housing shortage, and decreases when there is no housing surplus. How might you represent housing availability?

Also, of course, houses occupy land. How might the housing construction rate depend on the fraction of land occupied in Boomtown? (*Note:* The formulation should be similar to the effect of land availability on business construction.) Assume that when the fraction of land occupied in Boomtown rises, it is easier to build houses than business structures on the land that remains unoccupied. (Some suggestions may be provided from Figures 16.7 and 16.8 of Chapter 16.)

In addition, assume that housing availability influences migration into and out of Boomtown. How should this be formulated?

Once you have worked out equations to add housing to the model, adjust the parameters and initial conditions so that the model is in equilibrium in 1930. If the model does not remain in equilibrium, reexamine your initial values, parameters, and table functions.

Now, step-up the business and housing construction fractions in 1935 to represent the new highway. Choose a long enough value for LENGTH to permit housing, population, and business structures to reach equilibrium. Discuss your results. Does the model-generated behavior match the reference mode? What advice would you offer Mayor Bentubo on the basis of your work on the model so far?

Exercise 2: Policy Tests

Once you obtain model behavior characterized by rapid growth and decline in the number of jobs and rapid growth of businesses and population followed by slow decline, you may want to try out some policies to overcome the declining trends. You may analyze any number of policies. The following illustrate some of the possibilities:

a. One approach to the problems of Boomtown might be to encourage new housing construction. You can test this policy with a step increase in the housing construction fraction. Simulate and discuss the results.

b. You might examine the effects of a subsidized employment program by using a STEP function to increase the number of jobs per business. Simulate and discuss the results.

c. You might try increasing the housing demolition rate to represent a slum clearing policy. This can be done by using a STEP function to decrease the average life of housing. Simulate and discuss your results.

Exercise 3: (Optional) The Validity of the Model

  a. The model assumes that the land area available for the growth of a city is limited. Is that a good assumption? If not, what other limiting factors may restrict growth of a city?

  b. Exercise 1e suggested making the business construction rate more sensitive to land fraction occupied than the housing construction rate. Is that a good assumption? Rerun the model making housing construction and business construction equally sensitive to the land fraction occupied. How do the results differ?

  c. How well does your model represent a real city? Point out shortcomings of your model and limitations of your analysis.

CHAPTER 21

COMMODITY CYCLES: A STUDY OF HOGS*

Economic and environmental systems are subject to numerous and unexpected external shocks and disturbances. A drought, for example, can cause a sudden reduction in the harvest of corn. A strike can lead to changes in the availability of a product. A rapidly spreading disease can quickly reduce the size of a herd of cattle. And a chemical spill can affect the ecology of a coastal area.

One question to ask about an economic or environmental system is how it responds to shocks such as these. Following the disturbance, does the system smoothly return to equilibrium? If it reaches equilibrium, is the new equilibrium value the same as the equilibrium value before the disturbance? Does the system amplify disturbances?

Many systems tend to oscillate in response to external disturbances. The national economy, for example, shows cycles of high and low Gross National Product (GNP) and employment. And animal populations (such as rabbits and lynx) often exhibit oscillations.

In this chapter, a classic example of cyclical behavior will be explored— the production and consumption of pork products in the United States. While the pork example is interesting in itself, it is also important because it focuses on some general questions about the nature of oscillations and the stability of systems in response to disturbances.

The data in Figure 21.1 show the pig crop and slaughter rate in the United States from 1925 to 1960. As can be seen, there are cycles of high and low production, and the average period from peak to peak is about four years. The exercises that follow seek an explanation for why these cycles occur.

---

*Information in this chapter draws on material from Dennis L. Meadows, *Dynamics of Commodity Production Cycles* (Cambridge, Mass.: MIT Press, 1970).

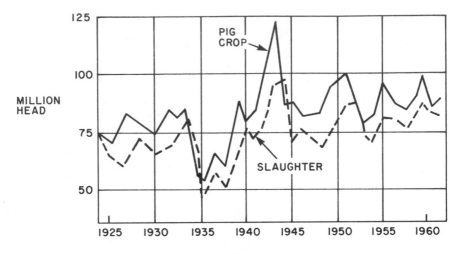

Figure 21.1 Pig crop and slaughter rate in the United States, 1925–1960

The production and consumption of pork products involve three groups of actors: farmers, butchers, and consumers. Butchers buy hogs from farmers, "dress" the hogs to obtain pork products, and then sell the pork products to consumers.

In order to understand the cycles that take place in the production and consumption of pork products, it will be helpful to build two models—one focusing on the process of breeding and fattening hogs on the farm, and the other focusing on the sale of pork products. Then the two models can be combined to generate cyclical behavior. It is easiest to begin with a model of the sale of pork products, then turn to the breeding of hogs, and finally combine the two models into one larger model and analyze the behavior the combined model generates.

Exercise 1: The Sale and Consumption of Pork

The main element in the model of the sale of pork is the inventory of pork maintained by butchers. When hogs are slaughtered on the farm, the pork products that are produced are added to the inventory kept by butchers; when pork is sold, that pork is taken from the inventory. In general, the amount of pork people buy depends on the price of pork; the price of pork, in turn, depends on the size of the pork inventory. When the inventory is high, prices fall; when the inventory is low, prices rise. (It may be helpful to sketch a rough causal-loop diagram at this point, before continuing with the exercise.)

For the initial model of the sale of pork, assume that the number of hogs slaughtered each month is an exogenous constant—4 million hogs per month. Hogs weigh about 250 pounds each, and the "dressed yield" of pork is about

60 percent. (This means that each hog produces 0.60*250 = 150 pounds of pork.) Thus a slaughter rate of 4 million hogs per month corresponds to a pork production rate of 600 million pounds of pork each month.

On the average, each person in the United States normally consumes about 3 pounds of pork each month. If a constant population of 200 million people is assumed, the total amount of pork normally consumed each month is 3*200 million = 600 million pounds of pork per month. But when the price is high relative to the normal price of pork, people usually consume somewhat less than 3 pounds each month; when the price is low, they consume somewhat more.

The price of pork depends on the price of hogs, and the price of hogs depends on supply and demand. Assume that butchers try to keep on hand about two weeks' worth of the amount of pork products they usually sell. When the available inventory of pork runs low, relative to their normal inventory, butchers are willing to pay higher prices to get hogs. When the butchers' inventory of pork is high, butchers are less willing to buy and the price of hogs falls.

Assume that the normal price of hogs at slaughter is $0.30 per pound. This means that butchers normally pay $0.30/0.6 = $0.50 for a pound of pork, since the dressed yield of pork is 0.60. Assume that butchers charge consumers a $0.50 per pound markup on pork. Thus when the price of hogs is at its normal value ($0.30/pound), the price of pork to consumers is ($0.30/0.6) + $0.50 = $1.00 per pound.

a. Draw a causal-loop diagram and flow diagram of the sale and consumption of pork.

b. Now write equations for the model. It is easiest to represent the pork consumed per person as the product of a "normal value" (3 pounds per person per month) and a TABLE function multiplier that reflects the influence of pork price on consumption. Similarly, it is easier to represent the hog price as the product of a "normal value" ($0.30/pound) and a TABLE function that reflects the influence of pork availability (relative to demand) on hog price.

c. In order to run the model, you will need to choose an initial value for the pork inventory. What value is necessary in order to place the system in equilibrium?

d. Once you have obtained a model that runs in equilibrium, try testing its response to exogenous disturbances. For example, how does the system respond to a year-long 10-percent reduction in the hog slaughter rate? (This might represent the impact of a drought or a disease. You can use two STEP functions to represent the year-long reduction in the slaughter rate.)

e. How does the system respond to a year-long 10-percent increase in the normal consumption of pork per person? (This might represent a change in buying habits, or a change in the availability or price of other kinds of meat.)

Exercise 2: The Breeding and Maturation of Hogs

In Exercise 1 we assumed the hog slaughter rate was an exogenous constant. Now let's develop a model of the breeding and maturation of hogs to simulate the slaughter rate. For this model, we will assume that the price of hogs is an exogenous constant. In Exercise 3, we will combine the two models to simulate both the hog price and the slaughter rate.

Farmers distinguish between two kinds of hogs: hogs for market and hogs for breeding stock. Hogs for market (which can be either male or female) are fattened for about a year after they are born, and then they are slaughtered. Females intended for market are not bred. Females to be bred (called sows) are raised separately as "breeding stock," and they are used entirely for breeding. They are not slaughtered for pork.[1]

Farmers change the size of their hog herd by adjusting the number of sows in their breeding stock. When the price of hogs is higher than normal, farmers generally desire to increase the size of the breeding stock, and when the price of hogs is lower than normal, they desire to decrease the size of the breeding stock.

a. Draw a flow diagram and write equations for a model of the breeding and maturation of hogs. Assume that each sow gives birth to 24 hogs per year (2 per month). Also, assume that hogs must be fattened for 12 months before they are ready for market. (The most difficult part of the model is the "breeding stock adjustment rate." One way to formulate this is to use a TABLE function to represent a "desired breeding stock size" as a function of hog price. This desired size can be expressed most easily as a percentage of the actual breeding stock size. The breeding stock adjustment rate can then be formulated as the difference between the "desired size" and the "actual size," divided by an adjustment time.)

Set the exogenous hog price at $0.30 per pound. You will also need to choose an initial value for the breeding stock and an initial value for the number of hogs. Choose values that will produce an equilibrium slaughter rate of 4 million hogs per month (the value we assumed in Exercise 1.)

b. Once you have obtained a model that runs in equilibrium try testing its response to a year-long 10 percent increase in the price of hogs. What happens? Why?

Exercise 3: Combining the Two Models

a. Now let's merge the models developed in Exercises 1 and 2. To do this, all that is necessary is to use the hog price from model 1 in place of the exogenous hog price in model 2, and the slaughter rate from model 2 in place of the exogenous slaughter rate in model 1. Run the model and examine the results. If you use the same initial values as those you chose in Exercises 1 and 2, the model should be in equilibrium.

b. Once you have obtained an equilibrium run, you can test the model's response to external disturbances. For example, test the model's response to a year-long 10-percent increase in the pork consumed per person. What happens? Can you explain why the cycles occur?

Two features of cyclical behavior are particularly important: period and damping. The period of a cycle is the time that elapses from one peak to the next. For the parameter values we have chosen, the hog model produces cycles with a period of about 48 months. Damping refers to the degree to which oscillations fade away over time. For the parameter values we have chosen, the hog model produces oscillations that are slightly damped. (Some models produce oscillations that grow over time. These are sometimes called "explosive" oscillations.)[2]

c. (Optional) See if you can determine which parameters in the hog model influence the period and damping. (This will take some investigation.) Once you have discovered which parameters influence the damping of the system, you may wish to explore some policies that might reduce the degree to which the system oscillates in response to external disturbances.

ENDNOTES

1. This is a somewhat simplified description of the process of breeding and fattening hogs. For a more detailed description, see Dennis L. Meadows, *Dynamics of Commodity Production Cycles* (Cambridge, Mass.: MIT Press, 1970).
2. One measure of the degree of damping is the "damping ratio," which is defined as "one minus the ratio of the amplitude of two successive peaks." If the oscillations are fading, the damping ratio will have a value between zero and one. If the oscillations are steady, the damping ratio will have a value of zero; if the oscillations are growing, the damping ratio will have a negative value.

PART VII

DEVELOPING MORE COMPLEX MODELS

OBJECTIVES

Part VII exposes the reader to problems closer to his or her day-to-day decision-making process by:

1. Introducing additional heuristic techniques for representing these problems as mathematical models;

2. Presenting less structured and more muddied problems;

3. Suggesting literature for readings about real-world applications.

CHAPTER 22

REPRESENTING DECISION PROCESSES*

Decisions made by individuals or groups arise in all situations. You decide (very quickly) to decrease the hot water in the shower. A number of people decide to migrate into the city. Decisions like these are made explicitly, based on combinations of rational and emotional factors. Other decisions are made implicitly, like the "decision" by coffee to cool or by yeast to bud. All rates of flow can be seen as resulting from explicit or implicit decisions. This chapter treats the representation of decision processes in equation form, going from simple representations to more complex ones. This discussion was begun in Chapters 15, 16, and 17.

CONTROLLING THE DAM WATER

The Smithfield Dam regulates the flow of the Rappanno River. Water accumulates in the reservoir behind the dam and is released from the dam into the Rappanno Valley below where it provides for irrigation of farmlands as well as fishing and other recreational uses of the river. Low water flow into the valley creates problems for the farmers while also decreasing the recreational benefits. But too high a river flow causes flooding and also wastes water that might later be in short supply during the dry season. Joan Perez, the controller of the dam, decides on a daily basis how much water to release from the reservoir into the river below.

In the examples and exercises presented next, thirteen different ways of writing equations to simulate Ms. Perez's decision to release water from the Smithfield Dam are discussed. These equations range from the very simple to some rather involved formulations. The earlier formulations are probably not

The authors deeply appreciate the assistance of Edward B. Roberts in the writing of this chapter.

realistic formulations of how Ms. Perez would actually react, but they do illustrate several general types of decision processes often encountered in system
dynamics models. The latter formulations are more realistic representations of
the types of decisions involved in controlling water levels in the Smithfield
Dam and the Rappanno Valley. Throughout, the purpose of these exercises is
to examine in more detail how decision processes can be expressed in equation
form.

EXAMPLE I: SIMPLE DECISION PROCESS

Write a DYNAMO program to test what happens to the reservoir and the
downstream flow in response to a step increase in water entering the reservoir
from upstream. Assume that Ms. Perez releases downstream exactly what has
entered the reservoir in the most recent time interval.

Figure 22.1 shows the causal-loop diagram for this example. Note that
there is no feedback loop in this decision process. Ms. Perez is running "open-
loop." Figure 22.2 shows the flow diagram for the example.

The following model represents the situation as described. (See Chapter
15 for a reminder on use of the STEP function.)

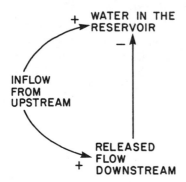

Figure 22.1 Causal-loop diagram of dam control (simple process)

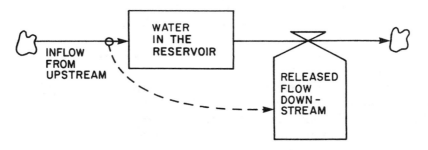

Figure 22.2 Flow diagram of dam control (simple process)

```
*           RESPONSE TO SIMPLE DECISION PROCESSES
NOTE
NOTE        -------------------------------------
NOTE
R           INFLOW.KL=NFLOW+STEP(0.2,5)
NOTE              UPSTREAM INFLOW (MILLION GALLONS/DAY)
NOTE
NOTE              20 PERCENT STEP INCREASE ABOVE NORMAL INFLOW
NOTE              AT TIME = 5 DAYS
NOTE
C           NFLOW=1
NOTE              NORMAL FLOW (MILLION GALLONS/DAY)
L           RES.K=RES.J+(DT)(INFLOW.JK-RLSFLW.JK)
N           RES=30
NOTE              WATER IN RESERVOIR (MILLION GALLONS)
R           RLSFLW.KL=INFLOW.JK
NOTE              RELEASED FLOW (MILLION GALLONS/DAY)
NOTE
NOTE              CONTROL STATEMENTS
NOTE
PLOT        INFLOW=I,RLSFLW=F/RES=R
SPEC        DT=1/PLTPER=1/LENGTH=30
RUN
```

Figure 22.3 depicts the graphical results for this DYNAMO simulation.
Note that the decision to release just what entered in the latest time interval

R   RLSFLW.KL = INFLOW.JK

results in precise tracking between reservoir input I and output F.

Question: Would this be a good water release policy to follow if the inflow to the reservoir varied significantly from day to day, perhaps due to rainfall variations?

Exercise 1: Delayed but Simple Decision Process

Assume that Ms. Perez can only get information on inflows from upstream from a measurement system that acts like a third-order delay with an average delay time of five days. (See Chapter 17 for a refresher on the DELAY3 function.) But she still releases downstream exactly what her most recent information indicates has entered the reservoir.

a. Draw the causal-loop diagram for this situation. How is it different from Figure 22.1? Is it a closed-loop or open-loop system?

b. Enter the model from Example I on your computer, changing RLSFLW and PLOT and adding DIINFL and MSDL, as shown below.

```
R           RLSFLW.KL = DIINFL.JK
R           DIINFL.KL = DELAY3(INFLOW.JK,MSDL)
```

```
INFLOW=I RLSFLW=F    RES=R

      0.8500           0.9500          1.0500          1.1500          1.2500  IF
      0.000           10.000          20.000          30.000          40.000   R
 0.0000 - - - - - - - - - - - - I - - - - - - - - - - R - - - - - - - IF
     .               .         I    .               R                . IF
     .               .         I    .               R                . IF
     .               .         I    .               R                . IF
     .               .         I    .               R                . IF
     .               .              .               R        I       . IF
     .               .              .               R        I       . IF
     .               .              .               R        I       . IF
     .               .              .               R        I       . IF
     .               .              .               R        I       . IF
10.000 - - - - - - - - - - - - - - - - - - - - - - - R - - - I - - - - IF
     .               .              .               R        I       . IF
     .               .              .               R        I       . IF
     .               .              .               R        I       . IF
     .               .              .               R        I       . IF
     .               .              .               R        I       . IF
     .               .              .               R        I       . IF
     .               .              .               R        I       . IF
     .               .              .               R        I       . IF
     .               .              .               R        I       . IF
20.000 - - - - - - - - - - - - - - - - - - - - - - - R - - - I - - - - IF
     .               .              .               R        I       . IF
     .               .              .               R        I       . IF
     .               .              .               R        I       . IF
     .               .              .               R        I       . IF
     .               .              .               R        I       . IF
     .               .              .               R        I       . IF
     .               .              .               R        I       . IF
     .               .              .               R        I       . IF
     .               .              .               R        I       . IF
30.000 - - - - - - - - - - - - - - - - - - - - - - - R - - - I - - - - IF
```

Figure 22.3 DYNAMO results of dam control (simple process)

NOTE	DELAYED INFORMATION ON INFLOW
NOTE	(MILLION GALLONS/DAY)
C	MSDL = 5
NOTE	MEASUREMENT DELAY (DAYS)
PLOT	INFLOW = I,DIINFL = D,RLSFLW = F/RES = R

Run the model as modified. What are the differences between your output and Figure 22.3? Why? How would this process for making decisions perform under conditions of high daily variations in reservoir inflow?

Exercise 2: Averaged, Delayed Information (Simple Decision)

Besides depending upon the inflow measurement system of Exercise 1, Ms. Perez decides to average the information she receives for another 10 days before acting upon it. (See Chapter 17 for information on the SMOOTH function.)

a. How does this change the causal-loop diagram of the situation? Is it a closed-loop or open-loop system?

b. Modify your model by adding averaged delayed information on inflow ADIINF as a SMOOTH equation. The initial condition equation needed for it is:

N   ADIINF = DIINFL

The water release decision should be changed to depend on ADIINF, and ADIINF added to the PLOT instruction for completeness. Run the model as modified. Comment on the behavior produced and the adequacy of the decision process used. Specifically, explain why the reservoir level has risen over the course of the run. How well would this decision rule perform if an unexpected dry spell stopped the flow of water into the reservoir?

c. Predict how the simulation would look if the time to smooth DIINFL were 20 rather than 10 days.

EXAMPLE II: MORE SIMPLE DECISION PROCESSES

In the three preceding model variations, the rate decision ended up in the form of an identity, A = B. Other simple decision processes will be shown in this example.

A consultant suggests to Perez that she should add stability to her water release decision by having her downstream flow depend partly on the dam's long-term constant inflow rate of 1,000,000 gallons per day and only partly on the recent inflow. She decides to rely half on the long-term inflow and half on her newer information.

The Exercise 2 model can be modified to test this decision process, by the following changes:

```
R           RLSFLW.KL = (W1)(LTINFL) + (W2)(ADIINF.K)
C           W1 = 0.50
NOTE            WEIGHTING #1 (DIMENSIONLESS)
C           LTINFL = 1
NOTE            LONG-TERM INFLOW (MILLION
NOTE            GALLONS/DAY)
N           W2 = 1 − W1
NOTE            WEIGHTING #2 (DIMENSIONLESS)
SPEC    DT = 1/PLTPER = 2/LENGTH = 60
```

The form of the equation for RLSFLW allows for weighting (or taking into account) two different input factors, here the constant LTINFL and the variable ADIINF. The SPEC changes produce results over a longer time frame. Figure 22.4 shows the DYNAMO plots of this simulation.

```
INFLOW=I RLSFLW=F DIINFL=D ADIINF=4     RES=R

    0.8500          0.9500          1.0500          1.1500          1.2500 IFD4
    29.000          31.000          33.000          35.000          37.000  R
0.0000 - - - - R - - - - - - - I - - - - - - - - - - - - - - - - - - IFD4
     .          R      .        I       .               .            . IFD4
     .          R      .        I       .               .            . IFD4
     .            R    .        F       .               .        I    . FD4
     .              R  .        F     D.                 .        I    . F4
     .                R       F   4.               D .        I    .
     .                 . R        F         4        .      D I    .
     .                   . R      .  F           4      DI    .
     .                     . R    .   F        .      4   I    . ID
     .                       . R  .    F       .      4 I    . ID
20.000 - - - - - - - - - - - - - R - - - - F - - - - - -4I - - - - ID
     .                     .        R  .    F       .        I    . ID4
     .                     .        R .    F       .        I    . ID4
     .                     .        R.    F       .        I    . ID4
     .                     .        .R    F       .        I    . ID4
     .                     .        . R   F       .        I    . ID4
     .                     .        .  R  F       .        I    . ID4
     .                     .        .   R F       .        I    . ID4
     .                     .        .    RF       .        I    . ID4
     .                     .        .    FR       .        I    . ID4
40.000 - - - - - - - - - - - - - - - - - F R - - - - - I - - - ID4
     .                     .        .    F  R  .        I    . ID4
     .                     .        .    F   R. .        I    . ID4
     .                     .        .    F    R.        I    . ID4
     .                     .        .    F    .R        I    . ID4
     .                     .        .    F     . R      I    . ID4
     .                     .        .    F     .   R I    . ID4
     .                     .        .    F     .    R I    . ID4
     .                     .        .    F     .     RI    . ID4
     .                     .        .    F     .     IR    . ID4
60.000 - - - - - - - - - - - - - - - - - F - - - - - - I R - - - ID4
```

Figure 22.4 Dam control using weighted inputs to decision

Notice that if Ms. Perez does take constant long-term information into account as indicated in the equations just shown, and if inflow does step up as shown in Figure 22.4, then there will be a continuing gap between inflow and release flow. If this situation were left unchecked, then the amount of water accumulating in the reservoir would become very large (in fact, since the present model does not have any simulation of ultimate reservoir capacity, it would eventually accumulate a nearly infinite amount of water). Similarly, under the equations just described, a step down in inflow rate would eventually lead to a complete depletion of the reservoir and eventually to negative water being stored behind the dam.

Exercise 3: Weighted Decision Processes

By inspection of the equation for RLSFLW in Example II, what would be the effect of setting Wl = 1? of setting W1 = 0?

Modify your Exercise 2 model as indicated above and test the implications of various values for W1. Take advantage of the DYNAMO rerun feature in doing these simulations.

Exercise 4: Variable Weightings

Another type of decision process depends on two or more inputs, but has the degree of dependency changing during the simulation run. In the example above assume that Ms. Perez wants to weight the long-term inflow data most heavily during the first and last seasons of each year and less heavily during the middle portion of the year.

The following revisions in the Example II version of the model should accommodate the desired change:

R	RLSFLW.KL = (W1.K)(LTINFL) + (W2.K)(ADIINF.K)
A	W1.K = TABLE(TW1,TIME.K,0,360,90)
T	TW1 = 1.0/0.7/0.3/0.4/0.75
NOTE	.TABLE FOR WEIGHTING #1
A	W2.K = 1 – W1.K

(Review Chapter 16 for the use of TABLE functions.) Note that each of the weights has now become an auxiliary equation, needed to help in the expression of the rate equation RLSFLW.

Modify your model from Example II to embody these changes, and run the test simulation with the step increase in INFLOW. Note that while the decision process is now taking more things into account (e.g., long-term and short-term inflows, as well as seasonality), it is not clear that any objective is being better accomplished. To achieve an objective the objective must be clearly formulated and appropriately reflected in a decision process.

GOALS AND DECISIONS

People as well as systems usually have goals, objectives, or purposes they are trying to achieve. Often the goals are not clearly identified, and frequently decisions or actions taken are inconsistent with the goals. This is particularly true when the situation involves multiple and complex feedback-loop interactions over time, making difficult the determination of what decision process will in fact accomplish the objective.

When the actual state-of-affairs is different from the goal, decisions are initiated to try to close the gap (or discrepancy). Figure 22.5 provides a causal-loop representation of this view of decision-making, as discussed initially in Chapter 15, beginning with Example V of that chapter.

The larger the goal, relative to the actual state, the bigger the gap or discrepancy, the more vigorous the decision or actions taken, the more change will be produced in the actual state, thereby reducing the prior gap. This decision process closes the loop of a negative feedback loop; it is "goal-seeking" in its attempt to bring the actual state closer to the goal.

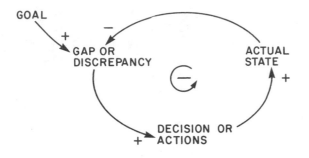

Figure 22.5 Diagram of decision processes

EXAMPLE III: INTRODUCING GOALS IN DECISION PROCESSES

After reflecting further on her job, Ms. Perez decides that what she really wants to do (her objective?) is to maintain a certain amount of water in her reservoir at all times. (Later we shall reflect on what the goal of the farmers downstream of Ms. Perez's dam might be.) To achieve this, she will release on a daily basis what the long-term inflow has been, adjusted by whatever difference exists between the amount in the reservoir behind the Smithfield Dam and the amount she would like to have.

Figure 22.6 pictures this situation in a causal-loop diagram. Compare this to Figure 22.1. Note that a negative feedback loop is now involved with the water release decision. Ms. Perez is now running "closed-loop," releasing less water whenever the reservoir drops below her target level and more when she has a surplus in storage.

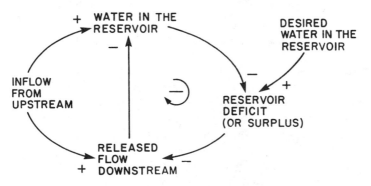

Figure 22.6 Diagram of dam control using goal for reservoir

The following model represents the situation as now described:

```
*           RESPONSE TO GOAL-ORIENTED DECISIONS
NOTE
NOTE        ------------------------------------
NOTE
R           INFLOW.KL=NFLOW+STEP(0.2,5)
NOTE                 UPSTREAM INFLOW (MILLION GALLONS/DAY)
NOTE
NOTE                 20 PERCENT STEP INCREASE ABOVE NORMAL INFLOW
NOTE                 AT TIME = 5 DAYS
NOTE
C           NFLOW=1
NOTE                 NORMAL FLOW (MILLION GALLONS/DAY)
L           RES.K=RES.J+(DT)(INFLOW.JK-RLSFLW.JK)
N           RES=30
NOTE                 WATER IN RESERVOIR (MILLION GALLONS)
R           RLSFLW.KL=LTINFL-DAFLW.K
NOTE                 RELEASED FLOW (MILLION GALLONS/DAY)
C           LTINFL=1
NOTE                 LONG-TERM INFLOW (MILLION GALLONS/DAY)
A           DAFLW.K=RESGAP.K/DT
NOTE                 DAILY ADJUSTMENT TO FLOW (MILLION GALLONS/DAY)
A           RESGAP.K=DESRES-RES.K
NOTE                 RESERVOIR GAP (MILLION GALLONS)
N           DESRES=30
NOTE                 DESIRED WATER IN RESERVOIR (MILLION GALLONS)
NOTE
NOTE                 CONTROL STATEMENTS
NOTE
PLOT        INFLOW=I,RLSFLW=F/RES=R,DESRES=D(28,32)
SPEC        DT=1/PLTPER=1/LENGTH=30
RUN
```

The goal for the reservoir is shown here as a constant amount DESRES of 30 million gallons. Whatever is the reservoir's deficit RESGAP becomes Ms. Perez's daily adjustment (RESGAP/DT) to the long-term flow in her release decision. (Incidentally, DAFLW[1] is an auxiliary equation based on another auxiliary RESGAP, all being used to help in the expression of a rate equation RLSFLW.)

Figure 22.7 shows the DYNAMO simulation results for this policy. Notice how closely Perez's objective of constant water in dam is maintained on a continuous basis, and how changes in inflow are immediately transmitted to changes in outflow. What causes the small constant difference between R and D in Figure 22.7 that shows up after the INFLOW change?

Exercise 5: Comparative Policy Analysis

a. Compare Figure 22.7 to Figure 22.3. To what extent are they similar? How are they different? Examine the equations for RLSFLW in Examples I and III to try to explain your observations.

```
INFLOW=I RLSFLW=F     RES=R DESRES=D

    0.8500            0.9500          1.0500          1.1500          1.2500  IF
    28.000            29.000          30.000          31.000          32.000  RD
0.0000 - - - - - - - - - - - I - - - R - - - - - - - - - - - - - - RD,IF
     .                .        I      R                .               . RD,IF
     .                .        I      R                .               . RD,IF
     .                .        I      R                .               . RD,IF
     .                .        I      R                .               . RD,IF
     .                .        F      R                .        I      . RD
     .                .            D  R                .        I      . IF
     .                .            D  R                .        I      . IF
     .                .            D  R                .        I      . IF
     .                .            D  R                .        I      . IF
10.000 - - - - - - - - - - - - - - D -R- - - - - - - - I - - - - IF
     .                .            D  R                .        I      . IF
     .                .            D  R                .        I      . IF
     .                .            D  R                .        I      . IF
     .                .            D  R                .        I      . IF
     .                .            D  R                .        I      . IF
     .                .            D  R                .        I      . IF
     .                .            D  R                .        I      . IF
     .                .            D  R                .        I      . IF
20.000 - - - - - - - - - - - - - - D -R- - - - - - - - I - - - - IF
     .                .            D  R                .        I      . IF
     .                .            D  R                .        I      . IF
     .                .            D  R                .        I      . IF
     .                .            D  R                .        I      . IF
     .                .            D  R                .        I      . IF
     .                .            D  R                .        I      . IF
     .                .            D  R                .        I      . IF
     .                .            D  R                .        I      . IF
30.000 - - - - - - - - - - - - - - D -R- - - - - - - - I - - - - IF
```

Figure 22.7 Constant goal for reservoir and immediate adjustment

b. From the perspective of the downstream users of the Rappanno River, is Ms. Perez's present water release policy a good one? How might the users feel if the inflow to the reservoir varied significantly from day to day, perhaps due to rainfall variations?

c. Ms. Perez currently adjusts the entire reservoir deficit (or surplus) in a single time interval by subtracting RESGAP.K/DT from LTINFL. Change the equation for DAFLW to allow her the flexibility of adjusting the reservoir gap over any time period. Use TAR for the time to adjust the reservoir. Run the model using values of 5, 10, and 20 days for TAR, comparing the results from the different perspectives of Ms. Perez and the river users.

MORE ON GOALS AND DECISIONS

The formulation, RESGAP.K/TAR or (DESRES-RES.K)/TAR, is the most frequently encountered cluster in the representation of decision processes: the desired (or goal) condition minus the actual condition, all divided by the time for attempting to correct the discrepancy. This was seen earlier in Chapter 15 where, for example, implicitly the "goal" for coffee that is cooling is the room temperature.

Another way of interpreting this formulation is that at each opportunity the decision takes into account a fractional part of the goal discrepancy. For example, Ms. Perez might say that on a daily basis she would like to correct for 20 percent of any reservoir deficit (or surplus). That could be reflected as:

A $\text{DAFLW.K} = (0.20)(\text{RESGAP.K})$

where 0.20 is the "fractional multiplier" of the reservoir gap.

Notice that adjusting the reservoir gap over a time period of 5 days is not the same thing as smoothing or averaging the reservoir gap for 5 days before computing DAFLW. By adjusting the reservoir gap over 5 days, the speed and amount of reaction is less each day. However, introducing a 5-day delay would lead to a response that is not less rapid on a per day basis, but lagged by five days. Hence, increasing TAR slows up the rate of release of water, and adding a 5-day smooth to the equation keeps the same magnitude of reaction, but delays the reaction by five days.

Exercise 6: Time Delays and Fractional Multipliers

Modify your model equation for DAFLW as indicated above and run the model. Compare the results with your earlier run in which TAR = 5 (Exercise 5(c)). Why are the results the same?

What results would be produced if Ms. Perez decided to adjust her daily release decision to eliminate 10 percent of the reservoir deficit? 5 percent?

Exercise 7: Introducing Variable Goals

Despite inflow variations Ms. Perez has been pursuing a goal of 30,000,000 gallons of water behind the Smithfield Dam, equivalent to 30 days of the long-term inflow to the dam of 1,000,000 gallons per day. This same constant goal might have been written as

N $\text{DESRES} = (30)(\text{LTINFL})$

where the N equation needs to be computed only at the beginning of the simulation to calculate the constant value of DESRES.

a. After further contemplation, Ms. Perez decides that she should instead be trying to maintain 30 days worth of water inflow in the reservoir, but that she should be using more recent information than the long-term inflow in setting this target. Write an equation for average inflow AVINFL, using 10 days as the averaging delay. (The initial condition for AVINFL is LTINFL.) Rewrite DESRES so that it is a variable, calculating 30 days worth of average inflow as Ms. Perez's goal. Rerun your model of Exer-

cise 6 (with 20 percent daily adjustment) using this new policy. How satisfactory is this new policy to Ms. Perez? to the river users?

b. Draw the causal-loop diagram of the situation under this new policy.
Compare it with Figure 22.6. Has any new closed loop been added to the
previous causal structure?

Exercise 8: More Goal Variation

Complaints from the dam users finally convince Ms. Perez that she does not
need to be so steadfast in keeping the reservoir at thirty days inflow. They
would like more water released during the growing and recreational seasons.
She agrees to take these seasonal considerations into account in setting her objectives for the reservoir.

The easy way to do this, she decides, is to vary the number of days of inflow that she is attempting to store, keeping it at 30 days at the beginning of
the year, allowing it to drop off during the middle of the year, and gradually
restoring her initial objectives near year-end.

To represent this decision process in the model, use the weighting function
W1 from Exercise 4, add a variable for the number of days of inflow desired in
the reservoir NDIDRS, and include these in the model from Exercise 7. (Do
not end up with more than one equation for each variable!)

A	DESRES.K = (NDIDRS.K)(AVINFL.K)
A	NDIDRS.K = (30)(W1.K)
NOTE	NUMBER OF DAYS OF INFLOW
NOTE	DESIRED IN RESERVOIR (DAYS)
A	W1.K = TABLE(TW1,TIME.K,0,360,90)
NOTE	WEIGHTING #1 (DIMENSIONLESS)
T	TW1 = 1.0/0.7/0.3/0.4/0.75
NOTE	TABLE FOR WEIGHTING #1

The equations generate a changing target for water to be stored behind the
dam throughout the year. Figure 22.8 shows some illustrative values for
NDIDRS, calculated from the equations.

TIME (Days)	W1 (Dimensionless)	NDIDRS (Days)
0	1.0	30
45	0.85	25.5
90	0.7	21
180	0.3	9
210	0.33	10
360	0.75	22.5

Figure 22.8 Illustrative values for variable seasonal goal

a. Test your understanding of TABLE functions (see Chapter 16) and of the equations above by recalculating some of the entries in Figure 22.8. Also calculate the values of NDIDRS at TIME = 99, 117, 270, and 350.

b. Run the model. You may want to change SPEC to allow a longer period for assessment of the policy results. (If you set LENGTH greater than 360, W1 will begin to produce inappropriate values beyond TIME = 360. See the discussion of the TABLE function in Chapter 16.)

MULTIPLE OBJECTIVES

In altering her policy to take into account the users' priorities relating to downstream flow at different times of the year, Ms. Perez acknowledged that multiple objectives often need consideration in the making of decisions. But the water release decision RLSFLW only implicitly reflects user desires by way of the seasonality multiplier W1.

Explicit identification of multiple objectives, with explicit reflection in a decision process, results in more complex but hopefully more effective decisions. The remainder of this chapter will treat the representation of multiple perspectives in a model's decision processes.

EXAMPLE IV: ALTERNATIVE GOALS IN THE DECISION PROCESS

Through their continuing protests, the Rappanno users succeed in getting evaluated an alternative set of criteria for operating the dam. They argue that the dam's purpose is to supply their river basin drainage area (the farmlands

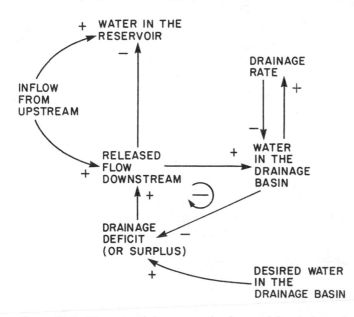

Figure 22.9 Diagram of dam control using goal for drainage basin

and water recreation sites below the dam) with the water it needs. They estimate that the drainage area normally holds 50,000,000 gallons of water, with
drainage out to the ocean normally occurring at the rate of 2 percent per day of
the water in the drainage area. Ms. Perez initially agrees to maintain the basin
at its level of 50,000,000 gallons.

Figure 22.9 pictures this new situation in a causal-loop diagram. Compare
this to Figure 22.6. Note that a different negative feedback loop is now involved with the water release decision, and the water in the reservoir is not part
of any feedback-loop process.

The model that follows represents the situation as now described.

```
*          RESPONSE TO ALTERNATIVE GOALS
NOTE
NOTE       ------------------------------------
NOTE
R          INFLOW.KL=NFLOW+STEP(0.2,5)
NOTE                UPSTREAM INFLOW (MILLION GALLONS/DAY)
NOTE
NOTE                20 PERCENT STEP INCREASE ABOVE NORMAL INFLOW
NOTE                AT TIME = 5 DAYS
NOTE
C          NFLOW=1
NOTE                NORMAL FLOW (MILLION GALLONS/DAY)
L          RES.K=RES.J+(DT)(INFLOW.JK-RLSFLW.JK)
N          RES=30
NOTE                WATER IN RESERVOIR (MILLION GALLONS)
R          RLSFLW.KL=LTINFL+DADRI.K
NOTE                RELEASED FLOW (MILLION GALLONS/DAY)
C          LTINFL=1
NOTE                LONG-TERM INFLOW (MILLION GALLONS/DAY)
A          DADRI.K=DRGAP.K/TADR
NOTE                DAILY ADJUSTMENT TO DRAINAGE INFLOW (MILLION GALLONS/DAY)
C          TADR=5
NOTE                TIME TO ADJUST WATER IN THE DRAINAGE BASIN (DAYS)
A          DRGAP.K=DESDRG-DRG.K
NOTE                DRAINAGE GAP (MILLION GALLONS)
N          DESDRG=50
NOTE                DESIRED WATER IN DRAINAGE BASIN (MILLION GALLONS)
L          DRG.K=DRG.J+(DT)(RLSFLW.JK-DRGRT.JK)
N          DRG=50
NOTE                WATER IN DRAINAGE BASIN (MILLION GALLONS)
R          DRGRT.KL=(0.02)(DRG.K)
NOTE                DRAINAGE RATE (MILLION GALLONS/DAY)
NOTE
NOTE                CONTROL STATEMENTS
NOTE
PLOT       INFLOW=I,RLSFLW=F,DRGRT=T/RES=R,DRG=G,DESDRG=S
SPEC       DT=1/PLTPER=2/LENGTH=60
RUN
```

The goal for the drainage basin is shown here as a constant amount
DESDRG of 50,000,000 gallons. Over a time period TADR (here equal to five
days), Ms. Perez's new release decision adds enough additional water to the
long-term inflow data to try to eliminate the drainage basin's water deficit
DRGAP.

Figure 22.10 shows the DYNAMO results for this policy. Notice how the
users' objective of constant water in the drainage area is maintained much better than shown in previous runs.

```
INFLOW=I RLSFLW=F  DRGRT=T    RES=R    DRG=G DESDRG=S

     0.8500           0.9500           1.0500           1.1500           1.2500  IFT
    10.000           20.000           30.000           40.000           50.000  RGS
 0.0000 - - - - - - - - - - - I - - - R - - - - - - - - - - - - - - - G GS,IFT
     .                .        I     R                .                    G GS,IFT
     .                .        I     R                .                    G GS,IFT
     .                .        F     R                .        I          G GS,FT
     .                .        F    .R                .        I          G GS,FT
     .                .        F    . R               .        I          G GS,FT
     .                .        F    . R               .        I          G GS,FT
     .                .        F    .  R              .        I          G GS,FT
     .                .        F    .   R             .        I          G GS,FT
     .                .        F    .   R             .        I          G GS,FT
20.000 - - - - - - - - - - - F - - - - .-R- - - - - - - - - - I - - - G GS,FT
     .                .        F    .    R             .        I          G GS,FT
     .                .        F    .     R            .        I          G GS,FT
     .                .        F    .      R           .        I          G GS,FT
     .                .        F    .      R           .        I          G GS,FT
     .                .        F    .       R          .        I          G GS,FT
     .                .        F    .        R         .        I          G GS,FT
     .                .        F    .        R         .        I          G GS,FT
     .                .        F    .         R        .        I          G GS,FT
     .                .        F    .          R       .        I          G GS,FT
40.000 - - - - - - - - - - - F - - - - . - - - -R- - - - - I - - - G GS,FT
     .                .        F    .           R     .        I          G GS,FT
     .                .        F    .           R     .        I          G GS,FT
     .                .        F    .            R    .        I          G GS,FT
     .                .        F    .             R   .        I          G GS,FT
     .                .        F    .             R   .        I          G GS,FT
     .                .        F    .             R.           I          G GS,FT
     .                .        F    .              R           I          G GS,FT
     .                .        F    .              R           I          G GS,FT
     .                .        F    .             .R           I          G GS,FT
60.000 - - - - - - - - - - - F - - - - - - - - - - - R - - I - - - G GS,FT
```

Figure 22.10 Constant goal for drainage basin and gradual adjustment

Exercise 9: More Comparative Policy Analysis

a. In Figure 22.10, note the behavior of RLSFLW and RES. Are these as you expected? Why or why not? How would the policy perform if IN-FLOW stepped down by 20 percent instead of up?

b. The users decide that if more water becomes available, more should be released to the drainage basin. In the model for Example IV, change DESDRG to an auxiliary equation, equal to 50 days of average inflow to the dam AVINFL, using 10 days as the averaging delay for AVINFL. Run the model, comparing the results with Figure 22.10. Be sure to examine the relative behaviors of RES and DRG under the two policies.

c. Compare the results of (b) with those generated in Exercise 7 from the viewpoints of the river users and of Ms. Perez.

Exercise 10: Multiple Goals in the Decision Process

Ms. Perez is concerned that the dam might run dry; the users are concerned that the drainage basin might run dry. A compromise is proposed under which Ms. Perez will explicitly gradually try to build up the reservoir's water storage toward her objective DESRES (i.e., 30 days of average inflow) while also trying to gradually build up the drainage basin's water level toward the user objectives DESDRG.

Change the Exercise 9 model by using the following equations from previous exercises as replacements or additions:

A $DESRES.K = (30)(AVINFL.K)$
A $RESGAP.K = DESRES.K - RES.K$
A $DAFLW.K = RESGAP.K/TAR$
C $TAR = 5$

Now write a new decision function for released flow downstream:

R $RLSFLW.KL = LTINFL + DADRI.K - DAFLW.K$

and add DESRES = D to the PLOT request.

a. Run the model. How well does this policy perform in achieving Ms. Perez's goals? the user's goals? Explain the observed behavior of the simulation.

b. As had been done in Example II, rewrite the RLSFLW equation to include W1 and W2 as weights, also adding an equation for W2 where $W2 = 1 - W1$. Let W1 be the multiplier of DADRI.K and W2 multiply DAFLW.K. Make several runs using different values of W1. Assess each simulation's adequacy in meeting the differing goals of Ms. Perez and the river users.

c. Draw a causal-loop diagram of the situation described in this exercise. How is its structure different from the prior causal structures examined in this chapter? What are the implications of the structure for possible behavior?

GOALS AS MULTIPLIERS IN DECISIONS

A favorite approach to representing complex decision situations in system dynamics models is to use multipliers to reflect goals or constraints that affect rates of flow. Just as the seasonal multiplier in Exercise 8 altered the number of days of water inflow desired in the reservoir, so too multipliers are used to have goals affect behaviors.

EXAMPLE V: USING MULTIPLIERS IN DECISION PROCESSES

Reexamine the model of the preceding exercise from this point of view. Ms. Perez's base release flow in this situation is the long-term inflow LTINFL. But she has agreed to increase that flow when the water in the drainage basin DRG is less than the desired amount DESDRG. If DRG is less than DESDRG, Perez will increase the outflow to become more than LTINFL, and she will reduce it when DRG exceeds DESDRG.

Figure 22.11 is a graph for a multiplier of LTINFL that has the conditions just described. The horizontal axis is the ratio of actual water in the drainage basin to desired. The vertical axis is the multiplier of LTINFL due to the drainage water situation. When the ratio is 1 (i.e., actual drainage water equals what is desired), the multiplier equals 1 (i.e., the release rate RLSFLW just equals the long-term inflow LTINFL). When drainage water is in excess and therefore the drainage ratio exceeds 1 (i.e., possible flooding or water wastage), the multiplier is less than 1 and RLSFLW becomes less than LTINFL. When the water in the drainage basin is less than desired, the drainage ratio is less than 1 and the graph shows a multiplier greater than 1 (i.e., RLSFLW becomes more than LTINFL).

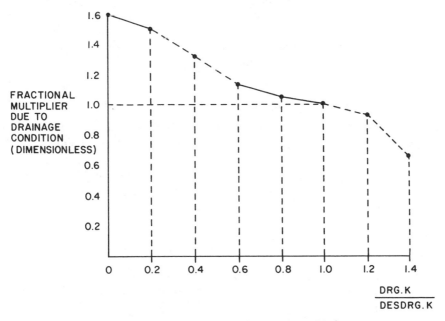

DRAINAGE RATIO (DIMENSIONLESS)

Figure 22.11 Relationship between drainage condition and release flow

Notice that the results shown in Figure 22.11 depend on the fact that:

LTINFL = DRGRT = DRG∗.02 = 1 million gallons/day

In the long run, drainage from the basin equals inflow. Therefore the model "seeks" an equilibrium where FMDD is equal to one. Notice that if LTINFL were to shift upward (perhaps due to a change in climate), then FMDD would seek a lower equilibrium. Eventually the dam would overflow and the drainage basin would be flooded with no real control possible (unless there was infinite capacity behind Smithfield Dam).

The equations for this formulation are:

A	DR.K = DRG.K/DESDRG.K
NOTE	DRAINAGE RATIO (DIMENSIONLESS)
A	FMDD.K = TABLE(TFMDD,DR.K,0,1.4,0.2)
NOTE	FRACTIONAL MULTIPLIER DUE TO
NOTE	DRAINAGE CONDITION (DIMENSIONLESS)
T	TFMDD = 1.6/1.5/1.3/1.1/1.05/1.0/0.9/0.6
NOTE	TABLE FOR FRACTIONAL MULTIPLIER
NOTE	DUE TO DRAINAGE CONDITION
R	RLSFLW.KL = (FMDD.K)(LTINFL)
NOTE	RELEASED FLOW (MILLION GALLONS/DAY)

With these equations inserted in the model used for Exercise 10 (only RLSFLW needs changing, as shown above; the other equations above can just be added), the results shown in Figure 22.12 are produced.

Exercise 11: Multiplier Approach vs. Explicit Goal Approach

Compare Figure 22.12 with your results for Exercise 9(b). Compare the equation structures used in the two models to represent RLSFLW. Both are responsive to the relative condition of water availability in the drainage basin. Depending on the specifics of a situation, you may choose either way of reflecting pressures on a decision process (e.g., the released flow) arising from consideration of a goal (e.g., the water in the drainage area).

Exercise 12: More Multipliers in Decision Processes

Using the approach developed in Example V, formulate a fractional multiplier due to the reservoir condition (and name it FMDR) that will multiply the long-term inflow rate LTINFL to affect Ms. Perez's release rate RLSFLW.

a. Develop a graph similar to Figure 22.11 for your multiplier. Will the multiplier FMDR be more or less than 1, when water in the reservoir RES is less than desired DESRES? Insert values in the table that seem reasonable to you.

```
INFLOW=I RLSFLW=F  DRGRT=T    RES=R    DRG=G DESDRG=S

    0.8500          0.9500          1.0500          1.1500          1.2500   IFT
    20.000          30.000          40.000          50.000          60.000   RGS
 0.0000 - - - - - - - R - - - I - - - - - - - - - - - G - - - - - - - - GS,IFT
    .                 R         I           .          G                 . GS,IFT
    .                 R         I           .          G                 . GS,IFT
    .                 R        TF           .          G S       I         .
    .                .R        T F          .          G   S     I         .
    .                . R       T  F         .          G       SI          .
    .                . R       T   F        .          G         I         . IS
    .                .  R      T    F       .          G         I S        .
    .                .   R      T    F      .          G         I  S       .
    .                .    R     T    F      .          G         I   S      .
 20.000 - - - - - - - - R - -T- -F- - - - - - - - - G - - - I - -S- -
    .                     R    T    F  .              .G        I    S    .
    .                     R    T     F .              .G        I     S   .
    .                      R   T     F .              .G        I     S  .
    .                       R   T    F .              .G        I     S.
    .                        R  T    F .              .G        I     S.
    .                         R T    F .              .G        I     S.
    .                         R T    F .              .G        I     S.
    .                          RT    F .              .G        I     S.
    .                          RT     F .             .G        I      S
 40.000 - - - - - - - - - - - RT- F - - - - - - - - -G- - - I - - - S
    .                          RT    F .              .G        I      S
    .                             T   F .             .G        I      S  TR
    .                             T   F .             .G        I      S  TR
    .                            TR F .               . G       I      S
    .                            TR F .               . G       I      S
    .                            T RF .               . G       I      S
    .                            T RF .               . G       I      S
    .                             T  F .              . G       I      S  FR
    .                              T F .              . G       I      S  FR
 60.000 - - - - - - - - - - - - - T FR- - - - - - - - G - - I - - - S
```

Figure 22.12 Drainage multiplier affecting the released flow

b. Following the approach taken in Example V, write the set of equations leading up to:

R   RLSFLW.KL = (FMDR.K)(LTINFL)

c. Substitute this new equation for RLSFLW and add the other equations you have written to the model of Example V. Run the model. Compare the results with those of Exercise 7, as well as with Figure 22.12.

d. Now, using your new equations as well as those produced in Example V, have Ms. Perez's released flow rate affected by both drainage and reservoir conditions.

R   RLSFLW.KL = (FMDD.K)(FMDR.K)(LTINFL)

Run the model. Compare the results with Figure 22.12, with the results of (c) above, as well as with the results of Exercise 10.

e. If you wish, using DYNAMO's rerun feature, modify TFMDD and/or TFMDR to test the effects of different degrees of responsiveness to the two goals.

SUMMARY

This chapter has exposed the student to examining, developing, and testing a wide variety of approaches for representing decisions. Although each formulation may not represent good water management practices for Ms. Perez, each form of decision function is appropriate to some situation and each is used often in system dynamics models. Usually the model builder will be free to choose from among several different ways of representing a decision depending on the situation being simulated. The alternative forms presented in this chapter are listed below as a guide to, but not as a limitation upon, efforts at representing decision processes.

Form of Decision Process	*Where Used in This Chapter*
Outflow = inflow	Example I
Outflow = delayed inflow	Exercise 1(b)
Outflow = averaged delayed inflow	Exercise 2(b)
Outflow = weighted sum of inflows, using constant fractional multipliers	Example II
Outflow = weighted sum of inflows, with variable weights coming from a table function	Exercise 4
Outflow = constant goal minus actual, immediately corrected	Example III
Outflow = constant goal minus actual, corrected over a time delay	Exercise 5(c) and Example IV
Outflow = constant fractional multiplier of constant goal minus actual	Exercise 6
Outflow = constant fractional multiplier of goal minus actual, where goal is a variable	
(a) goal is a constant multiplier times an average rate	Exercise 7 and Exercise 9(b)
(b) goal is a varying multiplier times an average rate	Exercise 8
Outflow = sum of responses to the conditions of multiple goals	Exercise 10(a)
Outflow = sum of weighted responses to the conditions of multiple goals	Exercise 10(b)

Outflow = base flow, times fractional
     multiplier reflecting goal
     condition   Exercise 5 and Exercise 12(b)

Outflow = base flow, times more than
     one fractional multiplier
     reflecting the conditions of
     multiple goals   Exercise 12(d)

ENDNOTE

1. You may wonder why DT appears in the denominator of the equation for
   DAFLW. Notice that if some time unit did not appear in the equation, then the
   units on the right-hand side of the equation would not match the units on the left-
   hand side of the equation. That is, the units of RESGAP is in millions of gallons
   and the units of DAFLW is in millions of gallons per day. However, this explanation in terms of matching units masks a more important point. Since this formulation is supposed to represent "instantaneous" adjustment, leaving out the DT
   would yield an odd result. Notice that if DT were changed from 1 day to 2 days
   (and if DT were not in the equation for DAFLW), then effectively, the compensation would be half as fast as in a truly "instantaneous" adjustment. Similarly, if
   DT were changed from 1 day to .5 day, the compensation due to reservoir gap
   would be twice as fast as the desired "instantaneous" adjustment. By not including DT in the equation for DAFLW, the alleged "instantaneous" adjustment
   could be made slower or faster just by changing the value of DT. As discussed previously, DT should be chosen so that the desired response of the model will occur
   regardless of how small a DT is chosen.

CHAPTER 23

NATURAL GAS: A NATURAL RESOURCE SIMULATION*

INTRODUCTION: THE PROBLEM

Natural gas is one of the principal sources of energy in the United States. It is a colorless gas, found in large underground reservoirs in gaseous form or dissolved in liquid. The formation of natural gas through the decomposition of organic matter requires hundreds of thousands of years. Thus, for practical purposes, natural gas can be viewed as a finite resource that may eventually be exhausted.

Natural gas is widely used for household heating. Over the decades from 1940 to 1970, the use of the natural gas increased at a rate of almost 7 percent per year. In 1940, about 3 trillion cubic feet were consumed. By 1970, the usage rate had increased to about 22 trillion cubic feet per year. In the early 1970s, however, the use of natural gas reached a peak, and by 1975, the usage rate had fallen to about 20 trillion cubic feet per year.

The total amount of natural gas that has been discovered but not yet consumed is called the level of "proven natural gas reserves." Until the mid-1960s, the amount of natural gas discovered each year exceeded the amount consumed, and thus the level of proven natural gas reserves increased from 1900 to 1960. But in the late 1960s, discoveries began to fall below the usage rate, and reserves have begun to decline. Gas shortages have appeared in various parts of the country.

No one knows the precise amount of natural gas remaining undiscovered, but many estimates suggest that perhaps one-half of the total supply of underground natural gas in the United States had been discovered by the mid-1960s.

---

*The model discussed in this chapter is based in part on Roger F. Naill, "The Discovery Life-Cycle of a Finite Resource: A Case Study of U.S. Natural Gas," from D. L. Meadows and D. H. Meadows, eds., *Toward Global Equilibrium: Collected Papers* (Cambridge, Mass.: Wright-Allen Press, 1973).

I THINK THEY'VE DISCOVERED GAS!!!

In the sections that follow, a simulation model is formulated to explain the historical pattern of natural gas discovery and usage from 1900 to 1980 and to examine the likely behavior of natural gas discovery and usage from 1980 to 2050. The reference mode for the model to be built is the behavior shown in Figure 23.1, and the time horizon is the period from 1900 to 2050. Once an adequate model is formulated, it can be used to examine various policies designed to influence the natural gas system.

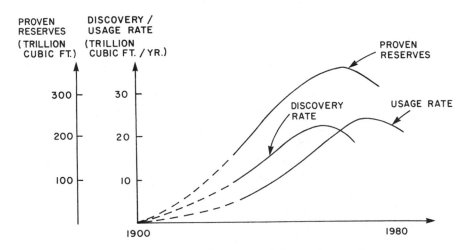

Figure 23.1 Historical data on the discovery and usage of natural gas

AN INITIAL MODEL

There are several ways to begin building a simulation model. Often, it is best to start by formulating a causal-loop diagram, and then prepare a flow diagram and equations. That is the approach taken, for example, in the Kaibab Plateau model developed in Chapter 18. Sometimes, however, it is helpful to begin by determining the principal levels and rates that ought to be included in the model. Then the feedback loops that influence the levels and rates can be added in steps. That is the approach taken in this chapter for building a model of the natural gas system.

Exercise 1 considers the levels and rates to include in the model of natural gas discovery and usage. Then, in the exercises that follow, some important feedback loops are added.

Exercise 1: The Discovery and Usage of Natural Gas

What are the principal levels and rates we should include in our model? To think about this, let's review the material contained in the Introduction of this chapter. According to that information, natural gas is a finite resource that eventually will be exhausted. Thus, whenever gas is discovered, the amount remaining to be discovered in the future is reduced. The amount of gas underground that has been discovered, but not yet used, is called the "level of proven gas reserves." Proven reserves are increased when gas is discovered and reduced when gas is used. This suggests that there are four main elements we should include in the model: undiscovered gas UDGAS, discovery rate DISCOV, proven gas reserves GASRV, and usage USAGE.

a. Draw a flow diagram indicating which of these elements are levels and which are rates.

Now in (b) through (e) write equations to simulate the behavior of the system over the period from 1900 to 2050.

b. First, write a level equation and initial value equation for undiscovered gas UDGAS. Assume that 1 quadrillion cubic feet of gas remained to be discovered in 1900. (1 quadrillion equals 1E15, or 1000 trillion. It is easier to measure gas in units of trillions of cubic feet. Thus the initial value of UDGAS in 1900 is 1000.)

c. Next formulate an equation for the natural gas discovery rate. The graph in Figure 23.1 clearly indicates that the discovery of natural gas increased over the period from 1900 to 1970 and has declined since 1970. For now, however, let's assume that the discovery rate remained constant over the period from 1900 to 2050, at 1 trillion cubic feet per year, its value in 1900. Later, we can add a more realistic representation of the discovery rate.

d. Now write an equation for the level of proven gas reserves GASRV. Assume an initial value of 6.4 trillion cubic feet.

e. Finally, consider the formulation for the natural gas usage rate USAGE. Like the discovery rate, it is clear from Figure 23.1 that the gas usage rate increased substantially over the period from 1900 to 1970. For now, however, let's assume that the usage rate remained constant at 0.32 trillion cubic feet per year over the period from 1900 to 2050. Later we can add a more realistic representation.

Exercise 2: Running the Initial Model

a. To simulate the model we have developed, we need to add the simulation specifications. Write the appropriate specifications, setting the solution interval DT = 0.25 years, the LENGTH = 2050 years, and the PLTPER = 3 years. Plot the undiscovered gas UDGAS and gas reserves GASRV. It is also necessary to write an initial value equation to set TIME = 1900 years.

b. Now, run the model and examine the output. Label the model run Exercise 2(b). What behavior mode does undiscovered gas UDGAS exhibit? Why? What behavior mode does the level of proven reserves GASRV exhibit? Why?

c. The behavior of the level of proven gas reserves GASRV depends on both the discovery rate DISCOV and the usage rate USAGE. Rerun the model, setting the discovery rate DISCOV = .32 trillion cubic feet per year. (You may want to choose a longer PLTPER, if you got bored while watching the output being printed in the previous question.) How does the behavior of the level of proven gas reserves GASRV differ from the behavior in Exercise 2(b)? Why?

d. (Optional) Rerun the model, setting the discovery rate DISCOV = 0. Label the run Exercise 2(d). How does the behavior of the level of proven gas reserves GASRV differ from the behavior in Exercise 2(b) and 2(c)?

NATURAL GAS DISCOVERY

Next formulate the feedback loops that influence the discovery and usage rates. Begin with the discovery rate. What might influence the rate at which gas is discovered? One factor is the adequacy of current reserves. If reserves are plentiful, there will be little reason for gas companies to invest in exploration for new gas. On the other hand, if reserves are low, gas companies will be under considerable pressure to find additional supplies. This section formulates a representation of the adequacy of gas reserves.

Exercise 3: The Adequacy of Gas Reserves

a. What is an adequate level of proven gas reserves? For example, in 1900, the level of gas reserves GASRV wa 6.4 trillion cubic feet. Was that an adequate supply? In order to answer this question it is necessary to compare the level of reserves with the rate at which gas was being used. How long would the gas reserves in 1900 have lasted if usage had continued at its 1900 rate and no new supplies were discovered? (Recall that in 1900 the usage rate USAGE was 0.32 trillion cubic feet per year.)

In general, to determine how long gas reserves will last if usage continues at its present rate, simply divide the level of gas reserves by the usage rate:

Amount of time	Level of gas	Usage rate
gas reserves will =	reserves	÷ (trillion cubic feet/year)
last (years)	(trillion cubic feet)	

There is, however, one complication. The usage of natural gas fluctuates to some extent from month to month (and also, of course, from day to day). For example, the usage of natural gas is generally much greater in January than in July. Thus it is more sensible to compute the amount of time present gas reserves will last by dividing the level of gas reserves by the usage rate averaged over a year.

Amount of time	Level of gas	Usage rate
gas reserves will =	reserves	÷ averaged over one year
last (years)	(trillion cubic feet)	(trillion cubic feet/year)

In DYNAMO, the average value of a variable can be computed by using the SMOOTH function. (For more information on the SMOOTH function, see Chapter 17.) The DYNAMO equation for the average usage rate AUSAGE can be written as follows:

A	AUSAGE.K = SMOOTH(USAGE.JK,UAT)
C	UAT = 1

AUSAGE	Average Usage Rate (trillion cubic feet/year)
USAGE	Usage Rate (trillion cubic feet/year)
UAT	Usage Averaging Time (years)

b. Call the amount of time gas reserves will last the "Reserve-Usage Ratio"[1] RUR. Copy the flow diagram you developed in Exercise 1(a), and add the reserve-usage ratio RUR and the average usage rate AUSAGE.

c. Now write an equation for the reserve-usage ratio RUR.

Exercise 4: Running the More Complete Model

a. Add the equations for the average usage rate AUSAGE and the reserve-usage ratio RUR to the model developed in Exercise 1. Also add the reserve-usage ratio RUR to the PLOT statement. (Set USAGE = .32 and DISCOV = 1 trillion cubic feet per year.) Now run the model and examine the output. Label the model run Exercise 4(a). What behavior does RUR exhibit? Why?

b. How does the behavior of RUR depend on the discovery rate DISCOV? Rerun the model, setting the discovery rate DISCOV = 0 trillion cubic feet per year. Label the run Exercise 4(b). What behavior does RUR exhibit? Why? When does GASRV reach zero?

c. Rerun the model again, this time setting the discovery rate DISCOV = 0.32 trillion cubic feet per year. Label the run Exercise 4(c). What behavior does RUR exhibit? Why?

Exercise 5: Investment in Exploration

Recall that the adequacy of gas reserves is one factor influencing the amount of exploration for natural gas. When reserves are plentiful, there is little exploration for new gas. But when reserves are low, gas companies invest a considerable effort in exploration.

Use the reserve-usage ratio developed in Exercise 3 as an indicator of the adequacy of gas reserves. Assume that gas companies prefer to maintain the level of gas reserves at about a twenty-year supply. When reserves fall below twenty years, companies increase the amount of revenue they invest in exploration for new gas. And when reserves rise above twenty years, companies decrease their investment in exploration.

a. Draw a causal-loop diagram showing the relationship between gas reserves, the reserve-usage ratio, investment in exploration, and the discovery rate. Is the loop positive or negative?

Now let's work out the elements of this feedback loop in more detail and then add them to the flow diagram developed in Exercise 3(b). To do this, we will need to examine the income that gas companies earn by selling gas, the portion of their income they invest in exploration for new gas, and the amount of gas discovered per dollar invested in exploration.

We will begin by looking at the income gas companies receive. The income received each year by selling natural gas is called the "sales revenue" SALREV (in dollars per year). The sales revenue SALREV is equal to the product of the gas usage rate and the price charged for gas PRICE. The price of gas in 1900 was about $0.00012 per cubic foot or $120 million per trillion cubic feet.

A   SALREV.K = USAGE.JK∗PRICE

SALREV	Sales Revenue (dollars/year)
USAGE	Usage Rate (trillion cubic feet/year)
PRICE	Price (dollars/trillion cubic feet)

Generally, gas companies invest a portion of their sales revenue in exploration for new gas. Call the amount invested (in dollars per year) the "investment in exploration" INVEXP. The investment in exploration INVEXP is equal to the product of the sales revenue SALREV (in dollars per year) and the fraction of sales revenue invested in exploration FSRIE.

A   INVEXP.K = SALREV.K∗FSRIE.K

INVEXP	Investment in Exploration (dollars/year)
SALREV	Sales Revenue (dollars/year)
FSRIE	Fraction of Sales Revenue Invested in Exploration (dimensionless)

What determines the fraction of sales revenue invested in exploration? According to the preceding discussion, gas companies adjust their investment in exploration in response to the level of gas reserves. When the reserve-usage ratio RUR is high, reserves are plentiful, and the fraction of sales revenue invested in exploration FSRIE is low. On the other hand, when the reserve-usage ratio RUR is low, reserves are inadequate and the fraction of sales revenue invested in exploration is high.

Assume that gas companies consider that the reserves level is adequate when the reserve-usage ratio RUR = 20 years. Call this value of RUR the "desired reserve-usage ratio" DRUR. Then we can define the relative reserve-usage ratio RRUR = RUR/DRUR, which is an indicator of the adequacy of current reserves. Reserves are adequate when RRUR = 1. Reserves are low when RRUR is less than 1, and reserves are high when RRUR is greater than 1.

b. For example, suppose the level of gas reserves GASRV = 6.4 trillion cubic feet, and the average usage rate AUSAGE = 0.64 trillion cubic feet per year. Then the reserve-usage ratio RUR = 10 years (RUR = GASRV/AUSAGE = 6.4trillion/0.64trillion/year = 10years). What is the value of the relative reserve-usage ratio RRUR? Are reserves too high or too low?

Now consider the relationship between the fraction of sales revenue invested in exploration FSRIE and the relative reserve-usage ratio RRUR. Assume that when RRUR = 1 (i.e., when RUR equals the desired reserve usage ratio), the fraction of sales revenue invested in exploration FSRIE = 0.35. When RRUR rises to 2 (i.e., when reserves are twice the desired level), FSRIE drops to 0. On the other hand, when RRUR drops to 0, FSRIE rises to its maximum value of 0.45.

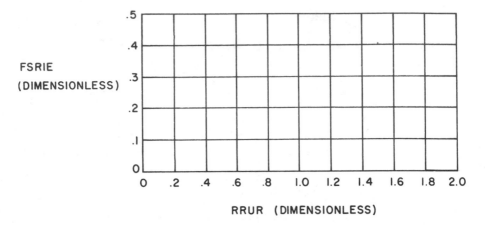

Figure 23.2 TABLE function relating RRUR to FSRIE

c. On a graph similar to Figure 23.2, draw a TABLE function showing the relationship between RRUR and FSRIE. Write the appropriate equations.

Now define a variable called the "efficiency of exploration" EFFEXP (measured in trillion cubic feet of gas discovered per dollar invested in exploration). Using EFFEXP, calculate the rate at which gas is discovered each year. The discovery rate (in trillion cubic feet per year) is equal to the product of the amount of revenue invested in exploration INVEXP (in dollars per year) and EFFEXP (in trillion cubic feet of gas per dollar invested). In 1900, the efficiency of exploration EFFEXP was about 67,000 cubic feet per dollar or 67E-9 trillion cubic feet per dollar.

R   DISCOV.KL = INVEXP.K*EFFEXP
C   EFFEXP = 67E – 9

DISCOV   Discovery Rate (trillion cubic feet/year)
INVEXP   Investment in Exploration (dollars/year)
EFFEXP   Efficiency of Exploration (trillion cubic feet/dollar)

d. Copy the flow diagram from Exercise 3(b) and add the PRICE, the sales revenue SALREV, the relative reserve-usage ratio RRUR, the fraction of sales revenue invested in exploration FSRIE, and the efficiency of exploration EFFEXP.

Exercise 6: Running the Model

a. Add the equations for PRICE, sales revenue SALREV, relative reserve-usage ratio RRUR, fraction of sales revenue invested in exploration FSRIE, and efficiency of exploration EFFEXP to the model developed in Exercise 3. Also add the fraction of sales revenue invested in exploration

FSRIE and the discovery rate DISCOV to the PLOT statement, choosing a plot scale of (0,.8) for FSRIE. Run the model and examine the output. Label the model run Exercise 6(a). What behavior do RUR and FSRIE exhibit? Why?

b. Why does FSRIE reach an equilibrium value of about .125? Does this value depend on the usage rate USAGE? To see, rerun the model, this time setting the usage rate USAGE = 1 trillion cubic feet per year. Label the run Exercise 6(b). How do the results differ?

c. (Optional) Does the equilibrium value of FSRIE depend on the desired reserve usage ratio DRUR? To see, rerun the model with USAGE = 0.32 trillion cubic feet per year and the desired reserve-usage ratio DRUR = 40 years. Label the run Exercise 6(c). Compare the results with those in Exercise 6(a).

A THIRD MODEL

In the version of the model worked out in Exercises 5 and 6, it was assumed that the usage of natural gas was constant. According to the data in Figure 23.1, however, the usage of natural gas increased exponentially over the period from 1900 to 1960. This section provides guidance for modifying the model developed in Exercise 2 to include a representation of this exponential growth.

The rapid growth in the usage of natural gas over the period from 1900 to 1960 had a number of causes. First, of course, the population of the United States grew substantially, which increased the amount of energy consumed. Second, the standard of living in the United States improved dramatically, increasing the amount of energy used per person. Third, and most important, natural gas proved to be an inexpensive and clean source of energy. Thus, once the technology of interstate pipelines was developed, natural gas replaced coal as a principal method of household heating.

It would be possible to build a model that included a representation of population growth, the rising standard of living, and the relative desirability of coal and natural gas, but for this purpose that would be unnecessarily complicated. Instead, assume that the demand for natural gas simply increases at a fixed rate of 7 percent per year. (We will later modify this assumption to reflect changes in the price of natural gas.)

One way to write the equations for the growth in usage of natural gas is to use the DYNAMO exponential function EXP. The EXP function is a shorthand way of generating exponential growth. It is generally used to introduce exponential growth in a model without precisely explaining its causes, or when fully explaining the causes would make a model unnecessarily complex. For example, consider the population of deer on the Kaibab Plateau. (See Chapter 18, Exercise 1.) The initial value of the deer population on the plateau in 1900 was 4000, and the fractional birth rate for deer FBRD was 0.5 per year.

In Chapter 18, we used a rate and level equation to represent deer births and the deer population because we were interested in explaining why the exponential growth in the deer population took place. If we had wanted to introduce growth in the deer population without explaining that it was caused by deer births, we could have used the following shorthand form:

$$DEER = DEERN*EXP(0.5*(TIME.K - 1900))$$

(For more information on the exponential function, see the *DYNAMO User's Manual* or the DYNAMO guide for your machine.)

Assume that the usage of gas grows at a rate USAGE = .07, and the initial usage of natural gas in 1900 was USAGEN = 0.32 trillion cubic feet per year. Then the equations should be as follows:

R	USAGE.KL = USAGEN*EXP(USAGEG(TIME.K −
X	1900))
C	USAGEN = .32
C	USAGEG = .07

USAGE	Usage Rate (trillion cubic feet/year)
USAGEN	Usage Rate, Initial Value (trillion cubic feet/year)
USAGEG	Usage Growth Rate (1/year)

In this formulation, the growth in the usage of natural gas is said to be "exogenous," or from outside the system, because we have simply introduced it as an assumption without including a representation of its causes.

Exercise 7: Running the Model

a. Add the equations for exponential growth in the usage of natural gas to the model developed in Exercise 5. Also add the usage rate USAGE to the PLOT statement. Choose a PLOT scale of (0,40) for DISCOV and USAGE, (0,400) for GASRV, (0,1000) for UDGAS, and (0,80) for RUR. Run the model and examine the results. Label the run Exercise 7(a).

b. Describe the behavior of the level of undiscovered gas UDGAS. When does the level of undiscovered gas UDGAS reach zero? What happens next? Does that seem realistic? (You may want to rerun the model, changing the PLOT scale for UDGAS.)

c. Do you think any important feedback loops have been omitted from the model developed so far? If so, what are they?

d. (Optional) When usage is growing exponentially, reserves do not last as long as the reserve-usage ratio indicates. To see this, rerun the model, setting the efficiency of exploration EFFEXP = 0. Label the run Exercise 7(d). (This will cause the discovery rate DISCOV to be 0. Why?) How long does it take the level of gas reserves GASRV to reach 0? What do you conclude?

e. (Optional) Rerun the model, setting the usage growth rate USAGEG = 0.04. Label the run Exercise 7(e). How do the results differ from the results in Exercise 7(a)? How long does it take usage to double in each case? When does the level of undiscovered gas UDGAS reach 0?

f. (Optional) Compare the behavior of the fraction of sales revenue invested FSRIE in Exercise 7(a) with the behavior of FSRIE in Exercise 7(e). What differences are there? Rerun the model setting the usage growth rate USAGEG = 0.02. Label the run Exercise 7(f). How do the results differ? Why does the fraction of sales revenue invested change when the usage growth rate changes?

Exercise 8: Variable Efficiency of Exploration

As seen in Exercise 7, the model developed so far has an obvious flaw: The discovery rate DISCOV continues to grow even after the level of undiscovered gas UDGAS is exhausted. This suggests that an important feedback loop has been omitted from the model, relating the discovery rate and the level of undiscovered gas.

Recall that Exercise 5 defined the efficiency of exploration EFFEXP (measured in trillion cubic feet of gas discovered per dollar invested in exploration). Exercise 5 assumed that the efficiency of exploration was constant: EFFEXP = 67E-9 trillion cubic feet per dollar. That is probably not a very good assumption. It is more reasonable to assume that as the level of undiscovered gas falls, it becomes more and more difficult to find the gas that remains. The gas that is easiest to find is probably discovered first. Gas that occurs in deep wells, offshore, or in unusual formations can be discovered only with additional ingenuity and expense. And when there is no gas left to find, none can be discovered, no matter how much revenue is invested in exploration.

a. Draw a causal-loop diagram showing the relationship between the level of undiscovered gas, the efficiency of exploration, and the discovery rate. Is the loop positive or negative?

b. Copy the flow diagram for the model developed in Exercise 5(d), and add the necessary elements to represent the feedback loop worked out in Exercise 8(a). (*Hint:* It will be helpful to define an auxiliary variable called the "fraction of undiscovered gas remaining" FUDGR.)

Now write equations for this new loop. The main task is deciding on a function showing the relationship between the fraction of undiscovered gas remaining FUDGR and the efficiency of exploration EFFEXP. We know that when FUDGR = 1, EFFEXP = 67E-9 trillion cubic feet per dollar. And when FUDGR = 0 (that is, when no undiscovered gas remains), the efficiency of exploration EFFEXP = 0. But what values should EFFEXP take when FUDGR is between 1 and 0?

Historical data might be used to determine values of EFFEXP for the period from 1900 to 1980. But since the level of undiscovered gas in the United States has not yet been fully exhausted, it is impossible to know with any certainty what the efficiency of exploration might be when the fraction of undiscovered gas remaining FUDGR falls below its present value. Hence, some reasonable guesses about the shape of the function have to be made, and later the model can be run using different values to see whether the model is sensitive to the exact values chosen.

One way to write the equation for EFFEXP is to express EFFEXP as the product of two terms: the efficiency of exploration in 1900, and an "efficiency of exploration multiplier," which falls from 1 to 0 as FUDGR declines.

One possible relationship between FUDGR and the efficiency of exploration multiplier EFFEXM is shown in Figure 23.3. The values in the graph suggest that the efficiency of exploration multiplier EFFEXM can be represented as equal to FUDGR squared. For example, when FUDGR = 0.8, EFFEXM = 0.64; and when FUDGR = 0.5, EFFEXM = 0.25.

Ordinarily, it would be appropriate to express a functional relationship like the one shown in Figure 23.3 as a DYNAMO table function. In this case, however, a table function has an important disadvantage. The TABLE func-

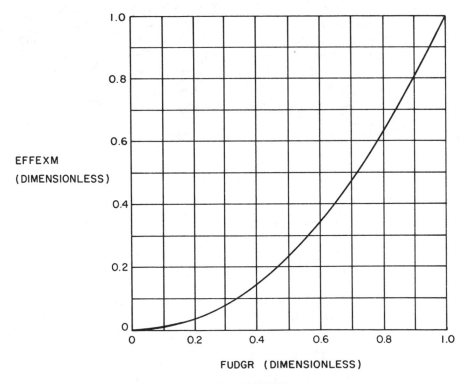

Figure 23.3 Function relating FUDGR to EFFEXM

tion operates by linearly interpolating between the points provided in the table. Thus, for example, if a TABLE were constructed from Figure 23.3 using an x-axis interval of 0.2, the TABLE would be linear over the range 0.0 to 0.2; linear over the range 0.2 to 0.4; and so forth. Unfortunately, much of the important behavior in the natural gas model occurs when FUDGR falls below 0.2; and, if a TABLE function were used, this would have the effect of establishing a linear relationship between FUDGR and EFFEXM during this part of the model run. This problem could be avoided by choosing extremely fine intervals on the x-axis, but it is simpler to define the relationship between FUDGR and EFFEXM analytically: EFFEXM.K = FUDGR.K∗FUDGR.K (or, EFFEXM equals FUDGR squared).

Exercise 9: Running the Model

a. Add the equations developed in Exercise 8 to the model worked out in Exercise 7. Run the model and examine the results. Label the run Exercise 9(a). What behavior does the natural gas discovery rate DISCOV exhibit? Why? Compare the model run with the run in Exercise 7(a). How do the results differ?

b. Rerun the model several times, choosing different values for the usage growth rate. Label the runs Exercise 9(b). Compare the results with those in Exercise 9(a).

c. Describe the behavior of the level of gas reserves GASRV. Does it seem realistic? Do you think any important feedback loops have been omitted from the model? If so, what are they?

d. (Optional) While it is plausible to assume that EFFEXM is proportional to FUDGR, there is no reason why the relationship should necessarily involve the square of FUDGR. Try running the model using other powers of FUDGR. (For example, try EFFEXM equals FUDGR cubed, or FUDGR to the fourth power.) Another way to express the relationship between EFFEXM and FUDGR is:

EFFEXM.K = EXP(S∗LOGN(FUDGR.K))

where S is the power to which FUDGR is to be raised. This permits non-integer values for S. (It is easy to show that S represents the elasticity of the efficiency of exploration, with respect to the fraction of undiscovered gas remaining.)

e. (Optional) Rerun the model several times, choosing different values for the initial level of undiscovered gas UDGASN. Label the runs Exercise 9(e). How do the results differ?

f. (Optional) Examine the model runs in Exercises 9(d) and 9(e). What determines the point at which the discovery rate peaks?

PRICE AND THE DEMAND FOR NATURAL GAS

The model worked out so far has one additional defect. The natural gas usage rate continues to climb exponentially, even after the level of gas reserves has been exhausted. This suggests that something else is missing from the model, something that reduces usage as gas becomes unavailable.

One important factor influencing usage is price. So far, the price of natural gas has been assumed to be constant. The following sections examine some of the determinants of the price of gas, and then consider the influence of a rising price on the natural gas usage rate.

Exercise 10: The Cost of Natural Gas

The price of natural gas is largely determined by two things: The cost of natural gas discovery (in dollars per trillion cubic feet) and the cost of natural gas production (i.e., the cost of pumping gas out of the ground). Consider the cost of discovery first.

The cost of discovery DCOST (in dollars per trillion cubic feet) is related to the efficiency of exploration EFFEXP. Recall that the efficiency of exploration EFFEXP indicates the amount of gas that can be discovered per dollar invested in exploration. Thus the cost of discovering a trillion cubic feet of gas can be determined by computing 1/EFFEXP.

a. We observed in Exercise 9 that the efficiency of exploration EFFEXP falls as the fraction of undiscovered gas remaining FUDGR declines. This implies that the cost of discovery DCOST rises as FUDGR declines. For example, when FUDGR = 1, EFFEXP = EFFEXM*EFFEXN = 1*67E-9 = 67E-9 trillion cubic feet per dollar, and so the cost of discovering a trillion cubic feet of gas DCOST = 1/EFFEXP = 1/67E-9 = $15,000,000. When FUDGR = 0.707, (the square root of ½), EFFEXP = 0.5*67E-9 = 33.5E-9, and DCOST = 1/EFFEXP = 1/33.5E-9 = $30,000,000. To make sure you understand the formulation for discovery cost DCOST, compute DCOST when FUDGR = 0.50.

b. The second factor determining the price of natural gas is the cost of pumping the gas out of the ground. This cost depends on such things as the depth of the gas underground, the nature of the rock formations in which the gas is contained, and so forth. For simplicity, let's assume that the cost of bringing gas out of the ground is proportional to the cost of discovery. That is probably a reasonable assumption, since gas that is more expensive to find is probably more expensive to pump out as well. We will assume that the production cost PCOST is equal to 3 times the cost of discovery.

We can then express the total cost of natural gas per trillion cubic feet TCOST as the sum of the discovery cost DCOST and the production cost

PCOST. Write DYNAMO equations for the discovery cost DCOST, the production cost PCOST, and the total cost TCOST.

c. To make sure you understand the formulation for total cost, it will be helpful to work out an example. If in 1960 the fraction of undiscovered gas remaining FUDGR was 0.707, what was the total cost per trillion cubic feet of the gas discovered that year?

d. What was the total cost per trillion cubic feet of the gas *used* in 1960? At first glance, it might seem that the answer to this question is just the same as the answer to the question about the cost of gas *discovered* in 1960. But that is not quite correct. Recall that natural gas is not used immediately after it is discovered. Instead, gas discoveries are added to the level of proven reserves until they are used.

   How much time elapses between the time gas is discovered and the time it is used? Suppose that in 1960 the gas usage rate was 12 trillion cubic feet per year and the level of gas reserves was 240 trillion cubic feet. If usage continued at its 1960 rate and no new gas was discovered, when would the level of gas reserves be exhausted? Why?

e. If some new gas were discovered in 1960 and usage continued at its 1960 rate, when do you think the gas discovered in 1960 would be used? Why?

Recall that we have already defined a variable called the reserve-usage ratio RUR, which indicates how long reserves will last if usage continues at its current rate and no new gas is discovered. If gas is discovered, the reserve-usage ratio indicates the length of time that will elapse between discovery and usage of the gas.[2]

According to Exercise 10(e) the reserve-usage ratio RUR in 1960 was 240 trillion/12 trillion/year = 20 years. Thus the cost of the gas used in 1960 was the cost of the gas discovered 20 years earlier. And, in general, the cost of the natural gas used in a particular year is the cost of the gas discovered RUR years before.

How can we write a DYNAMO equation to express this time-delay relationship? One way to obtain a delayed value of an auxiliary variable is to use DYNAMO's DLINF3 function. The DLINF3 function is a third-order information delay. (For more on DLINF3, see Chapter Seventeen.) Call the cost of the gas used in a particular year the delayed total cost DTCOST. Then the equations for DTCOST can be written as follows:

A   DTCOST.K = DLINF3(TCOST.K, RUR.K)

DTCOST   Delayed Total Cost (dollars/trillion cubic foot)
TCOST   Total Cost (dollars/trillion cubic foot)
RUR   Reserve-Usage Ratio (years)

f. Copy the flow diagram from Exercise 8(b) and add the discovery cost DCOST, the production cost PCOST, the total cost TCOST, and the delayed total cost DTCOST.

Exercise 11: The Price of Natural Gas

We have worked out equations for the cost of natural gas because the cost is one important element in determining the price of gas, but the availability of gas also plays a role in determining the price. When the reserve-usage ratio is low, indicating that gas is in short supply, gas companies raise the price, which lowers the rate at which gas is used. When gas is readily available, gas companies lower the price, which increases the usage rate.

a. Draw a causal-loop diagram showing the relationship between gas reserves, price, and the usage rate. Is the loop positive or negative?

Assume that gas companies base the price of natural gas on the cost of the gas they sell; that is, on the delayed total cost DTCOST. When the supply of gas is adequate, they set PRICE = 2*DTCOST. (Recall from Exercise 5(b) that reserves are adequate when the reserve-usage ratio RUR equals the desired reserve-usage ratio of 20 years. We defined the relative reserve-usage ratio RRUR = RUR/DRUR. Thus reserves are adequate when RRUR = 1.)[3]
When the supply of natural gas is short (i.e., when RRUR is less than 1), gas companies set the price above 2*DTCOST. For example, when RRUR = .4, PRICE = 5*DTCOST. On the other hand, when there is excess gas, gas companies set the price below 2*DTCOST. For example, when RRUR = 1.4, PRICE = 1.2*DTCOST.

b. Draw a TABLE function showing the relationship between RRUR and PRICE. (*Note:* It will be helpful to define a variable called the "price multiplier" PM and to represent the effects of RRUR on price. See Figure 23.4.) Write equations for PM and PRICE.

How does one represent the influence of the price of gas on the usage rate? Recall that Exercise 7 assumed that the usage of natural gas increases at a rate of 7 percent per year. We might think of that as the potential usage rate—the usage rate that would occur if the price of gas remained at its 1900 value. Of course, price has not remained at its 1900 value; it has gone up. Usually, when the price of a product rises, people buy somewhat less of it. Thus we would expect that as the price of gas rises, the usage rate would fall below the potential usage rate.

One way to represent this process is to express the usage rate USAGE as the product of the potential usage rate PUSAGE and a "demand multiplier" DEMM, which falls as the price rises.

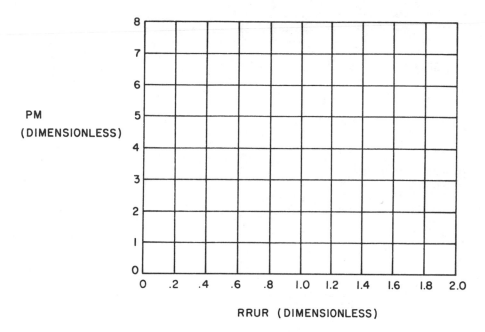

Figure 23.4 TABLE function relating RRUR and PM

R	USAGE.KL = DEMM.K*PUSAGE.K
A	PUSAGE.K = USAGEN*EXP(USAGEG(TIME.K-1900))
C	USAGEN = 0.32
C	USAGEG = 0.07

USAGE	Usage Rate (trillion cubic feet/year)
DEMM	Demand Multiplier (dimensionless)
PUSAGE	Potential Usage Rate (trillion cubic feet/year)
USAGEN	Usage Rate, 1900 value (trillion cubic feet/year)
USAGEG	Usage Growth (1/year)

Our only remaining problem is to work out a TABLE function for the demand multiplier DEMM. We need to know how fast the demand for gas falls as the price rises. For example, when the price of gas increases to twice the price in 1900, should we assume that demand falls to one-half the potential usage rate? Less than one-half? More than one-half?

One simple assumption is that the demand is inversely proportional to the price. It is helpful to define the price ratio PRATIO as the price of gas divided by the 1900 price. Thus the demand multiplier might be written as one over the price ratio. This relationship is graphed in Figure 23.5.

While it would be possible to express the relationship shown in Figure 23.5 as a DYNAMO table function, there is a problem. Over the course of the period, from 1900 to 2050, the price of natural gas will probably rise to far

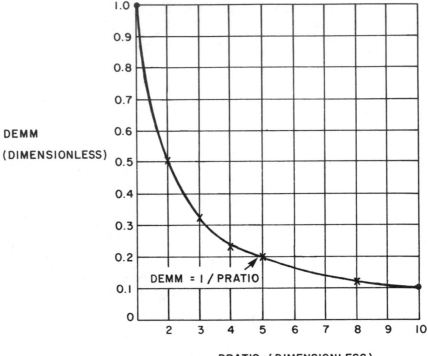

Figure 23.5 Function relating the price ratio and the demand multiplier

more than ten times its 1900 value. It may, in fact, rise to 100 or even 200 times its 1900 value. This problem could be taken into account by formulating an extremely large TABLE function, incorporating values of PRATIO from 1 to 200, but it is simpler to formulate the relationship analytically: DEMM.K = 1/PRATIO.K.

   c. Copy the flow diagram from Exercise 10f and add a representation of the price of natural gas PRICE, the price multiplier PM, and the demand multiplier DEMM.

Exercise 12: Running the Model

Add the equations developed in Exercises 10 and 11 to the model worked out in Exercise 8, and add the PRICE to the PLOT statement. There is also one interesting technical addition that needs to be made. Look again at the causal-loop diagram in Exercise 11(a). Generally, all closed feedback loops in a system dynamics model must contain at least one level. If you trace the variables around this loop, however, it looks at first as if all the variables are rates or auxiliaries. There is, in fact, a level in the loop, but it is a bit hidden. It is contained in the SMOOTH function which determines the average usage rate AUSAGE. Because AUSAGE involves the only level in the loop, it is necessary to give it an initial value. In this case, it is easy to use the same initial value as the initial usage rate: 0.32 trillion cubic feet/year.

a. Run the model and examine the results. Label the run Exercise 12(a). What behavior does the usage rate USAGE exhibit? How does it differ from the behavior in Exercise 9(a)?

b. Rerun the model several times, choosing different values for the initial level of undiscovered gas UDGAS. Label the runs Exercise 12(b). How do the results differ?

c. Rerun the model several times, choosing different values for the usage growth rate USAGEG. Label the runs Exercise 12(c). How do the results differ?

d. (Optional) Rerun the model several times, choosing different powers in the function relating EFFEXM and FUDGR. (See Exercise 9(d).) Label the runs Exercise 12(d).

e. (Optional) Examine the model runs in Exercises 12(b) through 12(d). What determines the point at which the discovery rate DISCOV peaks? What determines the point at which the usage rate USAGE peaks? (*Note:* This is a difficult question. It will probably be helpful to try Exercise 9(e) first.)

f. (Optional) While it is plausible to assume the DEMM is inversely related to PRATIO, there is no reason why the relationship should be precisely DEMM = 1/PRATIO. Try running the model using various alternative powers of PRATIO in the formulation. (For example, try DEMM = 1/(PRATIO*PRATIO).)

Another way to express the relationship between DEMM and PRATIO is:

$$\text{DEMM.K} = \text{EXP}(-\text{T}*\text{LOGN}(\text{PRATIO.K}))$$

where T is the power to which PRATIO is to be raised. (See Exercise 9(d) for a similar representation of EFFEXM.) It is possible to show that T represents the elasticity of demand for natural gas with respect to price.

EXPERIMENTS WITH THE MODEL

A number of questions can be asked about the model developed so far. For example, how well does the model behavior match the historical data for the period from 1900 to 1980? Are there adjustments that should be made in the parameter values we have chosen? Does the model behavior seem plausible for the period from 1980 to 2050? Have important factors been left out of the model?

To the extent there is confidence in the model, it can be used to examine the possible consequences of policies that might be proposed to ease the natural gas shortage. For example, what does the model indicate about the effects of imposing a tax on natural gas? What might happen if a new technology made it possible to mine gas which, up until now, had been inaccessible? The sections that follow consider how some of these questions might be addressed.

Exercise 13: The Model Behavior from 1900 to 1980

a. Reread the preceding introduction and compare Figure 23.1 with the model behavior you obtained in Exercise 12(a). How well do you think the model-generated behavior and the historical data match? In what ways is the model behavior similar to the data in Figure 23.1? In what ways is it different? Why do you think these differences occur? How important do you think these differences are? What additional data would you like to have to compare with the model's behavior from 1900 to 1980?

One of the main differences between the model behavior and the historical data concerns the peak in the usage rate. In the model output, the usage rate peaks at a considerably lower value than the discovery rate. But according to the historical data, usage and discovery peak at about the same levels.

One of the main reasons for this discrepancy is that the model does not include a representation of federal regulation of gas prices which began in 1960. Since 1960, the price of natural gas in the United States has been determined by a federal regulatory agency, the Federal Power Commission (FPC). The FPC has generally set the price of gas somewhat lower than the price that would otherwise have occurred.

Although it would require a good deal of additional model structure to represent price regulation in detail, it is possible to formulate a rough approximation of the regulated price without too much difficulty. Recall from Exercise 11 that the price of gas in the model is equal to the product of the delayed total cost DTCOST and a price multiplier PM. The price multiplier PM has a value of 2 when reserves are at the desired level. When reserves are low, PM increases, and when reserves are high, PM decreases.

The main effect of price regulation has been to hold the price of gas closer to the delayed total cost, particularly when reserves have been low. Thus, for example, when reserves have been at the desired level, the FPC has set the price of gas somewhat lower than 2 times the delayed total cost. And when reserves have fallen below the desired level, the FPC has not permitted the price to rise nearly as much as the model's price multiplier would indicate.

b. Figure 23.6 indicates how the price multiplier might be changed to represent the effect of price regulation.

Use the values in Figure 23.6 to add a representation of price regulation to the model. One way to represent the adoption of price regulation in 1960 is to use a DYNAMO CLIP function to switch from an "unregulated price multiplier" before 1960 to a "regulated price multiplier" after 1960. (The equation V = CLIP (A,B,C,D) means V = A if C>D; V = B if C≤D.) (In fact, the actual imposition of price regulation in 1960 occurred more gradually than this formulation indicates. The FPC did not cut the price of gas in a single jump. Instead, it held the price constant at about the 1960 level for several years. Then the FPC began to let prices drift upward to reflect rising costs. How would you include this complication in the model?)

c. Run the model with price regulation and examine the results. Label the run Exercise 13(c). How does the addition of price regulation influence the model behavior? Try using different values for the regulated price multiplier table. How does the model behavior change? What discrepancies remain between the model behavior and the historical data?

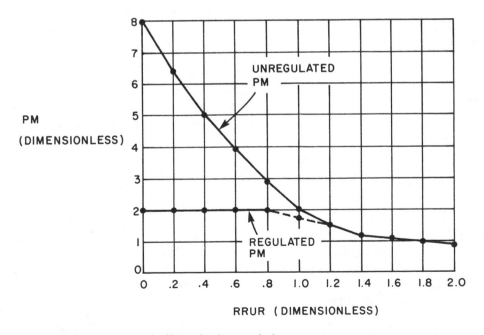

Figure 23.6 Hypothesized effect of price regulation

Exercise 14: Policy Tests

a. The model we have developed can be used to test the effects of many different policies that might be proposed to deal with the shortage of natural gas. But before using the model to examine policies, it is important to consider how much confidence we should place in the model. For example, do you think the model's behavior over the period from 1980 to 2050 seems plausible? Do you think the future course of natural gas usage and discovery from 1980 to 2050 will follow the general outline of the model behavior? Why or why not? What factors have been excluded from the model? How important do you think they are? What additional information would be helpful in improving the model?

b. The following list suggests some policies you might want to examine using the model. In comparing the effects of the policies on the model behavior, there are several things to consider. For example, what effects do the policies have on the natural gas usage and discovery rates in the short run (over the next 5 or 10 years)? What effects do they have on the usage and discovery rates in the long run (from 1980 until 2050)? What effects do the policies have on the price of natural gas, in the short and long run? What effects do they have on the profits earned by gas companies (as measured by the difference between price and total cost)?

   1. One policy you might consider is simply reducing the natural gas usage rate. Suppose a conservation policy is adopted in 1980 that cuts the usage of natural gas in half. What effects might that have? (*Note:* One way to represent a reduction in the usage rate is to multiply the usage rate by a "conservation factor" that has a value of 1 until 1980 and a value of 0.5 after 1980.)
   2. Another approach to limiting the use of natural gas might be to impose a tax. Suppose a 50-percent tax is imposed in 1980. What consequences might that have? (*Note:* A tax would increase the price that users of natural gas pay, but it would not increase the sales revenue received by gas companies.)
   3. Many observers of the natural gas industry have suggested that the natural gas shortage might be reduced by ending federal price regulation. Suppose regulation is ended in 1980. What effects might that have?
   4. Suppose price regulation is continued after 1980, but the regulated price is increased. What effects would different pricing schemes have?
   5. What might happen if a new technology becomes available in 1980 that doubles the efficiency of exploration?
   6. Suppose a new technology becomes available which makes it possible to mine 200 trillion cubic feet of gas from locations that previously had seemed entirely inaccessible. Assume that the efficiency of explo-

ration for this additional gas begins at 2E-9 trillion cubic feet per dollar and drops to 0 as the amount of gas remaining is exhausted. What effects might this have?

c. On the basis of your work with the natural gas model, what policies do you believe should be adopted? Why? Write a short paper discussing your conclusions.

ENDNOTES

1. In books and articles on natural gas and the energy crisis, this is usually called the "reserve-production ratio."

2. This assumes that when new gas is discovered, it is not used until all previously discovered gas has been used up. This is not a completely realistic assumption, but it is approximately valid. The formulation contains one other approximation as well. If you worked Exercise 7(d), you learned that the reserve-usage ratio RUR overestimates how long reserves will last, when usage is growing exponentially. How important do you think these approximations are?

3. A price equal to twice the total cost may seem extremely high, but it is necessary to take into account the time that elapses between an investment in exploration for gas and the sale of the gas that is discovered. One way to do this is to calculate the return on investment. The return on investment is the earnings produced by an investment, in percent per year, much like the interest rate for a savings account. Our model does not include enough detail to calculate the return on investment precisely, but we can work out a rough approximation to see what it means to set price equal to twice total cost.

   Suppose the Sunshine Gas Company invested $100 in exploration for gas in 1900. Since EFFEXP = 67E-9 trillion cubic feet per dollar that would result in 67E-7 trillion cubic feet of gas, and since RUR = 20 years, the gas would be sold in 1920. The total cost of the gas, in dollars per trillion cubic foot, would be DCOST + PCOST = $15 million + (3*$15 million) = $60 million per trillion cubic feet. The price would then be $120 million per trillion cubic feet.

   This would produce a total sale of 67E-7*$120 million = $800. Once the company paid the production cost (3*$15 million*67E-7 = $300), that would leave $500. Thus the initial investment of $100 in 1900 grew to $500 by 1920. This growth represents a compound interest rate of about 8 percent, i.e., the $100 investment earned 8 percent per year for 20 years. Thus the Sunshine Gas Company earned a bit more by investing its $100 in exploration for gas than it would have by putting its $100 in a bank. (The actual return on investment in gas exploration in the early 1900s was probably about 12 to 18 percent. So the model is not quite correct.)

CHAPTER 24

HEROIN ADDICTION AND ITS IMPACT ON THE COMMUNITY*

INTRODUCTION: THE PROBLEM

Heroin addiction is both a personal tragedy for the addict and a social problem for the community in which he lives. The addict is a victim of his drug and of society's attitude toward it and him. But the addict is also a criminal, not merely because the purchase and use of the drug are illegal, but because of the crimes the addict must commit to raise the money to support his habit. This chapter focuses on the relationships between heroin use and the heroin-related crime rate in a community.

The problem addressed here is briefly described in the newspaper article from *The Boston Globe*. The article reports the results of a study of heroin-related crimes in Detroit from 1970 to 1973. The study showed that when police efforts were successful in making large heroin "busts," confiscating large amounts of the drug, the crime rate went up rather than down. Some people, a majority perhaps, believe that police action in trying to control the supply of heroin helps to hold down crime rates, but this study indicates the opposite.

This chapter develops a view of heroin use and drug-related crime from the system dynamics perspective. The first purpose of the modeling effort is to understand how the structure of a heroin-and-crime system produces the kinds of behavior described in the article. The second purpose is to use the model constructed to analyze a few of the policy options available to a community facing a heroin problem.

---

*In addition to the newspaper article on the next pages, this project draws on material contained in Gilbert Levin, Edward B. Roberts, and Gary B. Hirsch, *The Persistent Poppy* (Cambridge, Mass. Ballinger Publishing Co., 1975). The authors deeply appreciate the assistance of George Richardson in the writing of this chapter.

When Heroin Supply Cut,
Crime Rises, Says Report*

Boston Globe, April 22, 1976
by Saul Friedman
Knight News Wire

Washington—The next time you hear of a big drug bust and the seizure of large quantities of heroin, don't go believing that your city's streets are necessarily any safer.

In fact, a new study of heroin traffic in Detroit has concluded that the tighter the heroin market, the more likely you will become a victim of a robbery or a burglary by an addict in need of a fix.

The study, which is to be released in a week by the privately funded Drug Abuse Council, is in the hands of Detroit officials. But it has implications beyond Detroit.

It supplies the first substantial statistical evidence challenging the conventional assumption that law enforcement campaigns to reduce the supplies of heroin will lead to a reduction of crime.

On the contrary, the study found that drug seizures lead only to higher heroin prices. And higher heroin prices result in an increase in "revenue-raising" crimes, such as robbery, burglary and theft.

The Detroit figures showed that crimes directly related to heroin climbed 3 percent, citywide, and much higher in poverty stricken areas—among whites as well as blacks—when the price of heroin rose by 10 percent.

"If the price of a bag of street heroin increased from $7 to $9 in any given month," the Drug Abuse Council said, "the number of revenue-raising crimes which occurred at the rate of about 11,000 a month would increase to almost 12,000 a month because of heroin alone."

Furthermore, "reports of specific offenses increased when heroin prices increased 10 percent. Citywide, the greatest increase seemed to be in three crimes often associated with heroin use in Detroit: Armed and unarmed robberies and burglaries of residences. Reports of armed robberies increased by 6.4 percent when heroin prices increased 10 percent."

Focusing on the crime of residential burglary, the study team of statisticians and economists found that "people living in poor neighborhoods were the most frequent victims" of the addict squeezed by higher heroin prices.

The study was based on an analysis of heroin prices and crime rates in Detroit for a 40-month period from June 1970 through September 1973. Drug prices were obtained from government narcotics agencies and the crime statistics came from the computers of the Detroit Police Department.

The Drug Abuse Council acknowledged that "any community that succeeds in eliminating or virtually eliminating its heroin supply could solve its heroin-related crime problem in fairly short order."

But while no community has been able to do so, many police departments, including the one in Detroit, have made large seizures and have temporarily reduced the heroin supply.

"Although conventional wisdom holds that such efforts lead to a reduction in heroin-related crime," the study said, "some people closer to the heroin scene, such as experienced law enforcement officials, believe the opposite occurs. In their judgment, marginally successful efforts at reducing heroin supplies result in more, not fewer, heroin-related crimes."

The Drug Abuse Council concluded that its study of Detroit heroin traffic and crime "suggests that the conventional wisdom is wrong and the people more familiar with the heroin scene are correct. That is, slight and temporary reductions in the supply and availability of heroin do not produce the reduction in heroin related crime policy-makers and the public alike want. Such efforts produce the increase in these crimes nobody wants."

A community heroin problem is not one problem, but a maze of many problems. A solution will not be found, but the work will form the beginnings of some understandings. Interested students can pursue the study further by reading Levin, Roberts, and Hirsch, *The Persistent Poppy* (1975).

Exercise 1: Problem Definition

a. The newspaper article identifies two dynamic patterns, one in which the heroin-related crime rate rises in response to police action, and one in which it falls. In the article, find evidence of these two patterns, and describe the situation in which each pattern occurs.

On the axes of Figure 24.1, the street heroin supply of a city is graphed as if it were in constant equilibrium for a time and then reduced by a significant seizure of the drug by a police bust.

b. Sketch the way the study claims the heroin-related crime rate would behave over time in response to this seizure. Begin your graphs as if the system were in equilibrium and continue it until the system returns to equilibrium.

c. Continue the graph of the heroin supply after the single bust until it, too, returns to some equilibrium.

d. On the same set of axes, sketch a graph of the price of an average bag of street heroin, showing how price responds to the seizure and how it returns to equilibrium.

e. Suggest some coordinates on the time axis, indicating how long you think it takes for these variables to change as you have shown.

f. On a similar set of axes, sketch a set of graphs illustrating how the article claims the system would behave if the police efforts were to succeed in "virtually eliminating" the heroin supply over a period of time.

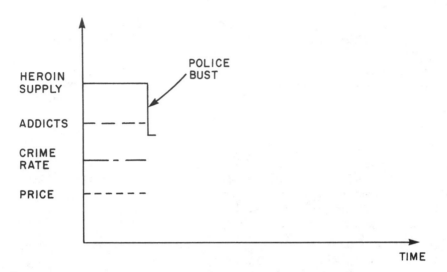

Figure 24.1 Key variables of heroin-and-crime system

g. Graph the heroin supply, the crime rate (crimes per month), the price of an average bag of heroin, the number of addicts in the community, and any other variables in the system you think are interesting or significant.

h. Suggest some coordinates on the time axis indicating how long you think it takes for these variables to behave as you have shown.

The graphs drawn—one set for an isolated bust and the other for persistently successful police efforts—form the reference behavior modes for the model to be developed. Whatever theory is proposed, whatever feedback structure is finally suggested, must be able to generate these two patterns of behavior over time.

AN INITIAL MODEL

Starting with what will become a piece of the final model, focus initially on what is probably the easiest reference mode: the response of the system to a single bust. Because portions of this system are easiest to conceptualize in rate/level terms, some of the model will be developed in quantitative terms without first drawing causal-loop diagrams. Such an approach is often helpful in developing the parts of a model that have a clear level-and-rate structure.

An available supply is like an inventory, an accumulation of inflows and outflows that should probably be modeled as a level. Figure 24.2 shows the level of available heroin "on the street" and its inflow and outflow rates in this system. To discover the inflow and outflow rates, think of where the heroin on the street comes from and where it goes. (It could disappear in two distinctly different ways.)

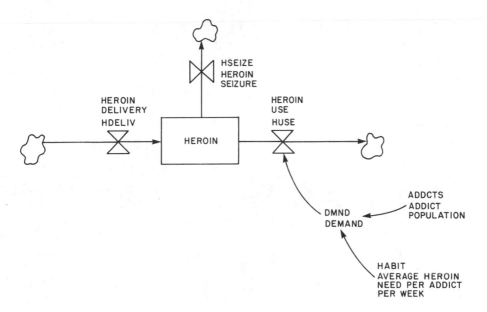

Figure 24.2 Determinants of the heroin use rate, HUSE

The rate at which heroin is used is a function of the "demand" DMND for heroin, as shown in the causal diagram in Figure 24.2. The diagram, which is partway between a causal-loop diagram and a formal flow diagram, implicitly defines what we mean by the word *demand*.

Exercise 2: Formulating the Model

a. Write an auxiliary equation for DMND that corresponds to the assumptions in the causal diagram.

b. Taking "weeks" as the time unit and "bags" as the unit of heroin (one bag is about $10 worth, or a single "hit"), what are the units of DMND, ADDCTS, and HABIT? Show that the units of these variables balance in your formulation for DMND.

In equilibrium, the heroin use rate HUSE should equal the demand DMND. But what if the heroin supply were very low, or perhaps even zero? HUSE would also have to be low or zero, no matter what DMND were. To handle such possible extremes, formulate HUSE as a product of DMND and a heroin availability factor HAF:

R $HUSE.KL = HAF.K * DMND.K$

HUSE   Heroin Use Rate (bags/week)
HAF   Heroin Availability Factor (dimensionless)
DMND   Demand (bags/week)

When HEROIN is zero, HAF will be zero, reducing the usage rate HUSE to zero as well. When the HEROIN supply is sufficient to take care of demand DMND, then HAF will be one, thereby setting the usage rate equal to demand. Formulate HAF as a table function of the supply/demand ratio SDR, defined to be HEROIN.K/DMND.K. Better yet, divide SDR by its normal value NSDR, and set HAF to be a function of SDR.K/NSDR. That way the (1,1) point in the table will represent "normal" conditions in which the supply/demand ratio is at its normal value and heroin usage is the same as heroin demand.

    c. Ignoring for the moment what the value of NSDR is, sketch on the graph (shown as Figure 24.3) a likely table function for HAF, and then write the DYNAMO equations for HAF and its table.

It is worth noting that by including a heroin availability factor of this sort, a negative feedback loop has been added that prevents the heroin level from becoming negative in response to abnormally high demand. (In many system dynamics models, levels that cannot logically go negative are linked to their outflows by negative loops—the outflow goes toward zero as the level goes toward zero.)

    d. The supply/demand ratio and its "normal" value NSDR have sensible meanings that one can discover by computing their units. Write the auxiliary equation for SDR, and then determine its units from the equation. Then think about what such a number would represent in the real system, and write your best guess.

The units of HEROIN/DMND are "weeks." Thus the supply/demand ratio SDR and its normal value NSDR represent lengths of time. A bit of

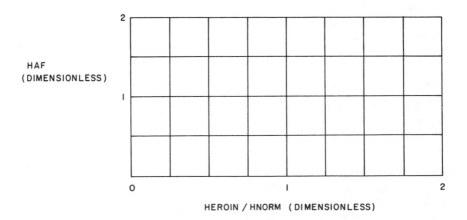

Figure 24.3 Heroin availability effect on use

thought shows that SDR is the length of time that the current supply of heroin would last if used up at the rate DMND. That is, if HEROIN = 20 units, and DMND = 5 units/week, then HEROIN/DMND = 4 weeks; the heroin supply would be used up at that rate in four weeks. That number of weeks is the "coverage" provided by that supply of heroin.

To pick a value for NSDR, imagine how many "weeks" of heroin supply is normally in the hands of street suppliers. Estimate 4 weeks, in which case set

C   NSDR = 4

Here the value for NSDR is being approximated. The article contains no such information. Perhaps 2 weeks is a better guess. It depends to some extent on what is meant by "street heroin," who has it, where it is stored, how close to the users it is, and so on. Think of NSRD as the average time between the delivery of a unit of heroin to the community area and the use of that unit by an addict. In any case, NSDR can be altered in the simulations to see if it has a significant effect on the dynamics of the system and to see if research needs to be done to get a more accurate figure.

At this point, the model has the structure shown in the flow diagram in Figure 24.4. The heroin use rate HUSE is determined.

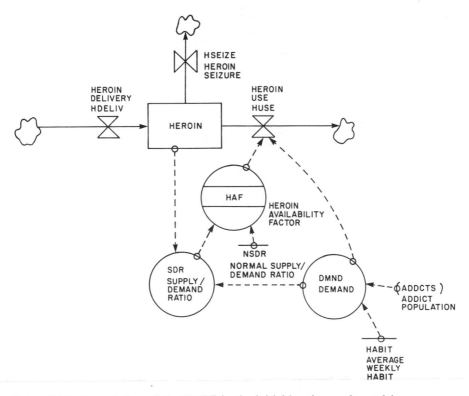

Figure 24.4 Formulation of the HUSE in the initial heroin supply model

Now formulate the heroin delivery rate HDELIV, which brings supplies of heroin into the community from outside. A network of suppliers exists that appears to be organized and efficient. Make a very simple formulation for the heroin delivery rate HDELIV so that the behavior of the simple structure can be seen in response to various police actions. Suppose that without police intervention the system is in equilibrium—the inflow to the heroin level equals the outflow. Assuming the supply of heroin is adequate to cover demand, the amount of heroin used per week equals the number of addicts multiplied by the average heroin need per addict per week,

ADDCTS*HABIT,

so formulate a constant heroin delivery rate:

R   HDELIV.KL = ADDCTS*HABIT

The DYNAMO timescript KL identifies HDELIV as a rate and suggests that it is varying over time, but here it is equal to a product of constants, so it, too, will be constant in the simulations.

To simulate the police busts described in the newspaper article, include an exogenous heroin seizure rate HSEIZE. ("Exogenous" means "determined outside the system boundary.") The following DYNAMO equation allows the testing of three kinds of police actions:

R   HSEIZE.KL = TEST1*PULSE(FSEIZE*HEROIN.K/DT,
X   TBEGIN, TBETWN) + TEST2*STEP(AMT,TBEGIN)
X   + TEST3*STEP(FSEIZE*HEROIN.K,TBEGIN)

A DYNAMO PULSE function is used to represent a single bust at TIME = TBEGIN, or a sequence of evenly spaced repeated busts separated by a time period of TBETWN. (See prior uses of PULSE in Chapter 17.) A STEP represents a sudden, one-time change in the amount of heroin the police are able to seize each week. Two types of STEP changes can be looked at with the above equation: TEST2 represents the initiation of continuous police action, which is able to seize a constant amount of heroin per week, while TEST3 represents the initiation of police action, which is able to seize a constant fraction of the available heroin supply per week. The constants TEST1, TEST2, and TEST3 are set equal to zero initially, using constant equations, and then changed in turn to 1 in DYNAMO's rerun mode to activate each test of simulated police action.

e. Write the complete set of DYNAMO equations for the simple, one-level heroin model we have constructed thus far. For your initial runs, use the following values for constants in the model:

ADDCTS   1000 people
NSDR   4 weeks

HABIT   20 bags/addict/week
FSEIZE   .05
TBEGIN   10 weeks
TBETWN   1000 weeks
AMT   1000 bags/week

DT   .5 weeks
LENGTH   80 weeks
PLTPER   2 weeks

In addition, set the constants TEST1, TEST2, and TEST3 equal to zero. For the initial value of the heroin level HEROIN, write an N equation that computes (from other quantities named in the model) the quantity of heroin necessary to place the model in equilibrium, assuming no police action yet.

Decide for yourself what variables you would like to see plotted, and write an appropriate PLOT statement.

Exercise 3: Running the Model

a. Now type your equations into the computer and produce an equilibrium run. (It should be dull—80 simulated weeks of no change. You may wish to change the LENGTH parameter to 20 weeks for the equilibrium run, in order to keep yourself from falling asleep while the machine is printing it out.) Label your equilibrium run Exercise 3(a).

If your model outputs are not in equilibrium, rethink the N equation you wrote for the HEROIN level and try again. Do not proceed with the other exercises until the model runs in equilibrium.

Once the model runs in equilibrium, perform three simulations in the re-run mode.

b. First, change the constant TEST1 to 1, and run the model. Label the run Exercise 3(b).

c. Then change the constant TEST2 to 1, and run the model again. Label the run Exercise 3(c).

d. Finally, change TEST3 in a similar fashion. Label the run Exercise 3(d).

Perhaps the most important ability in applying system dynamics to real problems is the ability to analyze and describe in real terms the behavior of a complex feedback system, using understandings gained from modeling the system and from simulation runs of the model. Therefore, consider the three types of police action you simulated. Note how the simple model behaved.

e. Now write a paragraph for each simulation describing in terms of the real system *how* the system behaves in response to the police action, and *why* it behaves that way. Then describe in a fourth paragraph the ways in which the model constructed so far is unrealistic. (At this point, the causal analysis may seem too simple to be worth four paragraphs, but the analytical skills you are trying to build here will be very useful as problems become more complex.)

f. (Optional) Simulate single, repeated police busts by running the model with TEST1 set equal to 1 and TBETWN (time between pulses) set equal to 20 weeks. Label the run Exercise 3(f). Explain the behavior as much as possible in terms of the real system we are simulating.

g. (Optional) Perform one or more of the tests of police action for different values of the constants in the model. You might want to vary NSDR and HABIT, as well as the constants describing the timing and extent of the police action, TBEGIN, TBETWN, FSEIZE, and AMT. Label the runs Exercise 3(g).

h. (Optional) Does the system always come back into the same equilibrium it started from? You may wish to try some simulations over longer time frames (change LENGTH and PLTPER accordingly). Label the runs Exercise 3(h). Explain the behavior of the model over longer time frames and compare it to how you feel the real system would behave under similar conditions.

A SIMPLE HEROIN AND CRIME MODEL

With the understandings gained from simulating the simple structure in Exercise 3, a more realistic heroin delivery rate and some feedback structure for the heroin-related crime rate can be added to the model. Both additions require adding the price of heroin to the model because price affects both the heroin supply and the need to commit crimes to support a drug addiction.

Assume that suppliers of heroin, wanting to make a profit, are sensitive to the street price obtainable for a unit, e.g., a bag of heroin. Thus the heroin delivery rate must be changed to depend on suppliers' perceptions of the current price of street heroin. First, the following exercises add price to the model, and then consider how to formulate its effects on supply.

Exercise 4: Formulating the More Complete Model

a. On what should price depend? For simplicity, convert Figure 24.4 to a causal-loop diagram showing the same variables and add price to the diagram. Add causal links to represent what you believe are the causal connections of price to the rest of the variables in the diagram. (Ignore for now its potential effects on the addict population.) Sign the arrows in the diagram.

The street price of heroin should be a function of the heroin supply and demand. If supply is low in comparison to demand, the price should be high, because pushers know that addicts will find it hard to buy heroin elsewhere, so they can raise their price without losing customers. To a certain degree, the reverse is also true. If supply is high in comparison to demand, the price may come down a bit, as addicts find it easier to locate several sources for the drug.

Thus it makes sense to set price in the model to be a function of SDR, the supply/demand ratio. SDR was shown to be a meaningful and useful concept in the system (recall that it has units of "weeks"), and it behaves nicely as a measure of the balance of supply and demand. When it is large, supply is in excess of demand; when the ratio is small, supply is inadequate to meet demand. Price can be set equal to some function of SDR. But now the question is: Should price be modeled as a level or as an auxiliary? If price responds rapidly to changes in the supply/demand situation, then it would be reasonable to model price as an auxiliary. If, on the other hand, price tends to be "sticky," responding more slowly to changes in SDR, then it would be more reasonable to represent price as a level.

Two considerations suggest that at this point the price should be computed in an auxiliary equation. First, street price probably does respond quite rapidly to changes in supplies available; sellers and pushers are directly connected with their markets. The time constant associated with the heroin level is NSDR = 4 weeks. Price probably reacts much faster than that, perhaps on the order of a few days. Second, the formulation as an auxiliary is simpler; the structure can be made more complicated later if it is decided that price really is a bit "sticky" or if price as a level might significantly alter the behavior of the system.

Therefore, let us write an auxiliary equation for price,

A   PRICE.K = PNORM*ESDRP.K

in the form where

PRICE = Heroin Price (dollars/bag)
PNORM = Normal Price (dollars/bag)
ESDRP = Effect of Supply/Demand Ratio on Price (dimensionless)

b. Sketch a table function for ESDRP, the effect of the supply/demand ratio on price. Let the input to the table (its horizontal axis) be SDR/NSDR, as a measure of the supply/demand ratio. (Note again that this is a "nice" fraction because it is clear what it means for the ratio to be zero (no heroin) and for the ratio to be one (HEROIN/DMND = NSDR, so the situation is "normal"). Dividing by NSDR is called "normalizing" the supply/demand ratio—a very useful practice in model formulation.) Plot two points precisely on the table, determine the slope and shape of the function and sketch a smooth curve meeting those requirements. Does price respond to low values of SDR roughly symmetrically to the way it responds to high values?

The equation for PRICE has been developed in the model so that the effect of heroin price on the heroin delivery rate HDELIV can be included. A high price will increase the delivery rate as suppliers try to bring more heroin into the community to make more money. A low price will probably decrease the delivery rate as suppliers find it more profitable to ship their heroin supplies to other markets (while they work in the community to entice new addicts in order to increase demand). Thus formulate the HDELIV rate as follows:

R   HDELIV.KL = (   )*EPHD.K

where EPHD represents the effect of price on heroin delivery. Ignoring for the moment what EPHD actually looks like, consider what should be put into the blank space in the equation.

In the initial model, HDELIV.KL = ADDCTS*HABIT is a constant. Here, being more realistic, the suppliers do not really know the number of addicts in the city, nor do they know the average habit of the addict population. What they are aware of is some rough estimate of the amount of heroin being consumed in the city over time.

c. With this observation in mind, suggest a variable that should go into the blank space above. As a check on the reasonableness of your suggestion, see if it has the correct units and if it has the proper value when the entire system is in equilibrium.

Assume that the nation's illegal heroin market machinery would act to keep the inflow of heroin to this community roughly equal to the outflow, with some modification (EPHD) representing the effect of price or profitability. Thus set HDELIV (roughly) equal to HUSE, the rate at which heroin is being used in the city, and then modify this by a price effect. However, no one in the real system actually knows HUSE because it is the current (instantaneous) rate at which heroin is being bought and taken by addicts. The most that could be perceived by people acting in the real system is some average of this usage rate. Therefore, base HDELIV on an average (or "smoothed") use rate AVUSE.

d. Write equations for HDELIV and AVUSE, including an effect of price on heroin delivery EPHD, which we will formulate as a table function.
    To complete this sector of the model, we must write an equation for EPHD, the effect of price on heroin delivery. Writing EPHD as a table function of PRICE.K/PNORM seems like a good idea, now that we have the notion of a reference or normal price PNORM. However, the variable called PRICE is the current street price of heroin, subject to potentially rapid change in response to changes in the heroin supply/demand ratio. The suppliers of heroin in the city cannot perceive that (instantaneous) price. As with the average use rate, the price that suppliers can perceive is

a longer-term trend or average. So we will first construct an equation for perceived price PPRICE as an average or smoothing of PRICE, and then formulate the effect of price on heroin delivery EPHD as a table function of PPRICE.K/PNORM.

e. Write an equation for PPRICE as an average of PRICE, using a smoothing time constant TPP (time to perceive price) of 4 weeks.

f. Now formulate EPHD as a table function of PPRICE.K/PNORM. Consider the slope, the shape, and at least two points in the table. Then draw an appropriate graph of EPHD and write the DYNAMO equations for it.

The supply side of the model is now complete. The model developed thus far is summarized at the end of the next short section.

It now remains to link the heroin system to the crime rate CRIMES, defined as heroin-related, revenue-raising crimes committed per week by the addict population. What governs how often an addict must commit a crime to raise the money needed to support his habit?

g. With just two additional constants (revenue per crime RPC and fraction of habit supported by crime FHSC), you should be able to write DYNAMO equations for the following auxiliaries:

MNEED	Money Needed per Addict per Week ($ per addict per week)
FREQPA	Frequency of Crimes per Addict (crimes per addict per week)
CRIMES	Crime Rate (crimes per week)

Check to see that the units of the variables in your equations make sense. If the units do not make sense or if both sides of an equation do not balance, try again, using the units as a guideline for correct formulations.

The structure now developed is shown in the flow diagram in Figure 24.5. The figure shows at this point that the addict population ADDCTS is still being considered as a constant—hardly a realistic view in the long term, but acceptable for the short term. Addicts do not stop being addicts, even for a short time, just because the price of their habit skyrockets. However, the addict population does change through the addiction of new people, or the departure of old addicts who leave the community through arrest, migration, or (a few) rehabilitation. None of those is a short-term process, so, for the moment, leave addicts as a constant, reserving until the third version of the model the more realistic structure representing the addict population as a level.

h. List the entire set of equations in the Heroin and Crime model, putting together the equations from Exercise 2 with those developed here in Exer-

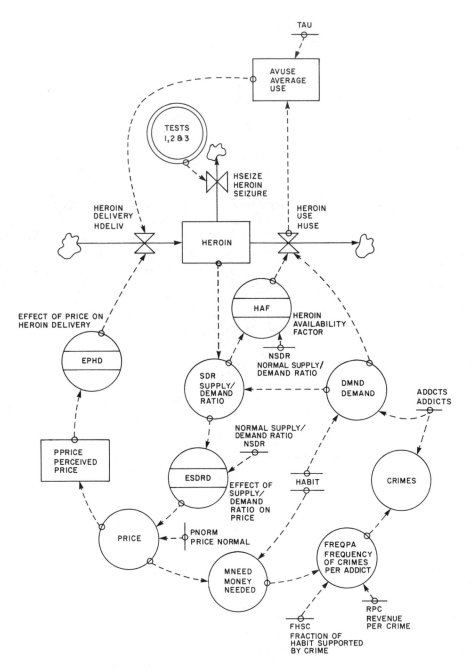

Figure 24.5 Flow diagram of the simple heroin and crime model

cise 4. Include constants and whatever NOTE statements you want for clarity. If you write out the "smoothing" or averaging equations for AVUSE and PPRICE rather than using DYNAMO's SMOOTH function, be sure to select initial values that put the model in equilibrium initially without disturbances from police action. Because of the way we "normalized" the table functions in the model, it should be in equilibrium almost automatically, given the equilibrium condition you developed for the structure developed in Exercise 2.

Exercise 5: Running the More Complete Model

a. Mentally trace through the feedback structure we have created, imagining its response to a single seizure of heroin (the PULSE in the first model). Write a description of what you imagine its response to be. Will the model we have constructed so far produce the first reference mode you drew? (See Exercises 1(b), 1(c), and 1(d).)

b. Perform the same sort of mental simulation, imagining a sustained police effort which is successful in permanently holding down the heroin supply (a STEP in the first model). How will the model we have constructed behave? Will it produce the second reference mode you drew? (See Exercise 1(f).)

c. Now submit the entire model to the computer in DYNAMO, edit it to remove any obvious typing errors, and produce a base run of the model in equilibrium. Once again, you may wish to change LENGTH to 20 weeks to avoid having to watch a long equilibrium run. If your model is not in equilibrium, try to find out why by checking initial values, constants, and table functions. Be sure, for example, that your tables are positioned at their "(1,1)" points for the initial values. Do not proceed until your model produces an equilibrium run. Label the run Exercise 5(c).

d. Once the model runs in equilibrium with no police action, perform the three simulations in rerun mode that we performed with the first model. Change, in separate runs, the constants TEST1, TEST2, and TEST3 to 1. Label the runs Exercise 5(d)(1), (2), and (3).

e. Describe the resulting behavior in each test, and, most importantly, explain how it follows from the structure we have assumed. Explain the behavior of the model as much as possible in terms of the real system, not in terms of model variables, table functions, or technical terms. Is the behavior realistic?

f. (Optional) Simulate a series of individual police busts spaced 20 weeks apart. Label the run Exercise 5(f). Explain the resulting behavior.

g. (Optional) Perform Tests 1, 2, and/or 3 with the table for the effect of price on heroin delivery EPTAB set equal to a string of 1s to simulate no feedback from price to heroin delivery. Label the run Exercise 5(g). Why does the model behave as it does? Which is more realistic, the runs with EPHD active, or the runs with EPHD deactivated (set identically to 1)? Why?

h. (Optional) Perform one or more runs with a steeper table function for PRICE. Try a less steep table. Label the runs Exercise 5(h). Can you predict the results before you see the computer output?

i. (Optional) What is the effect of changing revenue per crime RPC? Label the run Exercise 5(i). Can you predict the results before you see the computer output?

j. (Optional) Vary the averaging time for AVUSE or PPRICE. Label the runs Exercise 5(j). You can eliminate the averaging or delaying effect completely by setting the averaging time equal to the value of DT, for example. If you set TAU (time to average heroin use) equal to a very large number, e.g., 1E9 (10^9), you should see similarities to the runs of the first model. Why? What values of these time constants do you think are most realistic?

HEROIN AND CRIME IN THE LONG RUN

While in the short run it may be reasonable to view a city's heroin addict population as constant, a more realistic long-term view of a community's heroin and crime problem must consider the addict population a variable, not a constant. To do justice to the influences on heroin addiction, migration, and community responses to a heroin problem, a far more ambitious effort is needed than seems appropriate here. However, in a simple way the general feedback dynamics of an addict population can be captured with the structure diagrammed in Figure 24.6.

Figure 24.6 assumes that the addict population is influenced by the long-term trend or average supply of heroin. Plentiful supplies, relative to perceptions of supplies in other locales, lead to a growing addict population. Tight heroin supplies over the long term eventually lead to a reduction in the addict population. The structure assumes that price is not a factor directly, except as it follows the dynamics of long-term supply. The structure also ignores community programs that might remove addicts from the addict population, placing them in prisons or drug rehabilitation programs. The net addiction rate NETAR proposed in the diagram is a grouping together, or aggregation, of:

> Immigration resulting from perceptions that this city is a good place in which to be a heroin addict;
>
> Outmigration resulting from tight heroin supplies and police activity;

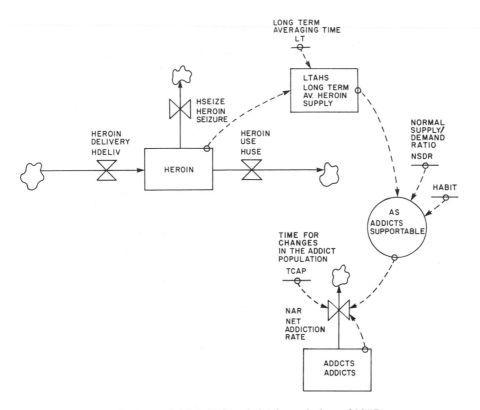

Figure 24.6 Flow diagram of ADDCTS and the formulation of NAR

Deaths, whether drug-related or from natural causes;

New addicts, usually increasing in number in times of plentiful heroin supply when pushers expand their markets;

Loss of addiction—rehabilitation—however difficult or rare.

The long-term average heroin supply LTAHS shown in Figure 24.6 is simply a smoothing of the heroin level in the previous models. Its smoothing time constant LT is likely to be much longer, however, than the averaging times assumed for AVUSE or PPRICE. One year, 52 weeks, is a first guess for the response time for sensing changes in the heroin supply situation. Similarly, the time to change the addict population TCAP is long, probably on the order of at least half a year (26 weeks). With these comments and Figure 24.6, write equations for this structural addition to the second model. The only subtlety is the equation for the addicts supportable on the long-term average heroin supply LTAHS, and that auxiliary equation is strongly suggested by the units of the quantities that combine in Figure 24.6 to compute it.

Exercise 6: Formulating the Long-Term Model

a. Write equations for the following variables shown in Figure 24.6: ADDICTS, NAR, ADDSUP, and LTAHS.

b. Now add to these equations the constants TCAP and LT, and add them all to the model developed in Exercise 4. To be sure you have an overview of the feedback structure of the resulting model, draw a complete rate-level-auxiliary diagram of the whole model.

Exercise 7: Running the Long-Term Model

a. Before running this expanded model, predict how it will respond to the three types of tests we have been using: a one-time seizure of 5 percent of the heroin supply, and the two different forms of permanent increase in the amount of heroin seized each week. Write your predictions and justify them briefly by pointing to feedback structure in the model that you believe will be responsible for the behavior you anticipate.

b. Now type this complete model into the computer, edit out the typing errors, and produce an equilibrium run for an addict population of 1000. Label the run Exercise 7(b). If your model does not run in equilibrium, follow the suggestions made previously for uncovering the problem(s). Again, do not proceed until your model runs in equilibrium.

c. Once the model runs in equilibrium with no police action, perform the three simulations of police activity by changing TEST1, TEST2, and TEST3 to 1 in separate runs. Label the runs Exercise 7(c)(1), (2), and (3).

d. Describe the resulting behavior in each test, and explain how it follows from the structure we have assumed. Does the model replicate the reference modes of the study? (See Exercises 1(b), (c), (d), and (f).) If your predictions in Exercise 7(a) were not borne out, pay special attention to explaining why the model behaved differently. At this point, you may not be sure quite why it behaved the way it did. It is perfectly acceptable to say so.

Exercise 8: Experiments with the Heroin and Crime Model

a. Simulate a sequence of police seizures of 5 percent of the heroin supply spaced 20 weeks apart. Label the run Exercise 8(a). Explain the resulting behavior in terms of the real system, using the structure of the model as a guide. Do you think the behavior is realistic?

b. Simulate TEST3 over a long time, e.g., 5 years (LENGTH $= 260$ weeks). Label the run Exercise 8(b). Describe what you think the community response to such a program of police action would be. Would it be the same in the first year as in the later years? What kinds of feedback (which we have not included) would operate in the real system, connecting police activity to community pressures?

c. Eliminate the feedback from perceived price PPRICE to the heroin delivery rate HDELIV by changing the table for EPHD to a string of 1s. Perform one or more of the tests of police action. Label the runs Exercise 8(c). In what ways is the behavior different from the runs with EPHD active? Why?

d. Weaken the effect EPHD by making the values in the table PRCTAB closer to 1, while keeping the same basic slope and shape as the original table. Perform the same runs as in Exercise 8(c) (label these new runs Exercise 8(d)) and compare the original runs, those in Exercise 8(c), and these. What do you think the real effect of price on the heroin delivery rate is? Why?

e. Explore the effects of the delay times TCAP and LT. Perform one or more of the basic tests of police action with TCAP twice its original value, then half its original value; then do the same for LT. Label the runs Exercise 8(e). What if both are different from their original values? Would you conclude from these runs that we should get more certain values of these constants from analysis of some sort of real data? Explain.

f. If the model in some of these runs did not behave as you predicted, and you still question why, try running the model deactivating some feedback loop that you think might be important (as in above). Label the run Exercise 8(f). You can test a wide variety of hypotheses this way and reach a firm understanding of the causes of model, as well as real system, behavior.

g. To pull together your work on the heroin and crime project, write one or more paragraphs which respond to the points made in the original newspaper article. Discuss the differences between short-term and long-term behavior, and explain why those differences exist in this system.

h. (Optional) Reformulate the action of the police in the final model to make police action *endogenously* determined (determined within the system, not from outside). One approach to this would be to write HSEIZE as a product of the heroin supply, some normal fraction seized per week (e.g., 5 percent as we have now), and an effect of pressure from the crime rate EPCR. EPCR could be a table function of the perceived crime rate, which could be formulated like perceived price as an average of the current crime rate CRIMES. Obtain an equilibrium run with FSEIZE equal to zero, and then "step" FSEIZE up to .05 to activate the new structure. Explore the resulting behavior of the model as previous exercises suggest.

i. (Optional) Add another rate affecting the addict population that represents (depending upon your interests) the flow of addicts into prisons (an arrest rate) or the flow of addicts into rehabilitation programs. Add a level to the model representing addicts in prison or addicts in rehabilitation (whichever is appropriate for the flow you chose). Provide two outflows from this new level: an outflow out of the system (some fraction of those in this new level) and a recidivism rate back into the addict population. Formulate these various rates as fractions according to the feedback effects you feel would operate in the real system, as we suggested previously for the effect of the perceived crime rate on police activity. In runs of this extended model, explore the potential effects of such community programs responding to a perceived heroin problem. What can you conclude about policy recommendations for a community with a heroin problem?

CHAPTER 25

BEYOND AN INTRODUCTION TO SIMULATION

The preceding chapters have provided an introduction to computer simulation using the system dynamics approach. This final chapter summarizes what has been accomplished and suggests additional literature beyond this introduction. The broad purpose of this text is to present a problem-solving technique that employs computer simulation and systems thinking as embodied in the system dynamics approach. The first section of the book introduces a problem-solving approach useful in day-to-day thinking without resort to computer simulation. The second section teaches the skills needed to implement insights gained through feedback thinking using computer simulation models.

SYSTEM DYNAMICS AS AN AID IN DAY-TO-DAY THINKING

The first four parts of this text argue that feedback thinking is an important way of looking at and making decisions concerning social, economic, environmental, managerial, and political problems. In essence, feedback thinking consists of three concept areas, each of which contributes to the system dynamics view of solving problems. These are:

1. *Thinking in terms of feedback loops.* This draws attention to the fundamental causes of the problem under investigation. Thinking in terms of feedback loops helps to clarify the problem's boundary; that is, what elements to include or exclude from the problem study. An intuitively appealing tool for helping to think in terms of feedback loops is causal-loop diagramming as taught in Part Two.

2. *Thinking over time.* This concept implies that feedback system problems are problems that show varying patterns of behavior over time. An early task involved in problem definition is sketching the behavior, over time, of

several important variables. This activity helps to define the problem's time frame, as well as provide a next stage analysis of the important variables for a given problem. Which variables to include or exclude initially are also determined by consideration of the perspective taken of the problem, its time frame, and potential policy choices. These skills are taught in Parts III and IV.

3. *Exploring how system structure, especially feedback loops and time delays, cause changes over time.* Working with computer simulation models in Parts V, VI, and VII shows how a wide variety of observed behaviors (exponential growth, exponential decay, oscillations, "s-shaped" growth) can all arise from precisely specified feedback-loop systems. However, merely taking the time to think in feedback terms can reveal that certain systems will probably be stable (they contain dominant negative feedback loops) or unstable (they contain positive growth loops). The student is urged to explore how structure generates behavior over time. This exploration can be either intuitive or aided by computer simulation.

Feedback thinking as a problem-solving tool is a complete skill that has been taught in the earlier chapters of this book. Using just the tools and insights gained from earlier chapters, problems arising in such diverse fields as history, economics, political science, biology, or environmental science can be viewed in terms of feedback dynamics.

Exercise 1: Feedback Thinking and Personal Habits

Most of us have one or more personal habits that we would like to see changed. For example, we might want to study harder, be more prompt for appointments, eat more regularly (or less), jog or exercise more frequently, or perhaps stop smoking. Often personal habits are difficult to change because we engage in a subtle form of self-reinforcing behaviors. Such reinforcing habits are often generated by feedback loops.

Describe, in feedback terms, a personal habit that you might like to change. Can you discover one or more feedback loops that might be reinforcing your bad habit? Does your feedback analysis give you any guidance for changing your behavior?

Exercise 2: Feedback Thinking and Current Events

The nation's economy, social programs, foreign trade, and foreign policy are all areas where feedback systems are probably at work. However, accounts of such events rarely discuss social, economic, and political problems in explicit feedback terms. You often have to "read between the lines" to infer where and how important feedback effects may be at work.

Find an article in the newspaper or a weekly news magazine that describes a social or economic problem that you believe contains important feedback effects. Explicitly describe the problem in question as a feedback problem, using causal-loop diagrams and sketches of several variables against time if possible. Include a copy of your reference article with the causal-loop diagrams. To what degree does the article fully describe a feedback system? What relationships are not explicitly stated in the article but inferred by your analysis?

SYSTEM DYNAMICS AS A COMPUTER MODELING TECHNIQUE

The final three parts of this text move beyond thinking about problems in feedback terms to introduce the use of the computer as an aid in simulating feedback problems. Examples from many different disciplines are used to demonstrate the applicability of this method to any feedback-related problem. Beyond the introductory materials of this book is a large and continuously building body of system dynamics literature reporting both research studies and practical applications. This section reviews some of this literature that is available in book form and suggests related student projects and advanced studies. Each book contains complete model descriptions, computer listings, and simulation results.

Urban Modeling

Urban Dynamics (Forrester, 1969) was one of the first major system dynamics works applied to problems outside of corporate management. The model focuses on the dynamics of growth, stagnation, and decline of a city. Principal considerations include business development, construction, employment, and migration. Policy simulations are aimed at recovery of the city, but show failure of most conventional proposals. The model is theoretical and is not based on any actual city's situation.

Organized as a textbook, with practical exercises at the end of each chapter, *Introduction to Urban Dynamics* (Alfeld and Graham, 1976) explains the interactions behind the growth of a city through the use of simulation models. Sequential evolution of ten urban models presents the assumptions, structure, behavior, and utility of models for urban policy analysis. These models are based on Forrester's *Urban Dynamics,* but the ideas are presented in a simpler manner.

Readings in Urban Dynamics: Volume I (Mass, 1974) and *Readings in Urban Dynamics: Volume II* (Schroeder, Sweeney, and Alfeld, 1975) treat the debate generated by Forrester's book. Additional model analysis and extensions are presented for the serious student of urban economic development and policy.

Exercise 3: Extensions of the Urban Model

Return to the model of urban growth that you constructed in Chapter 20. The usefulness of that model is limited by several simplifying assumptions. First, age structure of the population is ignored although changes in age structure have been observed to affect birth and death rates and the number in the labor force. Population must be divided into the various cohorts for understanding the changes in the labor force arising out of the changes in age structure.

Second, all jobs are assumed to be similar, permitting people to move freely from one job to another. In reality, most workers are trained to perform specific duties. A business executive cannot do construction work, and vice versa. Various categories of jobs and workers must therefore be explicitly represented in the model for more accurately portraying the real world.

Third, the city is treated as one homogeneous neighborhood. Most cities consist of many neighborhoods offering varying standards of living. A more detailed model of a city might incorporate submodels of various neighborhoods and the mechanisms of interaction among them.

Add to the sophistication of the urban model by including one of the classes of disaggregation just discussed. (For guidance in this project, you may wish to consult one or more of the references just described.)

World Issues

World Dynamics (Forrester, 1973), the first attempt to understand and model world problems from a dynamic feedback perspective, illustrates the quantification of the interrelationships among population, food, production, capital investment, natural resources, pollution, and quality of life. On an equation-by-equation basis, the book describes the less than fifty equations model, as well as discusses the general results obtained from the model runs. The model is simple enough for students who have completed this text to run and modify.

The Limits to Growth (Meadows et al., 1972) is a nontechnical presentation of problems associated with growth in world population and industrial output. The model upon which the book's discussion is based is similar to Forrester's *World Dynamics* model, but larger because of considerably more detail. This book can be read without any knowledge of system dynamics, and can be used effectively as a basis for provocative thinking about possible futures and economic and population policies.

Toward Global Equilibrium (Meadows and Meadows, 1973) is a collection of studies that identifies and analyzes specific issues connected with *The Limits to Growth* project. The book contains several small system dynamics models that can easily be entered into a computer and run with a DYNAMO compiler. Included are such topics as DDT and mercury impacts on the environment, natural resource dynamics, and population control in a primitive society.

The detailed technical report, including full model and supporting analysis, of *The Limits to Growth* book, is published in *Dynamics of Growth in a Finite World* (Meadows et al., 1974). The serious student of world modeling will want to study this book, as well as *Models of Doom* (Cole et al., 1973), a critical appraisal of the Meadows world model.

Exercise 4: The Boyd Critique

In an article in *Science* (Vol. 177, August 11, 1972), Professor Boyd presents one of the more interesting critiques of the *World Dynamics* model. Boyd argues that Forrester's assumptions represent a Malthusian viewpoint, in that an equilibrium between high death and birth rates will be reached due to the interaction of an exponential increase of population and capital with the finite resources of the world. He argues that Forrester's basic results could be deduced from his assumptions without the use of a computer simulation, and are, in fact, "intuitive" results. The opposing viewpoint is that taken by a "technological-optimist" who would argue that the increasing investment in technology will result in a limitless flow of goods, produced in nonpolluting facilities, which will allow the world system to reach an equilibrium with comfortably low birth and death rates.

Boyd adds an additional level variable called technology, *T,* to the *World Dynamics* equations. Technology, through its interactions with the original five levels, is modeled to represent the technological-optimist's viewpoint. Boyd's modified program is simulated, verifying the mentally predicted results of the hypothetical technologist-optimist.

Boyd argues that the use of a system dynamics model has failed to resolve the technological optimist—Malthusian controversy—and therefore is not useful as a policy tool.

The object of this modeling exercise is to formulate your own views concerning the technological optimist—Malthusian issue.

a. As a starting point, refer to Boyd's article and Forrester's *World Dynamics* book. Add the technology level per Boyd's equations and verify his results via computer simulation.

b. Next it would be useful to change the assumptions concerning technology such that Forrester's original results are obtained. This exploration will give you some indication as to the implicit inclusion of technology in the Forrester model and to the sensitivity of Boyd's assumptions to subsequent model behavior.

c. Formulate your own set of assumptions concerning the importance of technology. Make sure your assumptions do not violate physical laws. For example, it would appear that the Second Law of Thermodynamics would preclude PFTM going to zero. Also you might want to address the question of whether or not complete recycling is possible as indicated by Boyd's NRTM assumption. From a modeling point of view, is Boyd's method of incorporating technology into the model a sound one? If not, how would you change it?

d. Simulate the world system with your assumptions and evaluate your results. Compare your results to Forrester's and Boyd's, and assess your Malthusian ratio.

e. Could your results have been obtained without computer simulation?

Social and Human Service Issues

Since 1967, extensive system dynamics work has been performed in the areas of planning, organizing and delivery of health care. Several models of the factors affecting the delivery of human services including dental, mental health, and education are presented in *The Dynamics of Human Service Delivery* (Levin and Roberts, et al., 1976). The education model of the factors affecting student performance in the classroom can easily be entered and simulated using DYNAMO.

A number of specific social issues have received detailed modeling attention. *The Persistent Poppy: A Computer-Aided Search for Heroin Policy* (Levin, Roberts, and Hirsch, 1975) describes the application of system dynamics to the development of government policies on heroin addiction. Sections of the model presented in the book depict the growth of heroin addict populations, the effects of police enforcement, and the impact of each on the community, in much greater depth than shown in Chapter 24 of this text.

Exercise 5: Analyzing Model Boundary—The Flu Model Revisited

Reconsider the model of the flu constructed in Chapter 20. Imagine that a version of that model is constructed for the Commissioner of Health in a large midwestern state. The commissioner is interested in policies that might check or dampen the spread of flu across the state. Specifically, the commissioner is interested in investigating the effects of two policies: (1) selective closing of schools once the rate of sickness reaches some predetermined level within that district; and (2) selective immunization of persons within certain critical age brackets (such as persons aged 65 years and older).

How would the boundary of the model presented in Chapter 20 need to be changed to meet the policy needs of the commissioner? Describe the configuration of rates and levels that would need to be included within the model. Discuss any problems that you might foresee in formulating the rate equations for such a model.

Economics

System Simulation for Regional Analysis (Hamilton et al., 1969) is the first system dynamics application to a field outside of industrial management. This study uses simulation modeling as a means of synthesizing economic, environmental, and natural resources research on the Susquehanna River Basin. The study was funded by the Susquehanna River Basin Utility Group to determine the needs and economic consequences of additional dam construction.

Nathan Forrester's book, *The Life Cycle of Economic Development* (1973) focuses on Canada, although it is a generic study of the transition of an industrial economy from growth to equilibrium. The model concentrates on the distribution of factors and demands between different sectors of an economy.

Economic Cycles: An Analysis of Underlying Causes (Mass, 1975) reports on the early work on the National Economic Modeling Project, the continuing research effort of the MIT System Dynamics Group. This book suggests that labor hiring and termination policies underlie the short-term business cycle, while capital investment policies are primarily involved in generating economic cycles of much longer duration. By offering an understanding of the generic causes of business cycles, the book provides a basis for future investigation of alternative economic stabilization policies. The book describes, in detail, a simple simulation model that provides an example of system dynamics applied to economics.

No additional books have been written that report more current work on the National Economic Modeling Project. For such information and references to journal articles, see the latest issue of the *System Dynamics Newsletter,* available from the System Dynamics Group, MIT, 77 Massachusetts Avenue, Cambridge, MA 02139. The *Newsletter,* published annually, is an extensive review of system dynamics work being carried on around the world. The annotated bibliography of the *Newsletter* is an invaluable resource, including dissertations, journal articles, and books, covering all areas of system dynamics research, education, and application.

Dynamics of Commodity Production Cycles (Meadows, 1970) is a more narrowly focused economic study. The book contains a model that generates fluctuations in the supply and demand of agricultural and mineral commodities. The principal indicated application is to hog cycles with illustrations also of oscillations in mercury availability and price. Chapter 21 of this text is based on this study. *The Dynamics of the World Cocoa Market* (Weymar, 1968) is an earlier effort at modeling commodity fluctuations from a system dynamics perspective, but it also employs econometric methods. The serious student of economic modeling should find interesting a comparative assessment of these two books.

Another study with emphasis on one aspect of the economy is *Managing the Energy Transition* (Naill, 1977). The work was a specific follow-up of *The Limits to Growth* project and preceded the development of a number of energy policy models by the U.S. Department of Energy.

Exercise 6: Analyzing Model Boundary—The Natural Gas Model Revisited

The chief of long-range planning within the United States Department of Energy is interested in modifying the natural gas model developed in Chapter 23 to investigate several newly proposed questions and policy options. Specifically, he is interested in policies designed to: (1) increase the production of domestic gas by placing a tariff on imported natural gas; (2) tie the price of domestic gas to the price of imported oil via a form of modified price regulation; and (3) promote coal gasification (extracting natural gas from coal) by placing a tax on natural gas and using the proceeds from this tax to "seed" coal gasification projects.

Discuss how the boundary of the natural gas model might be expanded in order to answer each of these questions. To what degree should each of these questions be answered by adding exogenous variables (such as an assumed growth in OPEC prices) versus endogenous variables?

Managerial Problems

The first complete presentation of the system dynamics approach was *Industrial Dynamics* (Forrester, 1961). This book establishes the rationale of system dynamics, explains the methodology, and illustrates the initial application to industrial problems. The appendices contain technical information about system dynamics found nowhere else.

Managerial Applications of System Dynamics (Roberts, 1978) documents (including model listings) about forty applications of system dynamics in business. The first chapter is a concise exploration of the fundamental concepts of system dynamics and DYNAMO. Each part of the book begins with an overview containing references to related applications not included in the book, as well as to some applications that are not documented anywhere else in the literature.

Management System Dynamics (Coyle, 1977) talks about the design of managerial policies that help an organization perform satisfactorily over time in the face of external shocks and under the influence of internal processes. It then analyzes the problems and redesigns corporate policies to improve behavior. The book is an advanced managerial textbook that includes several examples drawn from industrial projects performed by the author and his associates.

Corporate Planning and Policy Design (Lyneis, 1980) is another advanced managerial text that focuses on building blocks of common corporate structures as a starting point for systems models. In addition to corporate planning, the book addresses production and employment instability and corporate growth. The book is unique in its use of a continuously expanding corporate model to cover all the topics treated.

One of the early applications of system dynamics to a managerial problem area is *The Dynamics of Research and Development* (Roberts, 1964). The book presents a broad descriptive theory of research and development projects in government and industry, and develops the corresponding system dynamics model. More than 1,000 computer-simulated project time-histories are presented to evaluate the R & D function.

General System Dynamics Textbooks

Principles of Systems (Forrester, 1968) is the classic, unfinished, introductory system dynamics text. It introduces the basic concepts of system structure, then shows by example how structure determines behavior. The book contains ten chapters of text followed by workbook problems and answers. Although the problems are framed in a corporate context, the principles are general to many fields.

A basic text, *Study Notes in System Dynamics* (Goodman, 1974) is especially useful in conjunction with *Principles of Systems*. Part I focuses on simple structures and describes elements of positive and negative feedback loops. Part II reinforces material in the first part with eight exercises that emphasize the relationships between structure and behavior. Part III contains advanced exercises on model conceptualization and analysis.

Richardson and Pugh's book, *Introduction to System Dynamics Modeling with DYNAMO* (1981), is an introductory book with a business focus. Even though the examples cover a wide range of applications, a single model addressing the problem of overruns in a large R & D project is traced throughout the book. Technical aspects of DYNAMO, including special features contained in DYNAMO II and III, are explained in this text.

In conjunction with any of the books mentioned, a modeler should have available the *DYNAMO User's Manual* (Pugh, 1976) in addition to the specific user's manual for the particular version of DYNAMO being used, such as *User Guide and Reference Manual for Micro-DYNAMO* (Pugh-Roberts Associates, 1982). Pugh's manual provides a complete description of DYNAMO, including the advanced forms of the language for DYNAMO II and DYNAMO III.

The literature reviewed here contains a wealth of both simple and more complex system dynamics models. Students defining a problem behavior in a field of special interest to them, are very likely to find a similar behavior discussed in one of these references. As in any area of life, learning by imitation is quite an appropriate approach. Incorporating a piece of a model developed by someone else is as valid a learning experience in this field as using a subroutine or procedure written by another computer programmer. This last chapter attempts to suggest to the student a myriad of modeling possibilities for developing expertise in system dynamics and computer simulation.

APPENDIX A

GLOSSARY OF SYSTEM DYNAMICS TERMS

Acronym   A word formed from the first or first few letters of a series of words. DYNAMO is an acronym formed from the words "DYNAmic MOdels."

Algorithm   A step-by-step procedure.

Arrow   An indication that one element has an influence on another element.

Auxiliary   A type of variable in the DYNAMO simulation language. Neither a flow nor an accumulation, auxiliaries are intermediate calculations used in a rate equation.

BASIC   A general-purpose computer language.

Behavior   *See* **Dynamic behavior**

Behavior of a system   The changes over time arising from the elements of a system.

Behavior pattern   The overall trends in the behavior of a system. Typical patterns are growth and decline, oscillation, and equilibrium-seeking or goal-seeking.

Boundary   Outer limits of what is being taken into account for problem being studied. Only those elements that most directly affect the behavior are included within the "system boundary."

Causal-loop analysis   Selecting from a situation description a perspective, time frame, problematic behavior, and policy choice(s) and from them constructing a causal-loop diagram that is consistent with the problem behavior.

Causal-loop diagram   An illustration using words or short phrases and arrows to show the pattern of influence between elements.

Closed loop   *See* **Feedback**

Closed system   *See* **Feedback**

Cloud   A symbol used in flow diagraming to indicate a source of something or a place "sink" where something terminates. (*See* Flow Diagram Symbols, p. 284.)

Computer model   A simulation model whose behavior over time is determined by a computer program. This is a special use of the term and refers in this book to a model written in the simulation language DYNAMO.

Computer simulation   Analyzing a problem by building a model consisting of instructions to a computer and using the model to imitate the behavior of the system causing the problem.

Constant   An element that does not change during a given simulation.

Delay   Passing of time. Each arrow in a causal-loop diagram represents a delay of some length.

DELAY1 function   A DYNAMO shorthand for representing a first-order delay.

DELAY3 function   A DYNAMO shorthand for representing third-order material delays.

DLINF3 function   A DYNAMO shorthand for representing third-order information delays.

Dimension of a problem   One of the four attributes of a problem: perspective, time frame, problematic behavior, or policy choice.

Dimensionless   Having no units associated with a quantity. For example, the ratio of two quantities with the same units is dimensionless.

Dimensions   Units in which a quantity is measured.

Doubling time   The length of time it takes for a quantity to double in size.

DT   Difference in Time, or delta (Δ) time.

Dynamic   Changing over time.

Dynamic behavior   The pattern of change in a system over time. Dynamic behavior is shown by the graphs of key system variables plotted against time.

Dynamic structures   The underlying pattern of causal relationships that determine the behavior of a system. Structure is often shown in causal-loop diagrams.

DYNAMO   A special-purpose computer language for performing certain types of computer simulation. An acronym for DYNAmic MOdels.

Equation   A statement in words and/or numbers, expressing a relationship where the elements on each side of the equal (=) sign are in fact precisely comparable quantities.

Equilibrium   The state of a system in which none of its level variables is changing.

Equilibrium value   The value of a variable at which the equilibrium state will exist. If a gap exists between a variable's current value and its equilibrium value, that gap will decrease over time. For example, the temperature of the coffee in a cup will come to the temperature of the room over a period of an hour or so.

Exogenous element   An element in a causal-loop diagram that affects another element, but is not affected by another element in the causal-loop diagram. This element is driven by some interaction other than the loops within the causal-loop diagram.

Exponential average   An averaging algorithm that weighs recent information more heavily than older information.

Exponential decay or growth   Decay or growth characterized by the change of the level variable being a constant fraction of the level variable itself. Exponential decay is the behavior often generated by a single negative feedback loop. Exponential growth is the behavior often generated by a positive feedback loop.

Feedback   The process in which an action taken by a person or thing will eventually affect that person or thing.

First-order delay   A rate that is delayed in a system dynamics model by flowing in and out of a single level.

Flow diagram   A diagram showing the relationships among variables in a system. More specific than a causal-loop diagram, a flow diagram indicates the types of variables as well as their interconnections.

FORTRAN   A widely used computer language for scientific calculations.

Function   A special algorithm built into DYNAMO, such as table functions; also indicates a dependency among variables, such as sales rate is a function of selling effort and price.

Goal   The end toward which effort is directed.

Goal gap   The difference between the goal and the actual. In a dynamic model this often represents the motivator or driving force underlying attempts at corrective action.

Initial Value   The beginning value of a level. Every level must have an initial value.

Interpolation   The estimation or calculation of a value by taking a weighted average of known values at neighboring points.

Lag   Delay.

Level   A type of variable that is the result of the accumulation of one or more flows (rates).

Linear relationship   Elements causally affecting one another in such a way that a change in the cause produces a proportional change in the result.

Link   A line drawn between two elements, indicating a causal relationship.

Maximum point   A point on a curve higher than points on either side.

Minimum point   A point on a curve lower than the points on either side.

Model   A simplification of real-world phenomena; the purpose of a model is to make the behavior of the real world more understandable.

Model-building   The process of defining a problem out of a situation, developing a causal-loop diagram, drawing a flowchart, defining relationships quantitatively, writing a DYNAMO model, running that model with several policy options, and analyzing the behavior of the model.

Negative loop   A causal loop whose behavior is characterized by oscillations, equilibrium, or goal-seeking behavior.

Net   The difference between the values of two variables having a special significance to each other, such as birth rate and death rate.

Nonlinear relationship   Elements causally affecting one another in such a way that the relationship cannot be written as a linear equation or graphed as a straight line.

Operator   A symbol that indicates a mathematical operation. Examples are + and −.

Oscillate   Behavior that is characteristic of movement back and forth or up and down over time.

Overshoot   To pass swiftly beyond, as if to miss the goal.

Parameter   A quantity, constant during a simulation, that may change from simulation to simulation depending on the application. DYNAMO constants and tables are sometimes referred to as the parameters of the simulation.

Perspective   Point of view of a concerned person.

Plot period   The time interval between successive entries in an output graph. In DYNAMO, the plot period is set by giving a constant PLTPER a value.

Policy choice   A strategy designed to solve a problem or reduce its severity. Several policy choices are usually tested in an analysis.

Positive loop   A causal loop whose behavior is usually characterized by continuous growth or decline.

Print period   The time interval between successive printing of output values. In DYNAMO, the print period is set by giving a constant PRTPER a value.

Problem   That aspect of a situation that behaves in an undesired fashion from the perspective of a concerned person.

Problem definition   The process of choosing the perspective, time frame, problematic behavior, and policy choice for a problem.

Problematic behavior   An undesired behavior pattern in a situation, usually shown as a time graph.

Pulse   A sudden increase or decrease in the value of a rate; also a DYNAMO function that produces this time behavior.

Rate   The change in a level variable over time; also the type of variable that is a flow.

Reference mode   Same as problematic behavior, emphasizing its use as a basis for comparison with behavior under various policy choices.

RUN   A computer term that instructs the computer to start or execute a program or model; also "a run" is the full execution itself of the program or model.

Sigmoidal growth   S-shaped growth characterized by exponential growth followed by a leveling off.

Simulation   A representation of a system in a simplified form to study its behavior over time. A problem is analyzed by imitating a system with a model.

Simultaneous equations   In DYNAMO, a set of either initial value equations or auxiliary equations that depend on each other at the same point in time. They therefore cannot be put in proper order for computation since, no matter what order, one equation is always before another equation whose value is required to calculate the first. Example:

$$A = B$$
$$B = A$$

Situation   A story describing some dynamic behavior from which a problem is defined in a causal-loop analysis.

Slope (of a variable on a graph)   The steepness of a curve that indicates the amount and speed of change of the variable being plotted. The more perpendicular the line of the variable to the horizontal axis of the graph, the more and faster that variable is changing. The more parallel the line of the variable is to the horizontal axis of the graph, the less and slower the variable is changing.

SMOOTH function   A DYNAMO function that performs a weighted averaging of past values, using an exponential average.

Statement identifier   A letter or word beginning in the first column of a DYNAMO equation that identifies the type of equation or statement that follows. These are listed on the last page of Chapter 14.

STEP function   A time behavior characterized by being constant until a time when it takes an immediate change and then remains constant again. Also the DYNAMO function that produces this time behavior.

Structure   *See* **Dynamic structure.**

Structured problem   One in which the elements and their interactions are clearly implied, if not stated.

Subscript   *See* **Timescript.**

System   A regularly interacting or interdependent group of elements forming a unified whole.

TABLE function   A DYNAMO algorithm for expressing a graphical relationship, usually nonlinear, between two or more variables.

Third-order delay   A rate that is delayed by flowing in and out of three levels.

Time frame   Length of time being considered in defining a problem.

Time graph   A graph where time is plotted along the horizontal axis; used in system dynamics to picture the dynamic behavior of systems.

Time interval   An amount of time between two events.

Timescripts   A shorthand way of representing a time interval. In system dynamics, J, K, and L are used to represent past time, the present, and the future, respectively.

Turning point   The point on a graph where the slope changes from increasing to decreasing or from decreasing to increasing. At the turning point itself, there is neither a decrease or increase in the value of the variable.

Undesirable behavior   A predicted or known pattern of change over time in the state of the system that "the concerned person" wishes to alter.

Unstructured problem   A situation in which the problem is not clearly defined and the elements are not obvious.

Variable   An element in a system whose value changes during the time period of interest.

APPENDIX B*

ANSWERS TO SELECTED EXERCISES

CHAPTER 2

Exercise 2

a. Difficult school work causes me to *study*.

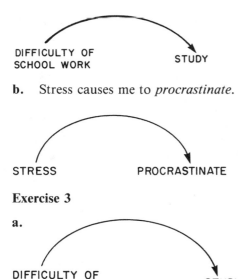

b. Stress causes me to *procrastinate*.

Exercise 3

a.

Difficult school work causes me to study. Studying causes my school work to be less difficult.

*Answers not included in this Appendix are available from Addison-Wesley on request.

455

b.

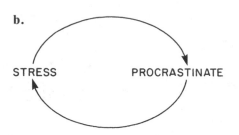

Stress causes me to procrastinate. Proctastination causes more stress.

Exercise 4

a. Loop:

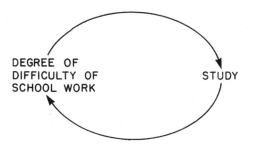

Pattern:

Or:
The *more* difficult school work is, the less I study, the *more* difficult school work becomes:

b. Loop:

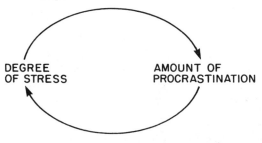

Pattern:

Exercise 5

a.   Electronic mail system
     How can individual rights to privacy be protected?
     Water system
     How can use of water be monitored so severe water shortages do not de-
     velop?

c.   How can my company profit from the use of an electronic information
     system?
     How can the most hydroelectric power be produced?

CHAPTER 3

Exercise 1

a.   The graph shows the population continuing to increase. The causal-loop
     diagram in Figure 3.5 is positive, indicating continuing change in the same
     direction, in this case, increasing.

b.   Yes, because our population system still seems to be dominated by a posi-
     tive feedback loop.

Exercise 4

a.

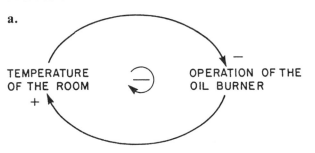

TEMPERATURE OF THE ROOM   —   OPERATION OF THE OIL BURNER
                        +

b.   Because you can follow the arrows around in a circle or closed loop. Each
     element in the system affects the other element, so that observed behavior
     can be explained.

c.   This is a system that maintains equilibrium. The *higher* the room tempera-
     ture, the *less* the oil burner operates, causing the room temperature to
     fall.

d.   Yes. An air conditioner will go off and on in response to the temperature of a room in the same way as the oil burner, but the signs on the arrowheads in the diagram will be reversed.

Exercise 6

a.

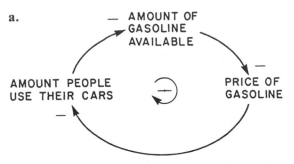

The *more* people use their cars the *less* gas available, the *less* gas available the *higher* the price, the *higher* the price the *less* people use their car. Self-regulating system.

Exercise 8

a.

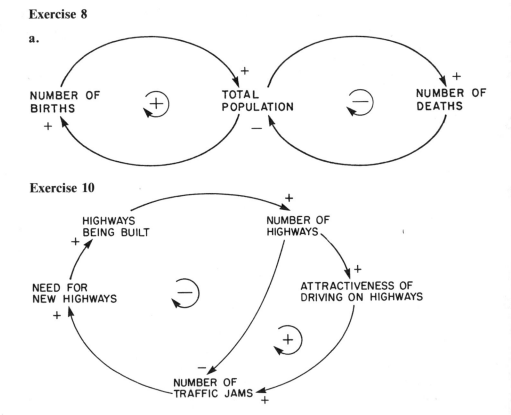

Exercise 10

c.   A negative and a positive loop linked together. The number of traffic jams tends to be the key to the system. If traffic *increases,* new highways are built causing *less* traffic jams but more attractiveness for driving which then causes *more* traffic. The only way traffic will continue to decrease is if the larger loop dominates, causing driving to continue to become less attractive.

Exercise 15

a.

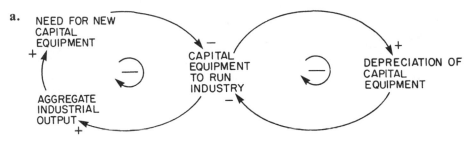

b.   Two negative loops linked together would be a system tending toward equilibrium (or perhaps oscillation).

Exercise 18

a.

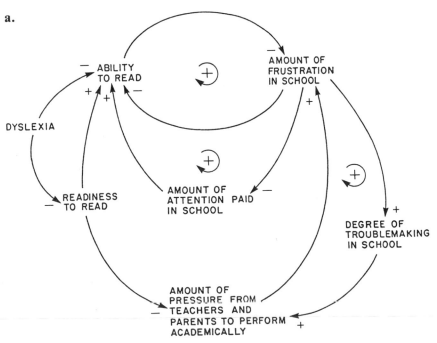

There are three positive feedback loops, which suggest once a student begins feeling school frustration the situation will become worse and worse.

b.   Yes. The principal's advice would work towards turning the positive loops around, so that the student's frustration is decreased and everything else improves.

CHAPTER 4

Exercise 2

a.   The student kept getting *more* upset and banged *harder,* causing the nail to slip *more.*

b.

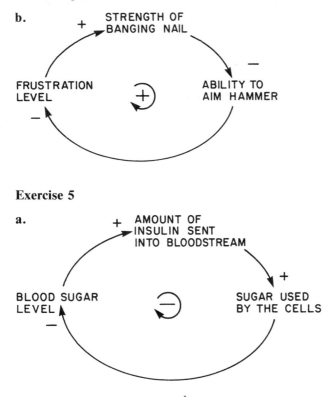

Exercise 9

a.

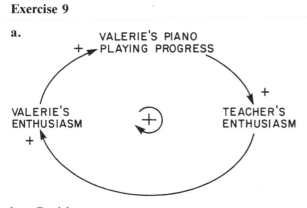

b. Positive

Exercise 12

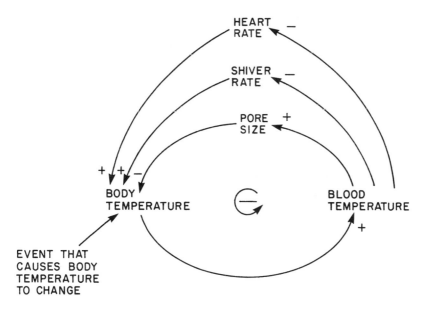

Exercise 16

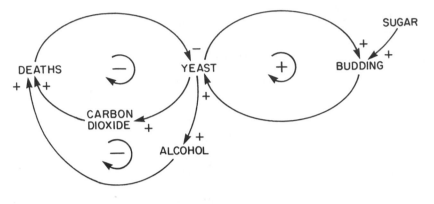

Exercise 20

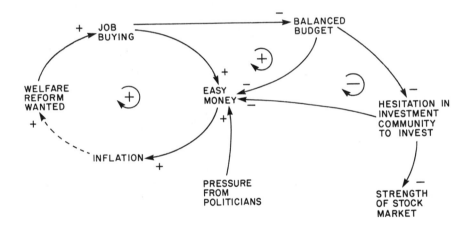

CHAPTER 5

Exercise 2

In 1850, Boomtown was practically nonexistent. Shortly thereafter, a few settlers came and discovered some deposits of precious minerals. A steady stream of people settled in Boomtown from 1855 to 1865. The most rapid rise in population occurred between 1865 and 1875, due maybe to the spreading rumor. As the mines were stripped and supplies dwindled, unemployment became a problem. After 1875, people started to leave in search of new fortunes. The population decreased steadily. By 1890, there were only a few people left in Boomtown.

Exercise 7

a.   Horizontal axis: Time, 1-day intervals, $1 \rightarrow 5$ days; Vertical axis: Number of yeast cells, 20-cell intervals, 0-120 cells

b.

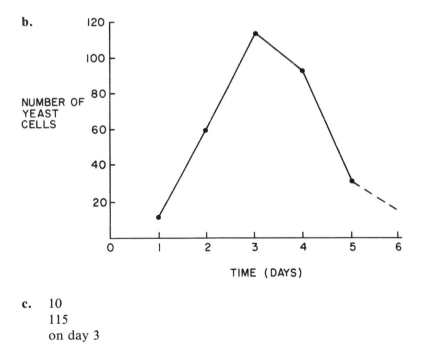

c.   10
115
on day 3

Exercise 9

a.

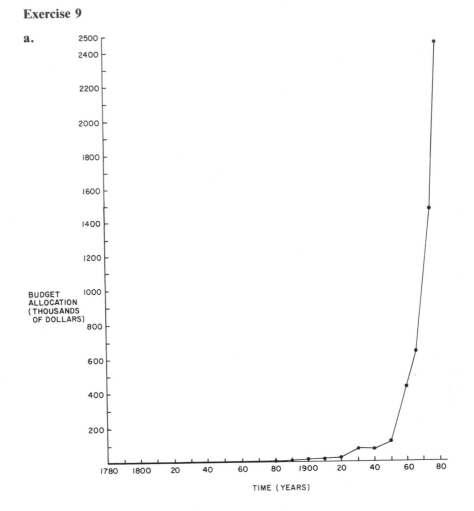

b. The general pattern of the graph gives us a fair idea of what the missing data should look like. We just need to extend the line between 1800 and 1882. This is called interpolation of the data. The trend of the budget is easy to predict. We can see that from 1780 until the middle of the 20th century, Littleton's school budget increases very slowly.

c. As population increases, so does the school budget. The population grows, however, in a more even rate throughout the 200 years, while the school budget increases very slowly until the 1950s, then shoots up dramatically throughout the second half of 20th century.

CHAPTER 6

Exercise 1

When a maximum or peak is reached, the net change in population goes from positive to negative, while for a minimum point, the net change goes from negative to positive.

Exercise 4

a.   Immigration and births dominated from 1850 through 1900; emigration and deaths from 1950 through 1970.
In balance—they balanced out.

b.   Emigration—movement to suburbs.

c.   It is clearly emigration from city to suburbs. Population decreasing in city, but increases the further you get from the city.

Exercise 8

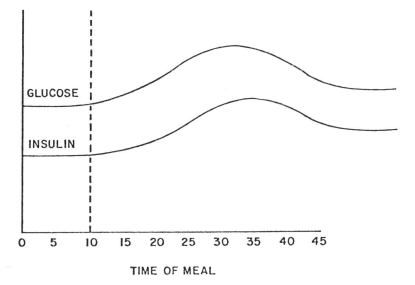

TIME OF MEAL

An increase in glucose level leads to an increase in the level of insulin which in turn decreases the glucose level.

Exercise 10

a. Until about 1930, the size of catch was nearly constant. Then it rose dramatically, peaking at about 1950, falling, then peaking again at 1960, after which it fell drastically until 1965 when it rose slightly and declined to pre-1930 levels.

b. Population increase→more fishing→insufficient breeding stock→smaller harvests. Population, price of fish, abundance of fish.

c.

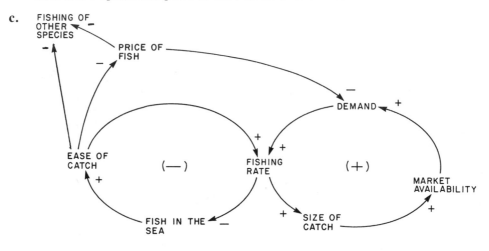

CHAPTER 7

Exercise 1

a.

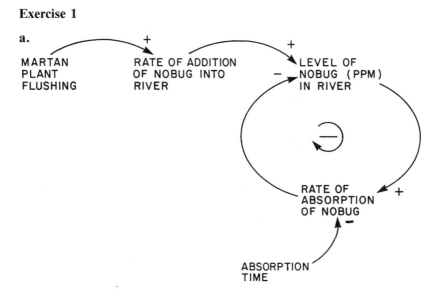

b. In terms of the causal loop, the only difference would be the different absorption time related to the half-life of Nobug in the river.

c. Several answers are possible. One might suggest flushing the plant into a holding tank and allowing Nobug to decompose for a week before dumping. Such a solution might be costly and would involve Martan officials who might need to be pressured by the Chamber of Commerce, state environmental agency, etc.

Exercise 3

a.

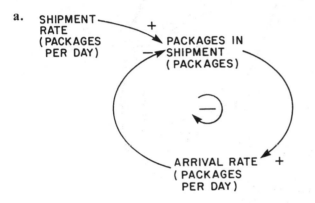

b. About 7 days, because this is roughly the time of peak arrivals and an "eyeball" estimate sees about as many packages arriving before as after April 7.

c. A primary reason might be different distances. Those closest to Growit would arrive earliest and those most distant would arrive latest. Freak scheduling "foul-ups" might explain deliveries of up to 13 or 14 days.

CHAPTER 8

Exercise 2

a. $1.80, based on a linear assumption. If exponential assumption is used, price should be $2.40.

b. The pattern of oil prices over the past 5 years. Its fluctuations and trends need to be studied. If the trend is exponential, a different strategy is used than if it is linear. The owner must be prepared for fluctuations as well to avoid a loss. And obviously, past data alone may be insufficient for projecting the future.

Exercise 4

b.

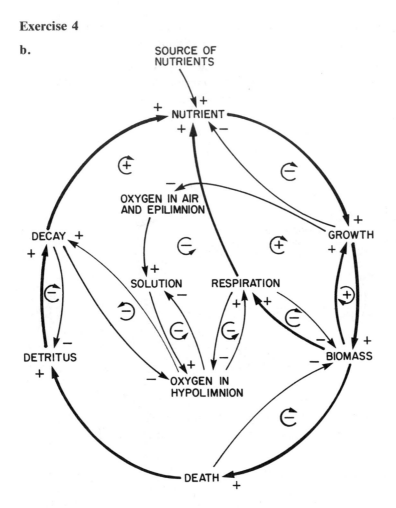

CHAPTER 9

Exercise 2

About 4 hours—half life of I-33; 15 years—time to develop an adequate
alternate power supply; 20 years—time for significant melting of the polar
ice cap; 1,000–12,000 years—accidental uncovering of nuclear waste.

Exercise 4

An insurance executive-policy: try to get the federal government's back-
ing to cover a low probability but very serious accident such that claims
based on the accident could exceed the resources of the insurance company.

Exercise 6

a.

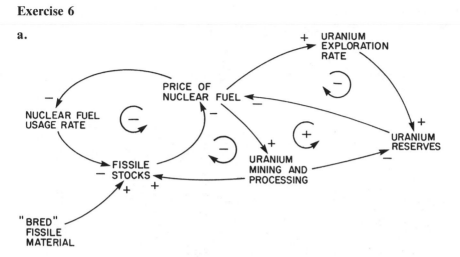

As fuel price rises, uranium exploration rate would gradually increase. But the extent of increased exploratory activity could well be far greater than the percentage price change, due to the significant effect of fuel price changes on uranium mining profitability. Drops in fuel price would generate similar reductions in exploration.

b. This system has even stronger likelihood of fluctuating. The positive feedback loop has the capacity of accentuating any oscillations in the three other negative feedback loops. But changes in uranium reserves are slow in developing. The time period for oscillations of this system might well be about ten to twenty years, or more. This reference mode is considerably different from the one expected in Exercise 5. Yet even twenty years are considerably shorter than the 50-year time horizon of interest to the long-range planner whose perspective is taken in the defined problem.

c. Long-term fluctuations in uranium reserves, as well as wider swinging nuclear fuel prices, might cause concern to a long-term planner in the Department of Energy, or to an owner of a uranium mine.

Exercise 11

a. and b.
public utility
200 years
accumulation of spent fuel rods
bury spent fuel rods in salt

Exercise 12

a.

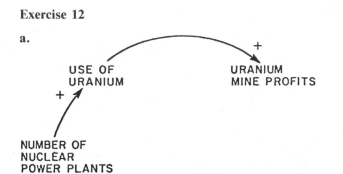

CHAPTER 10

Exercise 2

a. 2 years for an auto tire
2 months for a flashlight battery used one hour each day

b. 8 years for zinc
$U = R/8$
or
zinc/year Used = zinc Reserves$/8$ years

Exercise 5

a. A Detroit auto designer: build a lighter, more durable car.

b. *Problematic behavior:* increasing cost of raw materials,
increased sales losses to small
foreign import models because of
increased gas costs.
Time scale: 15 years

c.

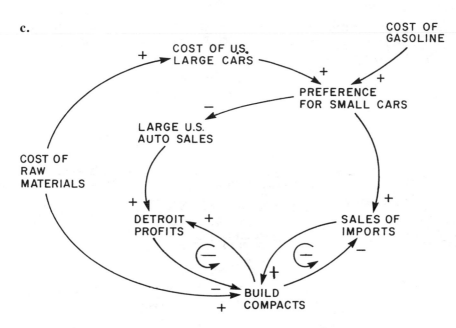

d. Compacts will sell better, leading to better profits.

e. Answer to this depends on (c).

CHAPTER 11

Exercise 1

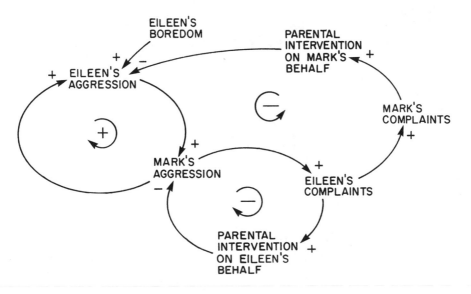

Given that both negative feedback loops are active, both children get punished. Mark may be happier due to a "balance of misery" but he's probably not elated!

Exercise 5

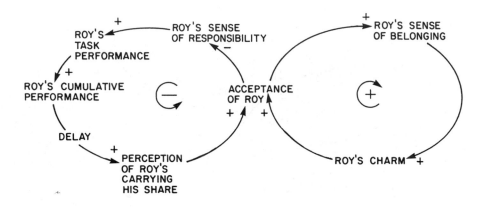

Exercise 8

a. *perspective:* Sam's
time frame: months
problematic behavior: Margaret becoming a fussy child
policy: leave Margaret to cry alone after checking her safety, dryness, . . .

b.

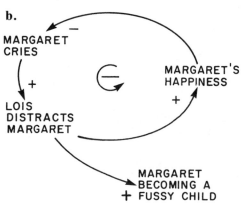

c. The more Lois distracts Margaret making her happy, the fussier Margaret becomes. This may be in turn feed back to reinforce Margaret's crying.

CHAPTER 12

Exercise 2

a. Yes. U.S. commentator on world affairs, concerned about oil dependence and national policy making.

b. 10 years? (not explicit)
The last one-third of the editorial points to problems in conversion including developments of RRs, ports, ships. The time to accept the necessity for conversion could be a few years.

c. Doubling of oil prices every year as production drops about 5% each year.

d. 1. Develop coal
2. Conservation
3. Develop renewable resources

e.

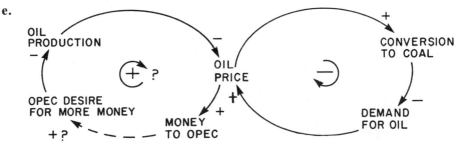

f. *Without policy:* see c
With policy: price rise slows down as shipping facilities and plant conversion proceed. Leveling occurs in about 15 years?

g. Time scale: about 10 years for conversion; about 20 years for conservation (in Abelson's opinion; perhaps much shorter, actually); about 50 years for renewable resources.

CHAPTER 13

Exercise 1

a.

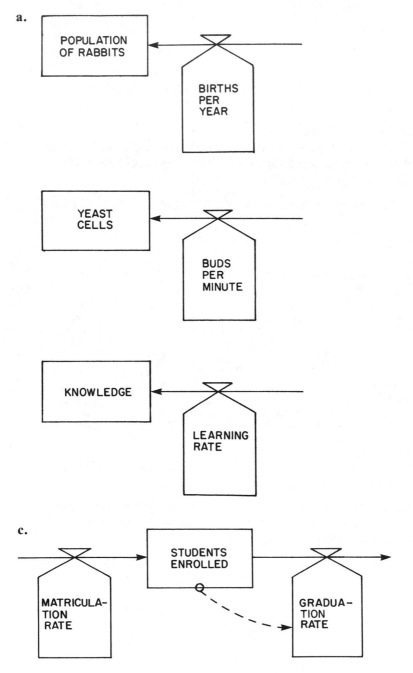

Exercise 4

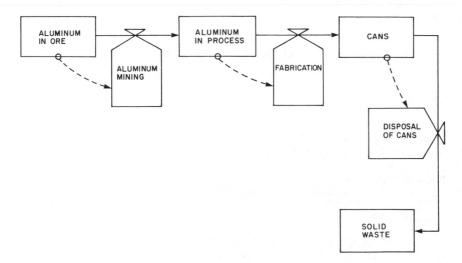

To represent recycling, a Recycling Rate should move aluminum from Solid Waste and bring it into Aluminum in Process.

Exercise 7

Population after first month:
    5510 + (1/12) (100) = 5518
Population in year 2520:
    5510 + (500) (100) = 55,510
Answer can be calculated without iterations.

CHAPTER 14

Exercise 4

a.   L   YEAST.K = YEAST.J + (DT)(BUDDNG.JK)

d.   correct

g.   L   YEAST.K = YEAST.J + DT*BUDDNG.JK

j.   L   POP.K = POP.J + (DT)(BIRTHS.JK)

Exercise 7

a.   YOUNG, MIDDLE, and AGING

b.   YOUNG – Y, MIDDLE = M, AGING = A

c.   one year

d.   50 years

e.   SPEC   DT = 1/LENGTH = 50/PLTPER = 1/PRTPER = 0
         PLOT   YOUNG = Y/MIDDLE = M/AGING = A

Exercise 12

```
* XANADU POPULATION MODEL
L POP.K=POP.J+DT*BIRTHS.JK
N POP=POPN
C POPN=5510
NOTE INITIAL POPULATION OF XANADU
R BIRTHS.KL=BIRCON
C BIRCON=100
N TIME=2020
NOTE STARTING TIME IS THE YEAR 2020
SPEC  DT=1/LENGTH=2040/PLTPER=1/PRTPER=1
PLOT  POP=P/BIRTHS=B
PRINT POP,BIRTHS
RUN
```

XANADU POPULATION MODEL

TIME E 00	POP E 00	BIRTHS E 00
2020.0	5510.0	100.00
2021.0	5610.0	100.00
2022.0	5710.0	100.00
2023.0	5810.0	100.00
2024.0	5910.0	100.00
2025.0	6010.0	100.00
2026.0	6110.0	100.00
2027.0	6210.0	100.00
2028.0	6310.0	100.00
2029.0	6410.0	100.00
2030.0	6510.0	100.00
2031.0	6610.0	100.00
2032.0	6710.0	100.00
2033.0	6810.0	100.00
2034.0	6910.0	100.00
2035.0	7010.0	100.00
2036.0	7110.0	100.00
2037.0	7210.0	100.00
2038.0	7310.0	100.00
2039.0	7410.0	100.00
2040.0	7510.0	100.00

```
XANADU POPULATION MODEL

      POP=P,BIRTHS=B

      4000.0         5000.0         6000.0         7000.0         8000.0   P
         0.00          50.00         100.00         150.00         200.00   B
2020.0  ------------------P-----B-------------------------
        .              .      P      B                  .              .
        .              .        P    B                  .              .
        .              .         P  B                   .              .
        .              .           PB                   .              .
        .              .            P                   .              . PB
        .              .           BP                   .              .
        .              .           B   P                .              .
        .              .           B     P              .              .
        .              .           B       P            .              .
2030.0  ------------------------------B-----P-------------------
        .              .           B       P            .              .
        .              .           B         P   .                     .
        .              .           B           P .                     .
        .              .           B            P.                     .
        .              .           B              P                    .
        .              .           B              .P                   .
        .              .           B              .   P                .
        .              .           B              .    P               .
        .              .           B              .      P             .
2040.0  ------------------------------B-------------------P------
```

Compare the output with your hand simulations. They should agree. Note the way the starting time of 2020 AD is specified. TIME is an internal level whose initial value is zero unless you specify the starting value as in this model. Note also that the LENGTH given is not the total length of the run but the value of TIME at which the model should stop running.

CHAPTER 15

Exercise 2

a.   YEAST=Y,BUDDNG=B

```
            0.00       200.00       400.00       600.00       800.00   Y
            0.000      20.000       40.000       60.000       80.000   B
   0.0000 -Y----------------------------------------------------- YB
          .Y             .            .            .           . YB
          .Y             .            .            .           . YB
          .Y             .            .            .           . YB
          .Y             .            .            .           . YB
          .Y             .            .            .           . YB
          .Y             .            .            .           . YB
          .Y             .            .            .           . YB
          .Y             .            .            .           . YB
          .Y             .            .            .           . YB
  10.000  --Y---------------------------------------------------- YB
          .  Y           .            .            .           . YB
          .  Y           .            .            .           . YB
          .  Y           .            .            .           . YB
          .  Y           .            .            .           . YB
          .   Y          .            .            .           . YB
          .   Y          .            .            .           . YB
          .   Y          .            .            .           . YB
          .   Y          .            .            .           . YB
          .    Y         .            .            .           . YB
  20.000  ----Y-------------------------------------------------- YB
          .    Y         .            .            .           . YB
          .     Y        .            .            .           . YB
          .     Y        .            .            .           . YB
          .      Y       .            .            .           . YB
          .       Y      .            .            .           . YB
          .       Y      .            .            .           . YB
          .        Y     .            .            .           . YB
          .         Y    .            .            .           . YB
          .          Y   .            .            .           . YB
  30.000  ----------Y-------------------------------------------- YB
          .           Y  .            .            .           . YB
          .            .Y .            .            .           . YB
          .            . Y .           .            .           . YB
          .            .  Y .          .            .           . YB
          .            .   Y .         .            .           . YB
          .            .     Y .       .            .           . YB
          .            .      Y .      .            .           . YB
          .            .       Y .     .            .           . YB
          .            .        .Y     .            .           . YB
  40.000  -------------------------------Y---------------------- YB
```

b.   YEAST GROWTH

TIME E 00	YEAST E 00	BUDDNG E 00
0.000	10.00	1.000
1.000	11.00	1.100
2.000	12.10	1.210
3.000	13.31	1.331
4.000	14.64	1.464
5.000	16.11	1.611
6.000	17.72	1.772
7.000	19.49	1.949
8.000	21.44	2.144
9.000	23.58	2.358
10.000	25.94	2.594
11.000	28.53	2.853
12.000	31.38	3.138
13.000	34.52	3.452
14.000	37.97	3.797
15.000	41.77	4.177
16.000	45.95	4.595
17.000	50.54	5.054
18.000	55.60	5.560
19.000	61.16	6.116
20.000	67.28	6.728
21.000	74.00	7.400
22.000	81.40	8.140
23.000	89.54	8.954
24.000	98.50	9.850
25.000	108.35	10.835
26.000	119.18	11.918
27.000	131.10	13.110
28.000	144.21	14.421
29.000	158.63	15.863
30.000	174.49	17.449
31.000	191.94	19.194
32.000	211.14	21.114
33.000	232.25	23.225
34.000	255.48	25.548
35.000	281.02	28.102
36.000	309.13	30.913
37.000	340.04	34.004
38.000	374.04	37.404
39.000	411.45	41.145
40.000	452.59	45.259

Time to double number of yeast cells = 7 hours since at 7 hours there are 19.49 cells and we started with 10 cells.

c.   The time to double is 7 hours at both the beginning and the end of the run.

Exercise 6

a. Two runs show the effect of changing the amount withdrawn per year. The first shows the effect with the same initial balance of $500. The account drops more and more rapidly towards zero as the interest earned falls further and further behind the withdrawals each year. (The first run for 40 years led to a negative balance at 19 years which would correspond to borrowing money.)

```
BANK ACCT. PT.II

        BAL=B,INT=I

           0.00        200.00       400.00       600.00       800.00  B
           0.000       20.000       40.000       60.000       80.000  I
 0.0000  ---------------------------------------B-------------------- BI
            .            .            .        B          .          . BI
            .            .            .        B          .          . BI
            .            .            .          B        .          . BI
            .            .            . B                 .          . BI
            .            .            .  B                .          . BI
            .            .            .B                  .          . BI
            .            .            B                   .          . BI
            .            .           B.                   .          . BI
            .            .          B  .                  .          . BI
 10.000  -------------------------B----------------------------------- BI
            .            .          B      .              .          . BI
            .            .        B        .              .          . BI
            .            .     B            .             .          . BI
            .            .   .B             .             .          . BI
            .            .  B.              .             .          . BI
            .            B   .              .             .          . BI
            .         B      .              .             .          . BI
            .      B         .              .             .          . BI
```

Clearly a larger initial balance is needed to provide $60 interest per year to match the withdrawals. Since the interest rate is 10%, that balance must be $600. The result is shown here to be an equilibrium balance.

```
BANK ACCT. PT.II

       BAL=B,INT=I

         0.00       200.00       400.00       600.00       800.00  B
         0.000       20.000       40.000       60.000       80.000  I
 0.0000 --------------------------------------------B------------ BI
           .           .            .            B            . BI
           .           .            .            B            . BI
           .           .            .            B            . BI
           .           .            .            B            . BI
           .           .            .            B            . BI
           .           .            .            B            . BI
           .           .            .            B            . BI
           .           .            .            B            . BI
           .           .            .            B            . BI
10.000  --------------------------------------------B------------ BI
           .           .            .            B            . BI
           .           .            .            B            . BI
           .           .            .            B            . BI
           .           .            .            B            . BI
           .           .            .            B            . BI
           .           .            .            B            . BI
           .           .            .            B            . BI
```

b. Again, the new interest rate of 8% will not provide $50/year but since .08*(BAL) should = $50, we get BAL = $50/.08 = $625. Runs are given here with the original balance of $500 and with the calculated balance of $625.

```
BANK ACCT. PT.II

       BAL=B,INT=I

         0.00       200.00       400.00       600.00       800.00  B
         0.000       20.000       40.000       60.000       80.000  I
 0.0000 -----------------------------------I-----B------------------
           .           .            I        B        .           .
           .           .            I.       B        .           .
           .           .             I .    B         .           .
           .           .             I . B            .           .
           .           .            I   . B           .           .
           .           .            I    .B           .           .
           .           .            I   .B            .           .
           .           .           I    B             .           .
           .           .          I    B.             .           .
10.000  ----------------------I---B------------------------------
           .           .      I   B    .              .           .
           .           .     I  B      .              .           .
           .           .   . I  B      .              .           .
           .           .    I   B      .              .           .
           .           .   I.  B       .              .           .
           .           .  I   B         .             .           .
           .           .  I B .         .             .           .
           .           . I B    .       .             .           .
```

```
BANK ACCT. PT.II

      BAL=B,INT=I

        0.00      200.00      400.00       600.00      800.00  B
        0.000     20.000      40.000        60.000      80.000  I
0.0000  ----------------------------------I-------B-----------
          .           .            .       I     . B          .
          .           .            .       I     . B          .
          .           .            .       I     . B          .
          .           .            .       I     . B          .
          .           .            .       I     . B          .
          .           .            .       I     . B          .
          .           .            .       I     . B          .
          .           .            .       I     . B          .
          .           .            .       I     . B          .
10.000  ----------------------------------I-------B-----------
          .           .            .       I     . B          .
          .           .            .       I     . B          .
          .           .            .       I     . B          .
          .           .            .       I     . B          .
          .           .            .       I     . B          .
          .           .            .       I     . B          .
          .           .            .       I     . B          .
          .           .            .       I     . B          .
```

Exercise 10

```
YEAST GROWTH

         YEAST=Y,BUDDNG=B,YEASTD=D

        0.000          5.000         10.000         15.000         20.000    Y
        0.0000         0.2500        0.5000         0.7500         1.0000    B
        0.0000         0.5000        1.0000         1.5000         2.0000    D
 0.0000 B-----------------------Y----------------------- YD
        B              .         Y .                  .           . YD
        B              .     Y     .                  .           . YD
        B              .   Y       .                  .           . YD
        B              . Y         .                  .           . YD
        B            . Y           .                  .           . YD
        B           .Y             .                  .           . YD
        B          Y.              .                  .           . YD
        B         Y .              .                  .           . YD
        B        Y  .              .                  .           . YD
10.000  B-------Y----------------------------------------- YD
        B        Y .               .                  .           . YD
        B       Y   .              .                  .           . YD
        B      Y     .             .                  .           . YD
        B     Y      .             .                  .           . YD
        B     Y      .             .                  .           . YD
        B    Y       .             .                  .           . YD
        B    Y       .             .                  .           . YD
        B    Y       .             .                  .           . YD
        B   Y        .             .                  .           . YD
20.000  B--Y--------------------------------------------- YD
        B   Y        .             .                  .           . YD
        B Y          .             .                  .           . YD
        B Y          .             .                  .           . YD
        B Y          .             .                  .           . YD
        B Y          .             .                  .           . YD
        B Y          .             .                  .           . YD
        BY           .             .                  .           . YD
        BY           .             .                  .           . YD
        BY           .             .                  .           . YD
30.000  BY----------------------------------------------- YD
```

```
YEAST  GROWTH

      YEAST=Y,BUDDNG=B,YEASTD=D

         0.000         5.000        10.000        15.000        20.000   Y
         0.0000        0.2500        0.5000        0.7500        1.0000   B
          .00000        .10000        .20000        .30000        .40000  D
 0.0000  B----------------------------Y-----D------------------
         B               .            Y.       D        .             .
         B               .            Y.         D      .             .
         B               .           Y .       D        .             .
         B               .           Y .   D            .             .
         B               .          Y   . D             .             .
         B               .          Y   . D             .             .
         B               .         Y     .D             .             .
         B               .         Y     D.             .             .
         B               .        Y      D              .             .
10.000   B------------------------Y---D-------------------------------
         B               .       Y      D.              .             .
         B               .       Y    D  .              .             .
         B.              .      Y     D  .              .             .
         B               .      Y    D   .              .             .
         B               .     Y    D    .              .             .
         B               .     Y   D     .              .             .
         B               .     Y   D     .              .             .
         B               .    Y   D      .              .             .
         B               .    Y   D      .              .             .
20.000   B-------------Y---D------------------------------------------
         B             . Y   D       .              .             .
         B             . Y    D      .              .             .
         B             .Y     D      .              .             .
         B             .Y    D       .              .             .
         B             .Y    D       .              .             .
         B             Y    D        .              .             .
         B             Y  D          .              .             .
         B             Y  D          .              .             .
         B             Y D           .              .             .
30.000   B---------Y--D------------------------------------------------
```

The behavior does still represent exponential decay. The time for the yeast
population to drop to half its initial size is, for a lifetime of 10 hours: 6
1/2 hours; and for a lifetime of 40 hours: 27 hours. (*Note:* The theoretical
relationship between the lifetime and the half-life time is half-life =
0.693*(lifetime) which is well approximated here; $\ln(2) = 0.693$ . .)

Exercise 12

a.

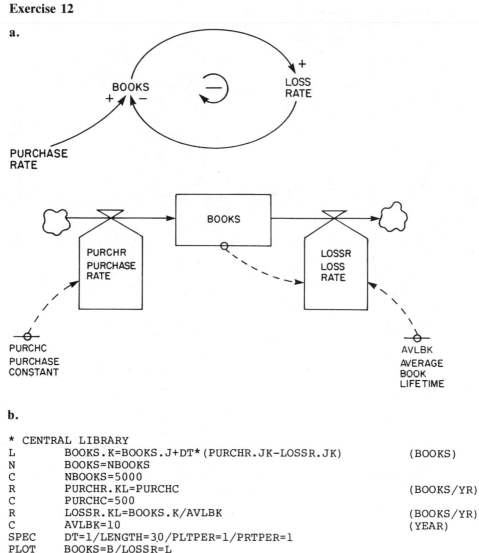

b.

```
*   CENTRAL LIBRARY
L       BOOKS.K=BOOKS.J+DT*(PURCHR.JK-LOSSR.JK)              (BOOKS)
N       BOOKS=NBOOKS
C       NBOOKS=5000
R       PURCHR.KL=PURCHC                                     (BOOKS/YR)
C       PURCHC=500
R       LOSSR.KL=BOOKS.K/AVLBK                               (BOOKS/YR)
C       AVLBK=10                                             (YEAR)
SPEC    DT=1/LENGTH=30/PLTPER=1/PRTPER=1
PLOT    BOOKS=B/LOSSR=L
PRINT   BOOKS,PURCHR,LOSSR
RUN
```

c. CENTRAL LIBRARY

```
        BOOKS=B,LOSSR=L

        0.0      2000.0      4000.0       6000.0       8000.0  B
        0.00     200.00      400.00       600.00       800.00  L
0.0000  ------------------------------------B------------------ BL
        .            .            .        B         .        . BL
        .            .            .        B         .        . BL
        .            .            .        B         .        . BL
        .            .            .        B         .        . BL
        .            .            .        B         .        . BL
        .            .            .        B         .        . BL
        .            .            .        B         .        . BL
        .            .            .        B         .        . BL
        .            .            .        B         .        . BL
10.000  ------------------------------------B------------------ BL
        .            .            .        B         .        . BL
        .            .            .        B         .        . BL
        .            .            .        B         .        . BL
        .            .            .        B         .        . BL
        .            .            .        B         .        . BL
        .            .            .        B         .        . BL
        .            .            .        B         .        . BL
        .            .            .        B         .        . BL
20.000  ------------------------------------B------------------ BL
        .            .            .        B         .        . BL
        .            .            .        B         .        . BL
        .            .            .        B         .        . BL
        .            .            .        B         .        . BL
        .            .            .        B         .        . BL
        .            .            .        B         .        . BL
        .            .            .        B         .        . BL
        .            .            .        B         .        . BL
        .            .            .        B         .        . BL
30.000  ------------------------------------B------------------ BL
```

Here is the expected output from the model run at equilibrium. While books come in and books are lost, there is no change in the number of books in the library over time.

```
CENTRAL LIBRARY

        BOOKS=B,LOSSR=L

     5000.0        5500.0        6000.0        6500.0        7000.0  B
     500.00        550.00        600.00        650.00        700.00  L
0.0000  -------------------------------------------------------B BL
        .             .             .             .        B    . BL
        .             .             .             .      B      . BL
        .             .             .             .  B.         . BL
        .             .             .         B       .         . BL
        .             .             .   B             .         . BL
        .             .           . B               .         . BL
        .             .         B.                    .         . BL
        .             .       B   .                   .         . BL
        .             .     B     .                   .         . BL
10.000  ------------------B--------------------------------------- BL
        .           .   B                           .         . BL
        .           . B                             .         . BL
        .         B                                 .         . BL
        .        B.                                 .         . BL
        .       B  .                                .         . BL
        .      B   .                                .         . BL
        .     B    .                                .         . BL
        .    B     .                                .         . BL
        .   B      .                                .         . BL
20.000  ------B--------------------------------------------------- BL
        .    B     .             .             .         . BL
        .    B     .             .             .         . BL
        .   B      .             .             .         . BL
        .   B      .             .             .         . BL
        .  B       .             .             .         . BL
        .  B       .             .             .         . BL
        .  B       .             .             .         . BL
        .  B       .             .             .         . BL
        . B        .             .             .         . BL
30.000  --B------------------------------------------------------- BL
```

That 5000 is the equilibrium amount is shown clearly in this run which begins with 7000 books and gradually comes towards the equilibrium number of 5000. At 30 years, there are 5084 books in the library.

Exercise 18

a.
```
      *  PUSHUPS
      L         PUSHUP.K=PUSHUP.J+DT*(IMPROV.JK)
      NOTE          NUMBER OF PUSHUPS IN A DAILY SESSION  (PUSHUPS)
      N         PUSHUP=NPUSH
      C         NPUSH=30
      A         AMTPR.K=TPP*PUSHUP.K                          (MIN)
      NOTE          AMOUNT OF PRACTICE IN MINUTES
      C         TPP=0.5                                   (MIN/PUSHUP)
      NOTE          TIME PER PUSHUP
      R         IMPROV.KL=(AMTPR.K-BASPR)*0.2          (PUSHUPS/MON)
      NOTE          UNITS FOR 0.2: PUSHUPS/MON/MIN
      C         BASPR=10                                      (MIN)
      NOTE          BASE PRACTICE AMOUNT
      SPEC      DT=1/LENGTH=18/PRTPER=1/PLTPER=1
      NOTE          UNITS FOR TIME: MONTHS
      PLOT      PUSHUP=P/IMPROV=I
      PRINT     PUSHUP,IMPROV,AMTPR
      RUN
```

b. Runs with differing numbers of pushups initially. Starting with 30 pushups:

PUSHUPS

```
          PUSHUP=P,IMPROV=I

         0.000        20.000       40.000       60.000       80.000  P
         0.0000       2.0000       4.0000       6.0000       8.0000  I
 0.0000 ------I-----------P------------------------------------
        .         I    .        P      .             .           .
        .         I    .        P      .             .           .
        .          I   .        P      .             .           .
        .           I  .         P     .             .           .
        .            I .          P .              .           .
        .             I.           P.             .           .
        .             I            P              .           .
        .            .I            .P             .           .
        .             . I          . P            .           .
 10.000 -----------------I-----------P--------------------
        .             .  I    .        P      .             .
        .             .    I  . P     .             .
        .             .     I .    P     .             .
        .             .      I.        P.             .
        .             .       .I          .P             .
        .             .        . I       .  P           .
        .             .        .  I      .    P         .
        .             .        .   I     .      P       .
        .             .        .    I    .        P     .
```

Jim is clearly getting ahead.
Starting with 20 pushups:

```
PUSHUPS

        PUSHUP=P,IMPROV=I

         0.000          10.000          20.000          30.000          40.000   P
         0.0000          0.2500          0.5000          0.7500          1.0000   I
0.0000  I----------------------------P----------------------------
        I               .            P              .               .
        I               .            P              .               .
        I               .            P              .               .
        I               .            P              .               .
        I               .            P              .               .
        I               .            P              .               .
        I               .            P              .               .
        I               .            P              .               .
10.000  I----------------------------P----------------------------
        I               .            P              .               .
        I               .            P              .               .
        I               .            P              .               .
        I               .            P              .               .
        I               .            P              .               .
        I               .            P              .               .
        I               .            P              .               .
        I               .            P              .               .
```

Jim is in equilibrium with his current practice pattern; that is, he will
maintain his current performance of 20 pushups indefinitely.
With 10 pushups:

```
PUSHUPS

        PUSHUP=P,IMPROV=I

         0.000           5.000          10.000          15.000          20.000   P
        -2.0000         -1.5000         -1.0000         -0.5000          0.0000   I
0.0000  ----------------------------P---------------------------- PI
        .               .          P .               .             . PI
        .               .        P   .               .             . PI
        .               .    P       .               .             . PI
        .              .P            .               .             . PI
        .          P    .            .               .             . PI
        .      P        .            .               .             . PI
        .P             .            .               .             . PI
```

Note the shortness of the run; at 8 months, the number of pushups Jim can do becomes negative—clearly nonsense.

The reason for the difference in Jim's performance with the differing number of initial pushups lies in the model equations. When Jim does 20 pushups, the amount of practice he gets is (0.5 min/pushup)*(number of pushups) = 10 minutes of practice. Since the base practice time is 10 minutes, this is the number he must do at a minimum if he is not to lose his current performance level. If he does fewer than 10, his performance will drop each month. If he does more than the 20 pushups required to maintain his performance, he can do more each day, taking more and more time on pushups until his free hours are filled. This natural limit is not imposed in the model but is surely recognized by the user!

CHAPTER 16

Exercise 2

```
*          BUILDING CONSTRUCTION
L          BUILD.K=BUILD.J+(DT)(CONST.JK)
N          BUILD=10
NOTE            INDUSTRIAL BUILDINGS (BUILDINGS)
R          CONST.KL=(BUILD.K)(CNSTF)
NOTE            CONSTRUCTION (BUILDINGS/YEAR)
C          CNSTF=0.1
NOTE            CONSTRUCTION FRACTION (1/YEAR)
NOTE
PLOT       BUILD=B/CONST=C
SPEC       DT=1/PLTPER=3/LENGTH=120
RUN
```

Exponential growth is caused by positive loop, in this case the feedback between construction and buildings.

```
BUILDING CONSTRUCTION

      BUILD=B,CONST=C

         0.0T        250.0T       500.0T       750.0T      1000.0T B
         0.00T        25.00T       50.00T       75.00T      100.00T C
0.0000 B------------------------------------------------------------ BC
       B              .            .            .          . BC
       B              .            .            .          . BC
       B              .            .            .          . BC
       B              .            .            .          . BC
       B              .            .            .          . BC
       B              .            .            .          . BC
       B              .            .            .          . BC
       B              .            .            .          . BC
       B              .            .            .          . BC
30.000 B------------------------------------------------------------ BC
       B              .            .            .          . BC
       B              .            .            .          . BC
       B              .            .            .          . BC
       B              .            .            .          . BC
       B              .            .            .          . BC
       B              .            .            .          . BC
       B              .            .            .          . BC
       B              .            .            .          . BC
       B              .            .            .          . BC
60.000 B------------------------------------------------------------ BC
       B              .            .            .          . BC
       B              .            .            .          . BC
       B              .            .            .          . BC
       B              .            .            .          . BC
      .B              .            .            .          . BC
      .B              .            .            .          . BC
      .B              .            .            .          . BC
      .B              .            .            .          . BC
      . B             .            .            .          . BC
90.000 ---B--------------------------------------------------------- BC
      . B             .            .            .          . BC
      .     B         .            .            .          . BC
      .       B       .            .            .          . BC
      .         B     .            .            .          . BC
      .            B. .            .            .          . BC
      .            . B             .            .          . BC
      .            .     B         .            .          . BC
      .            .            .B .            .          . BC
      .            .            .       B       .          . BC
      .            .            .            .       B-----BC
120.00 -----------------------------------------------------B--- BC
```

Exercise 5

A   CNSTF.K = TABLE(TCNSTF,FLANDO.K,0,1,0.2)
T   TCNSTF = 0.10/0.08/0.06/0.04/0.02/0.0

Exercise 8

```
* URBAN GROWTH MODEL
L         BUILD.K=BUILD.J+(DT)(CONST.JK-DEMO.JK)
N         BUILD=10
NOTE                INDUSTRIAL BUILDINGS (BUILDINGS)
R         DEMO.KL=BUILD.K/ALTB
NOTE                DEMOLITION OF IND. BUILDINGS (BUILDINGS/YEAR)
C         ALTB=50
NOTE                AVERAGE LIFETIME OF BUILDINGS (YEARS)
R         CONST.KL=(BUILD.K)(CNSTF.K)
NOTE                CONSTRUCTION (BUILDINGS/YEAR)
A         CNSTF.K=TABLE(TCNSTF,FLANDO.K,0,1,0.2)
NOTE                CONSTRUCTION FRACTION (1/YEAR)
T         TCNSTF=0.10/0.10/0.09/0.08/0.06/0.00
NOTE                TABLE FOR CONSTRUCTION FRACTION
A         FLANDO.K=(BUILD.K*AVAREA)/LAND
NOTE                FRACTION OF LAND OCCUPIED (DIMENSIONLESS)
C         AVAREA=1
NOTE                AVERAGE AREA PER BUILDING (ACRES/BUILDING)
C         LAND=1000
NOTE                LAND AVAILABLE FOR BUILDINGS (ACRES)
NOTE
NOTE                CONTROL STATEMENTS
NOTE
SPEC      DT=2/PLTPER=4/LENGTH=100/PRTPER=4
PLOT      BUILD=B/CONST=C,DEMO=D
PRINT BUILD,CONST,DEMO
RUN
```

A standard run for this model will provide a reference run for the changes made:

```
URBAN GROWTH MODEL

      BUILD=B,CONST=C,DEMO=D

        0.0          250.0         500.0         750.0        1000.0   B
      0.000         20.000        40.000        60.000        80.000   CD
0.0000 BC----------------------------------------------------------   BD
       DB             .             .             .             .  BC
       DB             .             .             .             .  BC
       DB             .             .             .             .  BC
       D B            .             .             .             .  BC
       .DBC           .             .             .             .
       .D BC          .             .             .             .
       .D    BC       .             .             .             .
       .D      BC     .             .             .             .
       . D       B C  .             .             .             .
40.000 --D------B--C----------------------------------------------
        . D          .B C           .             .             .
        . D          .    B C       .             .             .
        . D          .        B C   .             .             .
        . D          .            .  B            .             . BC
        .   D  .                   .    C    B.                 .
        .    D .                 C .           .     B          .
        .     D. C                 .           .          B     .
        .     DC                   .           .        B     .
        .     C.                   .           .        B   . CD
80.000 -----------C---------------------------------------B--- CD
        .     C.                   .           .        B . CD
        .     C.                   .           .        B . CD
        .     C.                   .           .        B . CD
        .     C.                   .           .        B . CD
        .     C.                   .           .        B . CD
```

When the negative feedback loop involving the demolition of buildings is effectively removed by increasing the average lifetime of buildings from 50 years to 100,000 years (1E5), the number of buildings increased more rapidly than before but was eventually limited by the end of the construction fraction table value as it approached zero.

```
URBAN GROWTH MODEL

        BUILD=B,CONST=C,DEMO=D

         0.0           250.0         500.0          750.0         1000.0   B
         0.000         20.000        40.000         60.000        80.000   CD
0.0000 BC-------------------------------------------------------- BD
       DB            .             .              .           .  BC
       DB            .             .              .           .  BC
       DBC           .             .              .           .
       D BC          .             .              .           .
       D   BC        .             .              .           .
       D    BC       .             .              .           .
       D       B C  .             .              .           .
       D        B C.             .              .           .
       D           .B C          .              .           .
40.000 D-----------------B--C------------------------------------
       D            .             .BC            .           .
       D            .             .        C    B .         .
       D            .    C        .              .      B    .
       D  C         .             .              .         B.
       DC           .             .              .         B
       C            .             .              .         B CD
       C            .             .              .         B CD
       C            .             .              .         B CD
       C            .             .              .         B CD
80.000 C--------------------------------------------------------B CD
       C            .             .              .         B CD
       C            .             .              .         B CD
       C            .             .              .         B CD
       C            .             .              .         B CD
       C            .             .              .         B CD
```

When the negative feedback loop involving that construction fraction
table was eliminated by making the construction fraction be a constant = 0.1/yr., the number of buildings explodes exponentially as seen in
this run:

```
URBAN GROWTH MODEL

      BUILD=B,CONST=C,DEMO=D

        0.000T        5.000T      10.000T      15.000T      20.000T B
        0.0           500.0       1000.0       1500.0       2000.0  CD
0.0000 B----------------------------------------------------------- BCD
       B              .            .            .             . BCD
       B              .            .            .             . BCD
       B              .            .            .             . BCD
       B              .            .            .             . BCD
       B              .            .            .             . BCD
       B              .            .            .             . BCD
       B              .            .            .             . BCD
       B              .            .            .             . BCD
       B              .            .            .             . BCD
40.000 B----------------------------------------------------------- BCD
       DB             .            .            .             . BC
       DB             .            .            .             . BC
       DB             .            .            .             . BC
       D B            .            .            .             . BC
       D B            .            .            .             . BC
       .D B           .            .            .             . BC
       .D   B         .            .            .             . BC
       .D     B       .            .            .             . BC
       .D         B   .            .            .             . BC
80.000 --D------B--------------------------------------------------- BC
       . D            B            .            .             . BC
       .  D           .    B       .            .             . BC
       .   D          .         B  .            .             . BC
       .     D        .            .    B       .             . BC
       .        D     .            .            .    B        . BC
```

Looking at the model equations, it is clear that we have two feedback loops, each proportional to the number of buildings, one positive, one negative. As long as the difference in these proportionality constants is positive, the number of buildings will expand exponentially as is seen when the average life of buildings is decreased from 50 to 20 years:

```
URBAN  GROWTH  MODEL

         BUILD=B,CONST=C,DEMO=D

         0.0           500.0         1000.0         1500.0         2000.0   B
         0.00          50.00         100.00         150.00         200.00   CD
0.0000  B---------------------------------------------------------------  BCD
        B                .             .             .            .  BCD
        B                .             .             .            .  BCD
        B                .             .             .            .  BCD
        DB               .             .             .            .  BC
        DB               .             .             .            .  BC
        DB               .             .             .            .  BC
        DB               .             .             .            .  BC
        .B               .             .             .            .  BCD
        .B               .             .             .            .  BCD
40.000  -DB------------------------------------------------------------  BC
        .DB              .             .             .            .  BC
        .DB              .             .             .            .  BC
        .D B             .             .             .            .  BC
        . DB             .             .             .            .  BC
        . D B            .             .             .            .  BC
        .  D B           .             .             .            .  BC
        .  D  B          .             .             .            .  BC
        .   D  B         .             .             .            .  BC
        .   D      B     .             .             .            .  BC
80.000  -----D-----B-------------------------------------------------  BC
        .        D    .B             .             .            .  BC
        .         D   .    B         .             .            .  BC
        .           D .        B        .           .            .  BC
        .             D           B.          .            .  BC
        .           . D             .    B      .            .  BC
```

If we choose a lifetime such that the number of buildings destroyed each year will equal the number constructed per year, the number of buildings should be constant. This is demonstrated with a choice of an average lifetime of 10 years:

```
URBAN GROWTH MODEL

       BUILD=B,CONST=C,DEMO=D

      0.000       5.000      10.000      15.000      20.000  B
      0.0000      0.5000      1.0000      1.5000      2.0000  CD
 0.0000 ------------------------------B------------------------- BCD
    .           .             B                .          . BCD
    .           .             B                .          . BCD
    .           .             B                .          . BCD
    .           .             B                .          . BCD
    .           .             B                .          . BCD
    .           .             B                .          . BCD
    .           .             B                .          . BCD
    .           .             B                .          . BCD
    .           .             B                .          . BCD
40.000 ------------------------------B------------------------- BCD
    .           .             B                .          . BCD
    .           .             B                .          . BCD
    .           .             B                .          . BCD
    .           .             B                .          . BCD
    .           .             B                .          . BCD
    .           .             B                .          . BCD
    .           .             B                .          . BCD
    .           .             B                .          . BCD
    .           .             B                .          . BCD
80.000 ------------------------------B------------------------- BCD
    .           .             B                .          . BCD
    .           .             B                .          . BCD
    .           .             B                .          . BCD
    .           .             B                .          . DCD
    .           .             B                .          . BCD
```

Exercise 11

a. Causal-loop diagram:

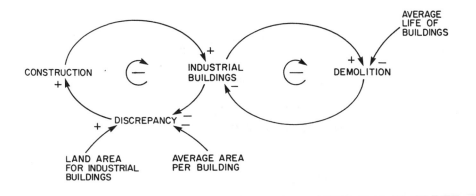

System dynamics flow diagram:

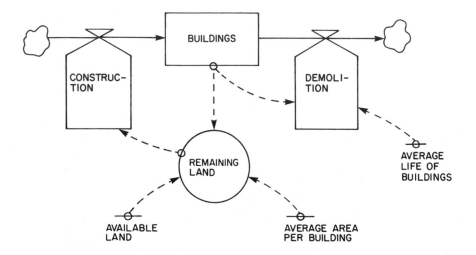

b.

```
* URBAN GROWTH MODEL
L       BUILD.K=BUILD.J+(DT)(CONST.JK-DEMO.JK)
N       BUILD=10
NOTE            INDUSTRIAL BUILDINGS (BUILDINGS)
R       DEMO.KL=BUILD.K/ALTB
NOTE            DEMOLITION OF IND. BUILDINGS (BUILDINGS/YEAR)
C       ALTB=50
NOTE            AVERAGE LIFETIME OF BUILDINGS (YEARS)
R       CONST.KL=CONPLR*BLDDIF.K
C       CONPLR=0.05
NOTE            CONSTRUCTION PER UNIT OF LAND REMAINING
A       BLDDIF.K=LAND-AVAREA*BUILD.K
NOTE            REMAINING LAND
C       AVAREA=1
NOTE            AVERAGE AREA PER BUILDING (ACRES/BUILDING)
C       LAND=1000
NOTE            LAND AVAILABLE FOR BUILDINGS (ACRES)
NOTE
NOTE            CONTROL STATEMENTS
NOTE
SPEC    DT=2/PLTPER=4/LENGTH=100/PRTPER=4
PLOT    BUILD=B/CONST=C,DEMO=D
PRINT BUILD,CONST,DEMO
RUN
```

c. A standard run of the model.

```
URBAN GROWTH MODEL

     BUILD=B,CONST=C,DEMO=D
      0.00         200.00        400.00        600.00        800.00  B
       0.000        20.000        40.000        60.000        80.000  CD
 0.0000 DB---------------------------------C--------------------
        . D            B             C             .             .
        .   D          .        B    .             .             . BC
        .     D        .     C    .  . B           .             .
        .       D      .   C       .      B        .             .
        .         D    .C           .        B .             .
        .         D    C            .           B             .
        .          D  C.            .            . B            .
        .          D C .            .            .  B           .
        .          D C .            .            .    B         .
 40.000 ---------D-C----------------------------------B-------
        .          DC  .            .            .     B         .
        .          DC  .            .            .      B        .
        .          DC  .            .            .      B        .
        .          DC  .            .            .      B        .
        .          DC  .            .            .      B        .
        .           C  .            .            .       B      . CD
        .           C  .            .            .       B      . CD
        .           C  .            .            .       B      . CD
        .           C  .            .            .       B      . CD
 80.000 ---------C------------------------------------B----- CD
        .           C  .            .            .       B      . CD
        .           C  .            .            .       B      . CD
        .           C  .            .            .       B      . CD
        .           C  .            .            .       B      . CD
        .           C  .            .            .       B      . CD
```

This model clearly does not exhibit S-shaped growth. Its behavior is a standard negative exponential approach to equilibrium where the demolition rate is equal to the construction rate.

It is not a very realistic model, especially at the beginning when the number of buildings built per year is very high at a time there are few buildings present—a strange model for industrial growth of a city! The entire model is now governed by two negative feedback loops so it can have no positive exponential component as do the earlier models. That is, there is no positive loop driving it into exponential growth at any time during the run.

d. The multiplier formulation is clearly superior as noted above.

CHAPTER 17

Exercise 1

a.

```
*  NOBUG
L         NOBUG.K=NOBUG.J+DT*(DUMP.JK-ABSORB.JK)              (PESTICIDE)
N         NOBUG=NOBUGN
C         NOBUGN=0
R         ABSORB.KL=NOBUG.K/NAT
C         NAT=2                                               (DAY)
R         DUMP.KL=60                                          (DAY)
SPEC      DT=0.25/LENGTH=18/PRTPER=1/PLTPER=1
PLOT      NOBUG=N/ABSORB=A
PRINT     NOBUG,ABSORB
RUN
```

The standard run for the model is:

```
NOBUG

        NOBUG=N,ABSORB=A

          0.00        50.00       100.00       150.00       200.00   N
          0.000       20.000      40.000       60.000       80.000   A
 0.0000  N-------------------------------------------------------- NA
             .            N    A         .                  .             .
             .            .          N       A        .             .
             .            .          N.        A      .             .
             .            .           .N          A   .             .
             .            .           .  N          A .             .
             .            .           .   N          A.             .
             .            .           .   N          A.             .
             .            .           .   N          A.             .
             .            .           .    N          A             .
10.000   -------------------------------------N------A------------
             .            .           .    N          A             .
             .            .           .    N          A             .
             .            .           .    N          A             .
             .            .           .    N          A             .
             .            .           .    N          A             .
             .            .           .    N          A             .
             .            .           .    N          A             .
```

The model exhibits negative exponential approach to an equilibrium value, namely 120 units of pesticide. This is the usual behavior of a simple system with one negative feedback loop.

b. With the normal absorption time increased from 2 to 4 days, the model behaves differently:

```
NOBUG

        NOBUG=N,ABSORB=A

      0.00          100.00        200.00         300.00        400.00   N
      0.000         20.000        40.000         60.000        80.000   A
0.0000 N--------------------------------------------------------------- NA
       .           NA      .              .              .             .
       .               N   A             .              .             .
       .             .     N   A         .              .             .
       .             .       N     A.    .              .             .
       .             .         N   . A   .              .             .
       .             .         N.      A .              .             .
       .             .           N        A  .          .             .
       .             .           .N          A  .       .             .
       .             .           .  N          A  .     .             .
10.000 --------------------------------N-----A----------------
       .             .             .  N        A  .                    .
       .             .             .  N        A  .                    .
       .             .             .    N      A.                      .
       .             .             .    N      A.                      .
       .             .             .    N      A.                      .
       .             .             .    N      A.                      .
       .             .             .    N        A                     .
       .             .             .      N      A                     .
```

However, the difference lies only in how rapidly the concentration of NOBUG comes to equilibrium and in the final equilibrium concentration of the pesticide. That equilibrium value has increased from 120 units to about 240 units. Note the same factor increase in equilibrium of NOBUG and the absorption time.

With NAT = 1, the material is absorbed very rapidly and the concentration is now much lower at equilibrium, as is seen in this run:

NOBUG

```
        NOBUG=N,ABSORB=A

        0.000        20.000       40.000       60.000       80.000   N
        0.000        20.000       40.000       60.000       80.000   A
 0.0000 N------------------------------------------------------- NA
                          .            .N             .          . NA
           .             .            .        N     .          . NA
           .             .            .           N.            . NA
           .             .            .            N            . NA
           .             .            .            N            . NA
           .             .            .            N            . NA
           .             .            .            N            . NA
           .             .            .            N            . NA
           .             .            .            N            . NA
           .             .            .            N----------- NA
 10.000 ---------------------------------------------N----------- NA
           .             .            .            N            . NA
           .             .            .            N            . NA
           .             .            .            N            . NA
           .             .            .            N            . NA
           .             .            .            N            . NA
           .             .            .            N            . NA
           .             .            .            N            . NA
           .             .            .            N            . NA
```

Exercise 4

```
a.   * NOBUG
     R          ABSORB.KL=DELAY1(DUMP.JK,NAT)              (PESTICIDE/DAY)
     C          NAT=2                                      (DAY)
     N          DUMP=NOBUGN/NAT
     C          NOBUGN=100                                 (PESTICIDE)
     R          DUMP.KL=(1/DT)*PULSE(420,1,PERIOD)         (DAY)
     C          PERIOD=7                                   (DAY)
     SPEC       DT=0.25/LENGTH=16.0/PRTPER=0.50/PLTPER=0.50
     PLOT       DUMP=D/ABSORB=A
     PRINT      DUMP,ABSORB
     RUN
```

b.

```
NOBUG

     DUMP=D,ABSORB=A

          0.0         500.0        1000.0       1500.0       2000.0  D
          0.00        100.00       200.00       300.00       400.00  A
 0.0000  D-----A--------------------------------------------------
         D     A         .            .            .            .
         .      A        .            .            .    D       .
         D              .           .A            .            .
         D              .     A      .            .            .
         D              .   A        .            .            .
         D            A.            .            .            .
         D           A   .          .            .            .
         D          A    .          .            .            .
         D       A       .          .            .            .
 5.0000  D---A----------------------------------------------------
         D A            .            .            .            .
         D A            .            .            .            .
         D A            .            .            .            .
         DA             .            .            .            .
         DA             .            .            .            .
         .A             .            .            .    D       .
         D              .          A.            .            .
         D              .   A       .            .            .
         D            .A            .            .            .
10.000   D--------A-----------------------------------------------
         D       A      .            .            .            .
         D       A      .            .            .            .
         D     A        .            .            .            .
         D  A           .            .            .            .
         D  A           .            .            .            .
         D A            .            .            .            .
         D A            .            .            .            .
         DA             .            .            .            .
         DA             .            .            .            .
15.000   -A---------------------------------------------D--------
         D              .          A.            .            .
         D              .     A      .            .            .
```

c.   With NAT = 4 days, the level of NOBUG does not fall as far between successive dumpings of NOBUG, as expected for a longer absorption time, as is seen in the listing:

NOBUG

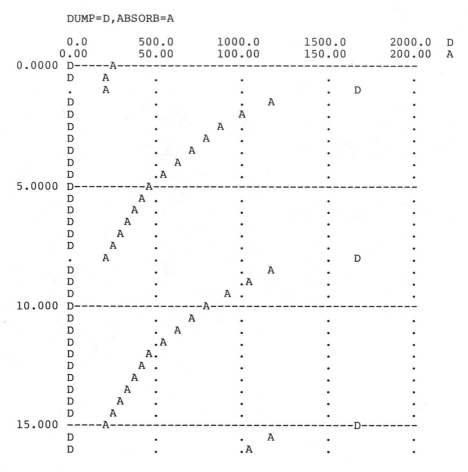

```
        DUMP=D,ABSORB=A

        0.0           500.0        1000.0       1500.0       2000.0  D
        0.00          50.00        100.00       150.00       200.00  A
0.0000  D-----A--------------------------------------------------------
        D     A       .            .            .            .
        .     A       .            .            .    D       .
        D             .            .    A       .            .
        D             .            .   A        .            .
        D             .        A   .            .            .
        D             .      A      .            .            .
        D             .    A        .            .            .
        D             .  A          .            .            .
        D            .A            .            .            .
5.0000  D----------A---------------------------------------------------
        D          A .            .            .            .
        D         A  .            .            .            .
        D        A   .            .            .            .
        D       A    .            .            .            .
        D      A     .            .            .            .
        .      A     .            .            .    D       .
        D             .            A            .            .
        D             .          .A            .            .
        D             .      A   .            .            .
10.000  D-----------------A--------------------------------------------
        D             .     A     .            .            .
        D             .    A      .            .            .
        D            .A          .            .            .
        D          A.           .            .            .
        D          A .          .            .            .
        D        A   .          .            .            .
        D        A   .          .            .            .
        D       A    .          .            .            .
        D      A     .          .            .            .
15.000  -----A----------------------------------------D---------
        D             .            .    A       .            .
        D             .          .A            .            .
```

Exercise 7

a.
```
    *     APARTMENTS
    L         APART.K=APART.J+DT*(COMPLR.JK)              (APARTMENTS)
    N         APART=APARTN
    C         APARTN=10000
    R         COMPLR.KL=DELAY1(INITR.JK,CT)              (APARTMENTS/YEAR)
    C         CT=4                                       (YEAR)
    R         INITR.KL=GAP.K/ADJT                        (APARTMENTS/YEAR)
    A         GAP.K=DESAPT.K-APART.K                     (APARTMENTS)
    C         ADJT=1                                     (YEAR)
    A         DESAPT.K=10000+STEP(5000,TINCD)            (APARTMENTS)
    C         TINCD=1                                    (YEAR)
    NOTE          TIME OF THE INCREASE IN DEMAND
    SPEC      DT=0.25/LENGTH=4.5/PRTPER=0.25/PLTPER=0.25
    PLOT      APART=A/INITR=I/COMPLR=C/GAP=G/DESAPT=D
    PRINT     APART,INITR,COMPLR,GAP,DESAPT
    RUN
```

c.   With a step increase in desired apartments at 1 year, the system begins a leisurely oscillation as is seen in the run for 9 years:

APARTMENTS

APART=A,INITR=I,COMPLR=C,GAP=G,DESAPT=D

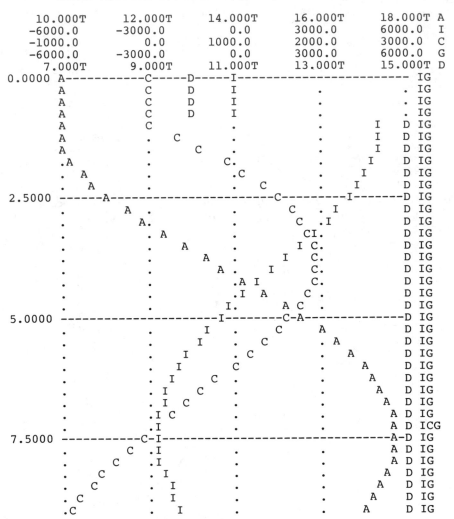

```
   10.000T        12.000T        14.000T        16.000T        18.000T A
   -6000.0        -3000.0            0.0         3000.0          6000.0 I
   -1000.0            0.0         1000.0         2000.0          3000.0 C
   -6000.0        -3000.0            0.0         3000.0          6000.0 G
    7.000T         9.000T        11.000T        13.000T         15.000T D
0.0000 A-----------C-----D-----I-----------------------------   IG
       A           C     D     I                    .           . IG
       A           C     D     I                    .           . IG
       A           C     D     I                    .           . IG
       A           C                 .              .       I    D IG
       A           .   C             .              .       I    D IG
       A           .     C           .              .       I    D IG
       .A          .           C.                   .      I     D IG
       .  A        .            .C                  .      I     D IG
       .   A       .            .      C            .     I      D IG
2.5000 ------A---------------------------------C---------I------D IG
       .      A         .              .         C   . I        D IG
       .        A.      .              .         C  .I          D IG
       .          . A         .              .   CI.           D IG
       .          .     A         .          .   I C.          D IG
       .          .          A    .      I       C.           D IG
       .          .            A .    I          C.           D IG
       .          .            .A I              C.           D IG
       .          .            .I  A         C .             D IG
       .          .            I.       A C  .              D IG
5.0000 ---------------------------------I--------C-A------------D IG
       .          .         I     .         C      A          D IG
       .          .         I     .    C        . A           D IG
       .          .      I      . C              .    A        D IG
       .          .     I       C                .   A         D IG
       .          .    I      C .                .   A         D IG
       .          .  . I    C                    .  A          D IG
       .          .  . I  C                      .  A          D IG
       .          .I C                           .  A D        IG
       .          .I                             .  A D        ICG
7.5000 -----------C-I--------------------------------------A-D IG
       .        C  .I                 .              .  A D     IG
       .      C    .I                 .              .  A D     IG
       .    C      . I                .              .  A        D IG
       .   C       . I                .              .  A        D IG
       . C         . I                .              .  A        D IG
       .C          .   I               .              .  A        D IG
```

d. This oscillatory behavior is caused by the delay in the negative feedback loop. The increase in number of apartments lags the demand enough so that the builders tend to have maximum building rates when the demand is dropping rapidly. This is typical behavior for a negative feedback loop with delay.

Exercise 10

a. Here is a working model with the desired equations with enough parameters defined as constants to permit doing all parts of this exercise:

```
*    EQUATIONS FOR THE EXPONENTIAL AVERAGE
L    AVG.K=AVG.J+DT*(ADJ.JK)
NOTE      AVERAGE SALES              (LOAVES)
N    AVG=NAVG
C    NAVG=100
R    ADJ.KL=GAP.K/AVGT
NOTE      ADJUSTMENT RATE            (LOAVES/DAY)
A    GAP.K=SALES.K-AVG.K
NOTE      GAP                        (LOAVES)
C    AVGT=5
NOTE      AVERAGING TIME             (DAYS)
A    SALES.K=INSAL+STEP(INCSAL,DINC)
C    INSAL=100
NOTE      INITIAL SALES              (LOAVES/DAY)
C    INCSAL=20
NOTE      INCREASE IN SALES          (LOAVES/DAY)
C    DINC=10
NOTE      DAY OF THE INCREASE
SPEC DT=1/LENGTH=30/PRTPER=1/PLTPER=1
PRINT SALES,GAP,AVG
PLOT  SALES=S,GAP=G,AVG=A
RUN
```

b. This run starts at equilibrium, but . . .

c. At day 10 the sales rise from 100 to 120 per day, thereby coming out of the
original equilibrium and coming to a new equilibrium as can be seen from
the output:

```
EQUATIONS FOR THE EXPONENTIAL AVERAGE

        SALES=S,GAP=G,AVG=A

        0.00         50.00       100.00      150.00      200.00  SGA
0.0000  G---------------------S------------------------- SA
        G            .          S            .          . SA
        G            .          S            .          . SA
        G            .          S            .          . SA
        G            .          S            .          . SA
        G            .          S            .          . SA
        G            .          S            .          . SA
        G            .          S            .          . SA
        G            .          S            .          . SA
        G            .          S            .          . SA
10.000  -----G-------------------A----S-------------------
        .   G        .          .A     S      .          .
        .    G       .          . A    S      .          .
        .   G        .          . A   S       .          .
        .   G        .          .  A S        .          .
        .   G        .          .  A S        .          .
        .G           .          .   AS        .          .
        .G           .          .   AS        .          .
        .G           .          .   AS        .          .
        .G           .          .   AS        .          .
20.000  -G-------------------------AS-------------------
        G            .          .   AS        .          .
        G            .          .   AS        .          .
        G            .          .   S         .          . SA
        G            .          .   S         .          . SA
        G            .          .   S         .          . SA
        G            .          .   S         .          . SA
        G            .          .   S         .          . SA
        G            .          .   S         .          . SA
        G            .          .   S         .          . SA
30.000  G-------------------------S------------------- SA
```

d. At day 10, of course, the average sales are 100. By day 16, they have risen to 114.8 and by day 17 to 115.8. Hence, the time for the average sales to rise half-way from 110 to 120 is between 6 and 7 days. This was for an averaging time of 5 days.

e. If the model is run with an averaging time of 10 days rather than an averaging time of 5 days, the results are:

```
EQUATIONS FOR THE EXPONENTIAL AVERAGE

          SALES=S,GAP=G,AVG=A

        0.00        50.00       100.00      150.00      200.00  SGA
0.0000  G---------------------S--------------------------- SA
        G            .           S              .          . SA
        G            .           S              .          . SA
        G            .           S              .          . SA
        G            .           S              .          . SA
        G            .           S              .          . SA
        G            .           S              .          . SA
        G            .           S              .          . SA
        G            .           S              .          . SA
        G            .           S              .          . SA
10.000  -----G-------------------A----S-------------------
        .    G       .           A    S          .          .
        .    G       .          .A    S          .          .
        .   G        .          .A    S          .          .
        .   G        .           . A  S          .          .
        .   G        .           . A  S          .          .
        .   G        .           . A  S          .          .
        . G          .           . A S           .          .
        . G          .           . A S           .          .
        . G          .           . A S           .          .
20.000  --G----------------------A-S-------------------
        . G          .           . A S           .          .
        .G           .           . A S           .          .
        .G           .           .    AS          .          .
        .G           .           .    AS          .          .
        .G           .           .    AS          .          .
        .G           .           .    AS          .          .
        .G           .           .    AS          .          .
        .G           .           .    AS          .          .
        .G           .           .    AS          .          .
30.000  -G-------------------------AS--------------------
```

The average sales rise more slowly to match actual sales as expected. The time to reach 115 (the half-way point) is about 13 days.

Exercise 13

a.   Here is a model with the original formulation.

```
*   STEVE'S ICE CREAM PARLOR
A   AD.K=FRSA*REV.K
NOTE     ADVERTISING                ($/MONTH)
C   FRSA=0.1
NOTE     FRACTION OF REVENUE SPENT ON ADVERTISING
A   SALES.K=SALESN+ADEFF*AD.K
NOTE     SALES                      (CONE/MON)
C   SALESN=1000
NOTE     NORMAL VALUE OF SALES      (CONES/MON)
C   ADEFF=5
NOTE     ADVERTISING EFFECTIVENESS
NOTE        (ADDITIONAL CONES PURCHASED/DOLLAR SPENT ON ADVERTISING)
A   REV.K=SALES.K*PRICE
NOTE     REVENUE                    (DOLLARS/MONTH)
C   PRICE=1
NOTE     PRICE                      (DOLLARS/CONE)
SPEC DT=0.10/LENGTH=20/PLTPER=1/PRTPER=0
PLOT   SALES=S,REV=R
RUN
```

b.   The micro-DYNAMO compiler on the APPLE gives the following message with that program:

> "Simultaneous active equations involving
> AD.REV.SALES
> 1 serious errors"
as expected from the text.

```
c.  *   STEVE'S ICE CREAM PARLOR
    A   AD.K=FRSA*AVGREV.K
    NOTE     ADVERTISING                (DOLLARS/MON)
    C   FRSA=0.1
    NOTE     FRACTION OF REVENUE SPENT ON ADVERTISING
    A   SALES.K=SALESN+STEP(SALESS,SALEST)+ADEFF*AD.K
    NOTE     SALES                      (CONES/MON)
    C   SALESN=1000
    NOTE     NORMAL VALUE OF SALES      (CONES/MON)
    C   SALESS=1000
    NOTE     SALES, STEP INCREASE (CONES/MON)
    C   SALEST=5
    NOTE     SALES STEP INCREASE TIME (MON)
    C   ADEFF=5
    NOTE     ADVERTISING EFFECTIVENESS
    NOTE     (ADDITIONAL CONES PURCHASED/DOLLAR SPENT ON ADVERTISING)
    A   REV.K=SALES.K*PRICE
    NOTE     REVENUE                    (DOLLARS/MON)
    C   PRICE=1
    NOTE     PRICE                      (DOLLARS/CONE)
    A   AVGREV.K=SMOOTH(REV.K,RAT)
    N   AVGREV=REVN
    NOTE     AVERAGE REVENUE            (DOLLARS/MON)
    C   REVN=2000
    NOTE     REVENUE, INITIAL VALUE
    C   RAT=1
    NOTE     REVENUE AVERAGING TIME   (MONTHS)
    SPEC DT=0.10/LENGTH=20/PLTPER=1/PRTPER=0
    PLOT   SALES=S,AVGREV=R
    RUN
```

```
STEVE'S ICE CREAM PARLOUR

        SALES=S,AVGREV=R

         0.0        1000.0       2000.0       3000.0       4000.0  SR
0.0000  ---------------------------S--------------------------- SR
          .            .          S                    .        . SR
          .            .          S                    .        . SR
          .            .          S                    .        . SR
          .            .          S                    .        . SR
          .            .          R               S             .
          .            .          .           R  .     S        .
          .            .          .             .   R      S    .
          .            .          .             .        R S    .
          .            .          .             .        RS .
10.000  -----------------------------------------------------RS-
          .            .          .             .            S. SR
          .            .          .             .            RS
          .            .          .             .            S SR
          .            .          .             .            S SR
          .            .          .             .            S SR
          .            .          .             .            S SR
          .            .          .             .            S SR
          .            .          .             .            S SR
          .            .          .             .            S SR
20.000  -----------------------------------------------------S SR
```

Run for Exercise 13(c).

d.

```
STEVE'S ICE CREAM PARLOUR

       SALES=S,AVGREV=R

        0.0       1000.0        2000.0       3000.0        4000.0  SR
0.0000 ----------------------------S---------------------------- SR
         .           .           S            .           . SR
         .           .           S            .           . SR
         .           .           S            .           . SR
         .           .           S            .           . SR
         .           .           R            S            .
         .           .           .       R        .   S       .
         .           .           .          R .      S        .
         .           .           .           .R       S       .
         .           .           .           . R      S       .
10.000 -------------------------------------------------R---S---
         .           .           .            .        R S  .
         .           .           .            .        R S .
         .           .           .            .        RS .
         .           .           .            .        RS.
         .           .           .            .        RS.
         .           .           .            .         S. SR
         .           .           .            .         S. SR
         .           .           .            .         RS
         .           .           .            .         RS
20.000 ----------------------------------------------------------RS

STEVE'S ICE CREAM PARLOUR

       SALES=S,AVGREV=R

        0.0       1000.0        2000.0       3000.0        4000.0  SR
0.0000 ----------------------------S---------------------------- SR
         .           .           S            .           . SR
         .           .           S            .           . SR
         .           .           S            .           . SR
         .           .           S            .           . SR
         .           .           R            S            .
         .           .           .            .      R    S    .
         .           .           .            .         R S.
         .           .           .            .          S.  SR
         .           .           .            .          S  SR
10.000 -------------------------------------------------------S SR
         .           .           .            .           S SR
         .           .           .            .           S SR
         .           .           .            .           S SR
         .           .           .            .           S SR
         .           .           .            .           S SR
         .           .           .            .           S SR
         .           .           .            .           S SR
         .           .           .            .           S SR
         .           .           .            .           S SR
20.000 -------------------------------------------------------S SR
```

Sales and average revenue rise and flatten out at 4000 in part (c) and both tests of part (d). With RAT = 2, it takes several weeks longer (twice as long?) for saturation to occur as with RAT = 1. With RAT = 0.5, saturation occurs in half as much time.

e.
```
STEVE'S ICE CREAM PARLOUR

        SALES=S,AVGREV=R

        0.0           500.0         1000.0        1500.0        2000.0  SR
0.0000  S--------------------------------------------------------- SR
        S             .             .             .           . SR
        S             .             .             .           . SR
        S             .             .             .           . SR
        S             .             .             .           . SR
        R             .             S             .           . SR
        .             .      R      .             S .           .
        .             .             .           R   .   S       .
        .             .             .             .   R     S   .
        .             .             .             .         R S .
10.000  -----------------------------------------------------R-S--
        .             .             .             .         RS.
        .             .             .             .         S.  SR
        .             .             .             .         RS
        .             .             .             .         S SR
        .             .             .             .         S SR
        .             .             .             .         S SR
        .             .             .             .         S SR
        .             .             .             .         S SR
        .             .             .             .         S SR
20.000  --------------------------------------------------------S SR
```
Half as many cones are sold in the same time period, with the same dynamic pattern.

f.

```
STEVE'S ICE CREAM PARLOUR

        SALES=S,AVGREV=R

        0.0           500.0         1000.0        1500.0        2000.0  SR
0.0000  S--------------------------------------------------------- SR
        S             .             .             .           . SR
        S             .             .             .           . SR
        S             .             .             .           . SR
        S             .             .             .           . SR
        R             .             S             .           .
        .             .      R      .       S     .           .
        .             .             .R       S    .           .
        .             .             .      R S     .          .
        .             .             .       RS     .          .
10.000  -----------------------------------RS----------------
        .             .             .       S     .          . SR
        .             .             .       S     .          . SR
        .             .             .       S     .          . SR
        .             .             .       S     .          . SR
        .             .             .       S     .          . SR
        .             .             .       S     .          . SR
        .             .             .       S     .          . SR
        .             .             .       S     .          . SR
        .             .             .       S     .          . SR
20.000  -----------------------------------S--------------- SR
```

```
STEVE'S ICE CREAM PARLOUR

     SALES=S,AVGREV=R

         0.000T       5.000T      10.000T      15.000T      20.000T SR
 0.0000  S--------------------------------------------------------- SR
         S                .            .             .        . SR
         S                .            .             .        . SR
         S                .            .             .        . SR
         S                .            .             .        . SR
         R S              .            .             .        .
         . R   S          .            .             .        .
         .     R S        .            .             .        .
         .        R  S .               .             .        .
         .          R S                .             .        .
 10.000  -------------R-S--------------------------------------------
         .              . R   S        .             .        .
         .              .     R S      .             .        .
         .              .       R   S .              .        .
         .              .         R S                .        .
         .              .            R S             .        .
         .              .            . R   S         .        .
         .              .            .     R S .     .        .
         .              .            .       R   S . .        .
         .              .            .         R S           .
 20.000  ------------------------------------------------R-S----------
```

```
STEVE'S ICE CREAM PARLOUR

     SALES=S,AVGREV=R

         0.0T         2000.0T      4000.0T      6000.0T      8000.0T SR
 0.0000  S--------------------------------------------------------- SR
         S                .            .             .        . SR
         S                .            .             .        . SR
         S                .            .             .        . SR
         S                .            .             .        . SR
         S                .            .             .        . SR
         S                .            .             .        . SR
         S                .            .             .        . SR
         S                .            .             .        . SR
         S                .            .             .        . SR
 10.000  S------------------------------------------------------- SR
         S                .            .             .        . SR
         RS               .            .             .        .
         .S               .            .             .        . SR
         .S               .            .             .        . SR
         . S              .            .             .        . SR
         .     RS         .            .             .        .
         .      R S       .            .             .        .
         .         R  S . .            .             .        .
         .            R.      S        .             .        .
 20.000  -------------------R--------S----------------------------
```

With FRSA = 0.05, the same pattern occurs as in (e), but saturation occurs sooner and at a lower value. With FRSA = 0.2 the sales and revenue do not saturate at all, but continue to grow linearly over time at a rate of increase of 1000 cones per month. With FRSA = 0.3, the sales and revenue begin to grow exponentially. For FRSA < 0.2, the negative loop dominates and the curves move toward equilibrium. For FRSA > 0.2, the positive loop dominates and the curves exponentially grow. For FRSA = 0.2, the loops balance each other and *equilibrium growth* (linear in time) occurs.

Exercise 15

d.   Here is the output from the original model (Exercise 7, Chapter 17):

```
APARTMENTS

        APART=A,INITR=I,COMPLR=C,GAP=G,DESAPT=D

        10.000T        12.000T        14.000T        16.000T        18.000T A
        -6000.0        -3000.0          0.0          3000.0         6000.0  I
        -2000.0        -1000.0          0.0          1000.0         2000.0  C
        -6000.0        -3000.0          0.0          3000.0         6000.0  G
         7.000T         9.000T        11.000T        13.000T        15.000T D
0.0000  A-----------------D-----I------------------------ ICG
        A              .      D      I              .    . ICG
        A              .      D      I              .    . ICG
        A              .      D      I              .    . ICG
        A              .             C              .   I    D IG
        A              .            .   C           .   I    D IG
        A              .            .      C        .   I    D IG
        .A             .            .           C.      I    D IG
        .     A        .            .          .C     I      D IG
        .        A     .            .          .     CI      D IG
2.5000  -------A--------------------------------------I-C-----D IG
        .            A  .            .             . I     C    D IG
        .             A.            .             .I      C D IG
        .            . A            .           I.       C D IG
        .            .    A        A             I  .       CD IG
        .            .         A   .          I  .          CD IG
        .            .            A .      I        .        CD IG
        .            .           .A I          .        CD IG
        .            .           .I  A          .       C D IG
        .            .          I.         A    .      C D IG
5.0000  -----------------------------I-----------A---------C----D IG
        .            .       I  .             A     C     D IG
        .            .       I  .            . A C      D IG
        .            .     I    .            . C A      D IG
        .            .   I      .           C     A     D IG
        .            . I        .        C  .      A     D IG
        .            . I        .      C     .      A     D IG
        .            .I         .    C       .      A   D IG
        .            .I         .  .C        .     A D IG
7.5000  -------------I---------C------------------------A-D IG
        .            .I          C  .          .      A D IG
        .            .I        C    .          .      A D IG
        .            . I  C         .          .     A    D IG
        .            .  IC         .          .      A    D IG
        .            . CI          .          .    A      D IG
        .            .C  I         .          .    A      D IG
        .            .C      I     .          .   A       D IG
        .            C        I    .          .  A      D IG
        .          C.          I   .          . A      D IG
10.000  -----------C---------I-----------------A---------D IG
        .          C.        I  .            A.         D IG
        .          C         I .            A   .        D IG
        .          C         I.          A     .        D IG
        .          .C        I       A           .        D IG
        .          .C        .I  A         .        D IG
        .          . C        . IA          .        D IG
```

Here is the effect of changing the delay from first to third order:

```
APARTMENTS - 3RD ORDER DELAY

        APART=A,INITR=I,COMPLR=C,GAP=G,DESAPT=D

        0.000T       10.000T       20.000T       30.000T      40.000T A
       -10.000T      -5.000T        0.000T        5.000T      10.000T ICG
         7.000T       9.000T       11.000T       13.000T      15.000T D
0.0000 -----------A-----D-----I-------------------------- ICG
             .       A      D     I                  .         . ICG
             .       A      D     I                  .         . ICG
             .       A      D     I                  .         . ICG
             .       A            C               I       D IG
             .       A            C               I       D IG
             .       A            C               I       D IG
             .       A            C               I       D IG
             .       A            C               I       D IG
             .       A          .C                I       D IG
2.5000 -----------A------------C----------I----------D IG
             .       A          .  C              I       D IG
             .       A          .  C            I.        D IG
             .      .A          .  C            I.        D IG
             .      .A          .  C          I .         D IG
             .      .A          .   C        I .          D IG
             .      . A         .     C  I   .            D IG
             .      . A         .     C I    .            D IG
             .      . A         .       I    .            D ICG
             .      . A         .     I C    .            D IG
5.0000 ------------A---------I----C----------D IG
             .          .    A     .I       C      .       D IG
             .          .      A I.         C      .       D IG
             .          .      A I .        C      .       D IG
             .          .      IA  .          C    .       D IG
             .          .    I   A .        C      .       D IG
             .          .    I     A .      C      .       D IG
             .          . I       A.        C      .       D IG
             .          I          A    C         .       D IG
             .      I .           .A    C         .       D IG
7.5000 ---------I----------------A--C---------------D IG
             .        I .         . A C           .       D IG
             .       I      .     .  A            .       D AC,IG
             .      I       .     .  CA           .       D IG
             .     I        .     .C A            .       D IG
             .     I        .      C   A          .       D IG
             .     I        .     C.   A          .       D IG
             .      I       .      C . A          .       D IG
             .      I       .        C  . A       .       D IG
             .     I        .        C    . A     .       D IG
10.000 --------I---------C-------A----------------D IG
             .        I  .      C       .A        .       D IG
             .      I. C              A           .       D IG
             .      .I C            A.            .       D IG
             .      . C I           A .           .       D IG
             .      .C     I  A  .                .       D IG
             .        C          AI  .            .       D IG
```

The oscillations are much larger with the third-order delay. This is made more dramatic by a plot of the first-order delay model with the same scales produced by the computer for the third-order delay:

```
APARTMENTS

          APART=A,INITR=I,COMPLR=C,GAP=G,DESAPT=D

          0.000T       10.000T        20.000T        30.000T        40.000T A
        -10.000T       -5.000T         0.000T         5.000T        10.000T ICG
          7.000T        9.000T        11.000T        13.000T        15.000T D
 0.0000 -------------A-----D-----I----------------------- ICG
          .          A     D     I                        .  ICG
          .          A     D     I                        .  ICG
          .          A     D     I                        .  ICG
          .          A           C           I            D IG
          .          A          .C           I            D IG
          .          A          .C           I            D IG
          .          A          . C          I.           D IG
          .         .A          . C          I.           D IG
          .         .A          . C       I .             D IG
 2.5000 -------------A-----------------C----I-------------D IG
          .         . A          . C      I .             D IG
          .         . A          . C   I                  D IG
          .          . A         . C I      .             D IG
          .          . A         .   I      .             D ICG
          .          . A         .  IC      .             D IG
          .          . A         . I C      .             D IG
          .           . A        . I  C     .             D IG
          .          . A         .I   C     .             D IG
          .          . A         I    C     .             D IG
 5.0000 -------------------A---I----C-------------------D IG
          .          .       A  I .    C     .             D IG
          .          .         AI .  C       .             D IG
          .          .         A  . C        .             D AIG
          .          .        IA  . C        .             D IG
          .          .        I A . C        .             D IG
          .          .       I  A .C         .             D IG
          .          .       I  A .C         .             D IG
          .          .       I  A .C         .             D IG
          .          .       I  A  C         .             D IG
 7.5000 ------------------I---A--C---------------------D IG
          .          .       I  A C.          .            D IG
          .          .       I  A C.          .            D IG
          .          .       I  A C.          .            D IG
          .          .       I   AC .         .            D IG
          .          .        I AC .          .            D IG
          .          .        IA C .          .            D IG
          .          .         A C .          .            D AIG
          .          .         A C .          .            D AIG
          .          .         AI  .          .            D ICG
10.000 ------------------A-CI----------------------D IG
          .          .         A CI .          .            D IG
          .          .         A   CI.          .            D IG
          .          .        A    CI.          .            D IG
          .          .        A    C I          .            D IG
          .          .        A    C .I          .            D IG
          .          .        A    C .I          .            D IG
```

The period of oscillation is about 6.3 years for the first-order delay and appears to be a little over 6 years for the third-order delay. Because of the slow response of the third-order delay, you would have to run the model for a longer time (maybe 30 years) to be sure of the period for the third-order delay.

The major difference is then in the amplitude of the oscillations (larger for the third-order delay). However, there is a significant difference in the development of the oscillations. For example, the completion rate starts off very steeply for the third-order delay but only slowly for the first-order delay. The peak in number of apartments is reached in 7.5 years with a first-order delay, but not until 9 years in the third-order delay run.

CHAPTER 18

Exercise 1

a.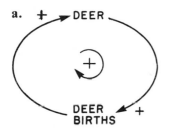

b.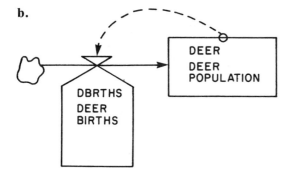

c. L   DEER.K = DEER.J + (DT)(DBRTHS.JK)
   N   DEER = DEERN
   C   DEERN = 4000

d. R   DBRTHS.KL = DEER.K*FBRD
   C   FBRD = 0.5

e.   N   TIME = 1900
     SPEC   DT = .1/LENGTH = 1950/PLTPER = 1
     PLOT   DEER = D,DBRTHS = B(0,1E6)
     RUN   KAIBAB

```
*   KAIBAB ONE(A)
NOTE
NOTE      **********   DEER POPULATION      **********
NOTE
L   DEER.K=DEER.J+(DT)(DBRTHS.JK)
N   DEER=DEERN
NOTE      DEER POPULATION (DEER)
C   DEERN=4000
NOTE      DEER POPULATION, INITIAL VALUE (DEER)
R   DBRTHS.KL=DEER.K*FBRD
NOTE      DEER BIRTHS (DEER/YEAR)
C   FBRD=0.5
NOTE      FRACTIONAL BIRTH RATE FOR DEER (1/YEARS)
NOTE
NOTE      **********   SIMULATION SPECIFICATIONS     **********
NOTE
N   TIME=1900
SPEC  DT=.1/LENGTH=1950/PLTPER=1
PLOT  DEER=D,DBRTHS=B(0,1E6)
RUN KAIBAB
```

First Kaibab Model

f.   Behavior mode is exponential growth.

```
KAIBAB ONE(A)

        DEER=D,DBRTHS=B

        0.0T        250.0T       500.0T       750.0T      1000.0T DB
1900.0  D------------------------------------------------------ DB
        D                 .            .            .        . DB
        BD                .            .            .        .
        BD                .            .            .        .
        .D                .            .            .        . DB
        .BD               .            .            .        .
        . B  D            .            .            .        .
        .   B   D         .            .            .        .
        .      B     D  .              .            .        .
        .          B    .    D         .            .        .
1910.0  -------------B------------D-----------------------------
        .                 .        B  .            B  .    D   .
        .                 .            .         B  .           .
        .                 .            .            .           .
        .                 .            .            .           .
        .                 .            .            .           .
        .                 .            .            .           .
        .                 .            .            .           .
1920.0  ------------------.------------.------------.-----------
        .                 .            .            .           .
        .                 .            .            .           .
        .                 .            .            .           .
        .                 .            .            .           .
        .                 .            .            .           .
        .                 .            .            .           .
        .                 .            .            .           .
1930.0  ------------------.------------.------------.-----------
        .                 .            .            .           .
        .                 .            .            .           .
        .                 .            .            .           .
        .                 .            .            .           .
        .                 .            .            .           .
        .                 .            .            .           .
        .                 .            .            .           .
1940.0  ------------------.------------.------------.-----------
        .                 .            .            .           .
        .                 .            .            .           .
        .                 .            .            .           .
        .                 .            .            .           .
        .                 .            .            .           .
        .                 .            .            .           .
        .                 .            .            .           .
1950.0  ------------------.------------.------------.-----------
```

Run for Exercise 1(f).

g.   KAIBAB ONE(A)

DEER=D,DBRTHS=B

```
        0.0T        250.0T      500.0T      750.0T     1000.0T DB
1900.0  D-------------------------------------------------- DB
        D              .           .           .        . DB
        D              .           .           .        . DB
        D              .           .           .        . DB
        BD             .           .           .        .
        BD             .           .           .        .
        BD             .           .           .        .
        B D            .           .           .        .
        .BD            .           .           .        .
        .B D           .           .           .        .
1910.0  -B--D-----------------------------------------------
        .B      D      .           .           .        .
        . B        D   .           .           .        .
        .  B         D .           .           .        .
        .   B         D            .           .        .
        .    B        .  D         .           .        .
        .     B       .       D  . .           .        .
        .      B      .              D         .        .
        .       B     .                   D    .        .
        .         B   .                       . D       .
        .             .   B                    .        .
1920.0  ---------------------------B-----------------------
        .             .           .   B        .        .
        :             .           .         . B         .
        .             .           .           .         .
        .             .           .           .         .
        .             .           .           .         .
        .             .           .           .         .
        .             .           .           .    -    .
        .             .           .           .         .
        .             .           .           .         .
1930.0  ---------------------------.-----------.-----------
        .             .           .           .         .
        .             .           .           .         .
        .             .           .           .         .
        .             .           .           .         .
        .             .           .           .         .
        .             .           .           .         .
        .             .           .           .         .
        .             .           .           .         .
        .             .           .           .         .
1940.0  ---------------------------.-----------.-----------
        .             .           .           .         .
        .             .           .           .         .
        .             .           .           .         .
        .             .           .           .         .
        .             .           .           .         .
        .             .           .           .         .
        .             .           .           .         .
        .             .           .           .         .
        .             .           .           .         .
1950.0  ---------------------------.-----------.-----------
```

Run for Exercise 1(g).

Exercise 4

a.

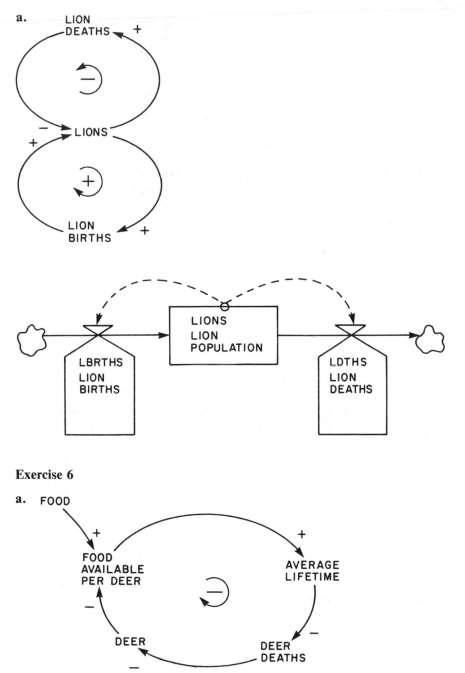

Exercise 6

a.

b.

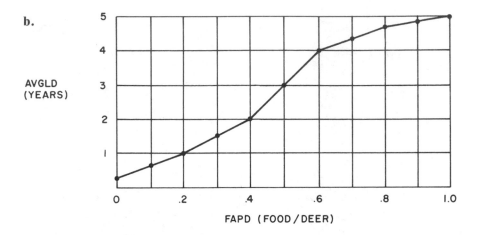

AVGLD
(YEARS)

FAPD (FOOD/DEER)

CHAPTER 19

No answers are included for this chapter.

CHAPTER 20

Exercise 1

a.   The Reference Mode

You should by now be familiar with the term "reference mode." The reference mode is the historical behavior of the problem you are modeling. The reference mode should cover as many aspects of the problem as possible, yet, you must be able to represent it in terms of known and identifiable components of the system. In our problem, businesses, land, people, jobs, and housing are the components whose historical behavior will serve as our reference mode.

By carefully reading the story of Boomtown, you should be able to draw the various components of the reference mode, as shown in Figure 1. The story suggests that everything was in balance (or dynamic equilibrium) in Boomtown before the construction of the highway connecting it to Zetroit in 1935. We also know that currently, both businesses and population are declining. We can safely assume that the business activity and population attained their highest levels sometime between now and the beginning of the boom.

Not much has been said about housing in the story, but the number of houses has probably been growing as business demand rose. The reference mode for businesses, population, and housing is shown in Figure 1 (a).

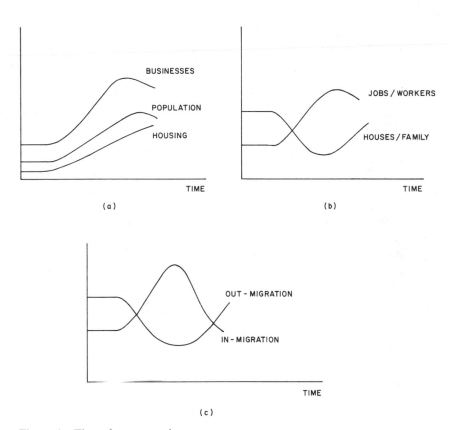

Figure 1. The reference mode

Throughout the growth phase, the number of jobs available to each worker clearly increased, as businesses grew and more business and construction jobs became available. After the peak, the number of jobs per worker declined to a point where the jobs available can hardly support the population. The reference mode for jobs per worker, is drawn in Figure 1 (b). (The graph representing houses/family is one possible scenario.)

Before the boom began, outmigration slightly exceeded immigration, so that the population of the town remained constant inspite of the net positive difference between births and deaths. During the growth phase, immigration rose while outmigration declined. After the peak, those trends reverse. Since the population of the town is currently decreasing, the outmigration exceeds the sum of net births and immigration. Figure 1 (c) illustrates the reference behavior of migration in our problem.

The reference mode serves two purposes: First, it gives us a clearer picture of the problem, and thus helps us understand how it occurred.

Second, it allows us to verify the logic of our model, which should be able to simulate various aspects of the reference mode.

b.   The Dynamic Hypothesis

Your view of how the behavior shown in the reference mode occurred is your dynamic hypothesis about the problem. At this point, the story and the reference mode you have drawn will be the basis of your dynamic hypothesis.

At the outset, it appears that the growth of Boomtown was limited by its land area, but the reasons for the decline after the peak are not immediately clear. From studying the turning points in the reference mode, we can guess that the interdependencies between the growths of businesses, population, and housing caused the problem. We will construct our model in a way so as to be able to understand the nature of those interdependencies.

It is a sound practice to start with a very simple model representing growth in a single variable, and systematically add to it additional structure representing growth in other variables. Building a model in phases has two advantages: it helps you understand behavior of various components of the model and it allows you to figure out the key elements responsible for creating the problem you are studying. Thus, by the time you have completed building the model, you are able to advance a precise and logically consistent hypothesis about the problem.

We will construct the model of growth in Boomtown in three phases. In each phase, we will study the model behavior, compare it with the reference mode, and observe the points of difference and similarity between the two. In the first phase, we will develop and test a model of the growth of businesses in an urban area with fixed land. Next, we will add a structure for population growth to the model and test it again to understand the business-population interaction. Finally, we will add a structure for housing construction and perform additional tests to understanding its implications.

CHAPTER 21

Exercise 1

a., b., and c.

One possible model of the sale and consumption of pork is shown in Figures 1, 2, and 3. The most difficult parts of the formulation are the two TABLE functions—one representing the relationship between the relative coverage RCOV (the number of weeks of pork product sales on hand, relative to what is normally available) and the hog price multiplier HPM,

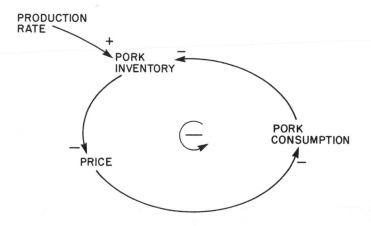

Figure 1. Causal-loop diagram for Exercise 1

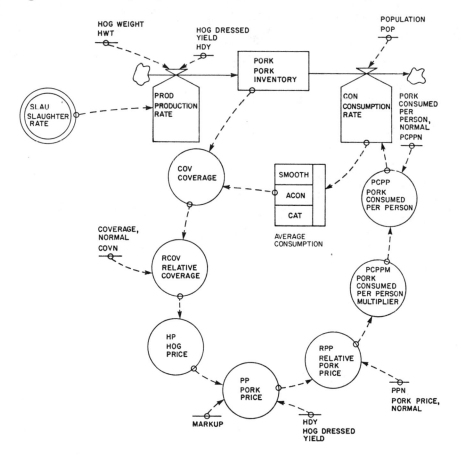

Figure 2. Flow diagram for Exercise 1

```
*   EXERCISE 1
NOTE
NOTE       *****       PORK INVENTORY      *****
NOTE
L  PORK.K=PORK.J+(DT)(PROD.JK-CON.JK)
NOTE       PORK (LBS)
N  PORK=PORKN
C  PORKN=300E6
NOTE       PORK, INITIAL VALUE (LBS)
A  PROD.K=SLAU.K*HWT*HDY
NOTE       PRODUCTION RATE (LBS/MONTH)
A  SLAU.K=SLAUN+STEP(STH,STTIME)+STEP(-STH,STTIME+12)
NOTE       SLAUGHTER RATE (HOGS/MONTH)
C  SLAUN=4E6
NOTE       SLAUGHTER RATE, INITIAL VALUE (HOGS/MONTH)
C  STH=0.4E6
NOTE       SLAUGHTER RATE TEST HEIGHT (HOGS/MONTH)
C  STTIME=200
NOTE       SLAUGHTER RATE TEST TIME (MONTHS)
C  HWT=250
NOTE       HOG WEIGHT (POUNDS/HOG)
C  HDY=.6
NOTE       HOG DRESSED YIELD (DIMENSIONLESS)
NOTE
NOTE       *****       PORK CONSUMPTION      *****
NOTE
R  CON.KL=POP*PCPP.K
NOTE       CONSUMPTION (LBS/MONTH)
C  POP=200E6
NOTE       POPULATION (PEOPLE)
A  PCPP.K=PCPPN*PCPPM.K*CTEST.K
NOTE       PORK CONSUMED PER PERSON (LBS/MONTH/PERSON)
C  PCPPN=3
NOTE       PORK CONSUMED PER PERSON, NORMAL (LBS/MONTH/PERSON)
A  PCPPM.K=TABHL(PCPPMT,RPP.K,0,3,.5)
T  PCPPMT=2/1.4/1/.66/.35/.15/0
NOTE       PORK CONSUMED PER PERSON MULTIPLIER (DIMENSIONLESS)
A  CTEST.K=1+STEP(CTH,CTTIME)+STEP(-CTH,CTTIME+12)
NOTE       CONSUMPTION TEST (DIMENSIONLESS)
C  CTH=.1
NOTE       CONSUMPTION TEST HEIGHT (DIMENSIONLESS)
C  CTTIME=200
NOTE       CONSUMPTION TEST TIME (MONTHS)
NOTE
NOTE       *****       PORK PRICE      *****
NOTE
A  RPP.K=PP.K/PPN
NOTE       RELATIVE PORK PRICE (DIMENSIONLESS)
C  PPN=1.00
NOTE       PORK PRICE NORMAL ($/LB)
A  PP.K=(HP.K/HDY)+MARKUP
NOTE       PORK PRICE ($/LB)
C  MARKUP=.50
NOTE       MARKUP ($/LB)
```

Figure 3. Model listening for Exercise 1—One possible approach

Figure 3. continued

```
NOTE
NOTE         *****      HOG PRICE       *****
NOTE
A   HP.K=HPN*HPM.K
NOTE       HOG PRICE ($/LB)
C   HPN=.30
NOTE       HOG PRICE NORMAL ($/LB)
A   HPM.K=TABHL(HPMT,RCOV.K,0,3,.5)
T   HPMT=5/2/1/.5/.25/.2/.2
NOTE       HOG PRICE MULTIPLIER (DIMENSIONLESS)
A   RCOV.K=COV.K/COVN
NOTE        RELATIVE COVERAGE (DIMENSIONLESS)
C   COVN=.5
NOTE        COVERAGE NORMAL (MONTHS)
A   COV.K=PORK.K/ACON.K
NOTE        COVERAGE (MONTHS)
A   ACON.K=SMOOTH(CON.JK,CAT)
NOTE        AVERAGE CONSUMPTION (LBS/MONTH)
C   CAT=3
NOTE        CONSUMPTION AVERAGING TIME (MONTHS)
N   ACON=PCPPN*POP
NOTE        AVERAGE CONSUMPTION, INITIAL VALUE (LBS/MONTH)
NOTE
NOTE         *****      SIMULATION SEPCIFICATIONS      *****
NOTE
C   DT=.1
C   LENGTH=10
C   PLTPER=2
PLOT   PORK=P/CON=C,PROD=R
PLOT   PP=P/HP=H/SLAU=S
RUN
```

and the other representing the relationship between the relative pork price RPP and the pork consumed per person multiplier PCPPM. (See Figures 4 and 5.) It is important to make sure that if the pork inventory drops to zero, the pork consumed per person also drops to zero. Otherwise, the pork inventory will become negative.

Let's check this for the values assumed in Figures 4 and 5. According to the TABLE in Figure 4, when RCOV = 0 (which occurs when PORK

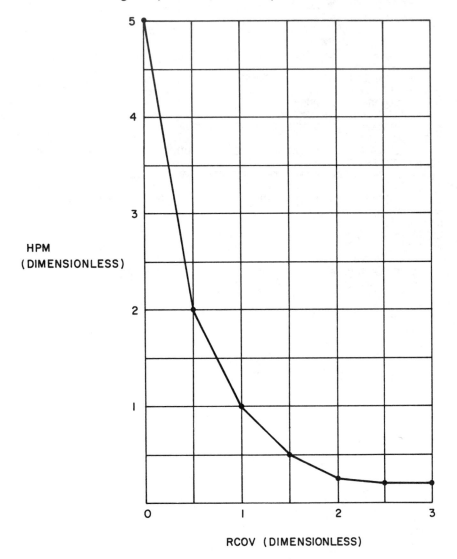

Figure 4. Assumed relationship between RCOV and HPM

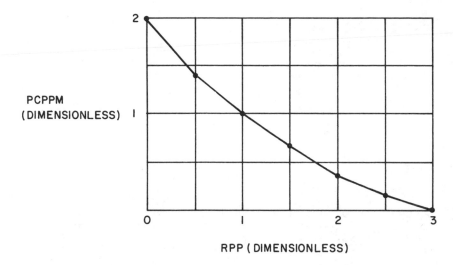

Figure 5. Possible relationship between RPP and PCPPM

= 0), HPM = 5. Thus the hog price HP = HPN∗HPM = $0.30∗ 5 =
$1.50. This means that the pork price PP = (HP/HDY) + MARKUP =
($1.50/0.6) + $0.50 = $2.50 + $0.50 = $3.00. And the relative pork price
RPP = PP/PPN = $3.00/$1.00 = 3. According to the TABLE function
in Figure 5, when the relative pork price RPP = 3, the pork consumed per
person multiplier PCPPM = 0, which is what we wanted.

   Another thing to check is the value of DT. It is important that DT be
less than one-third of the desired inventory coverage (in months). Thus,
since DCOV = .5 month, DT should be less than .133 month. A value of
0.1 is reasonable.

d.   Figure 6 shows an equilibrium run, and Figure 7 shows the model's response to a year-long 10% *increase* in the hog slaughter rate. Note that in
Figure 7 the increase in the slaughter rate causes the pork inventory
PORK to rise, which in turn causes the hog and pork prices to fall. This
causes the pork consumed per person PCPP to rise, until the consumption
rate CON equals the production rate PROD. (The reverse occurs when the
increase in the slaughter rate ends.) (*Note:* A 10% reduction would have
the opposite result.)

e.   Figure 8 shows the response of the system to a year-long 10% increase in
the pork consumed per person. Note that the increase in the consumption
rate causes the inventory to fall, which in turn causes the hog and pork
prices to rise. This causes the pork consumed per person to fall, until,
once again, the consumption rate CON equals the production rate PROD.

EXERCISE 1

```
        PORK=P,CON=C,PROD=R

        0.00M        100.00M       200.00M       300.00M       400.00M P
        0.00M        200.00M       400.00M       600.00M       800.00M CR
0.0000  ------------------------------------------P------------ PCR
         .            .             .             P          .  PCR
         .            .             .             P          .  PCR
         .            .             .             P          .  PCR
         .            .             .             P          .  PCR
         .            .             .             P          .  PCR
```

EXERCISE 1

```
        PP=P,HP=H,SLAU=S

        0.0000       0.5000        1.0000        1.5000        2.0000   P
        .00000       .10000        .20000        .30000        .40000   H
        0.0T         2000.0T       4000.0T       6000.0T       8000.0T  S
0.0000  -----------------------------P------------H------------ PS
         .            .             P             H          .  PS
         .            .             P             H          .  PS
         .            .             P             H          .  PS
         .            .             P             H          .  PS
         .            .             P             H          .  PS
```

Figure 6. Equilibrium run for Exercise 1

EXERCISE 1

```
        PORK=P,CON=C,PROD=R

       225.00M        275.00M        325.00M        375.00M        425.00M P
       580.00M        600.00M        620.00M        640.00M        660.00M CR
 0.0000 ------------C-----P--------------------------------- CR
            .        C     P     .              .          . CR
            .        C     P     .              .          . CR
            .        C     P     .              .          . CR
            .        C     P     .              .          . CR
            .        C     P     .              .          . CR
            .        C     P     .              .            R
            .        .           .        P  .  C            R
            .        .           .           . P    C        R
            .        .           .              .     P    C R
20.000 ---------------------------------------------------P---CR
            .        .           .              .          P   CR
            .        R           .              .          P   C.
            .        R        C  .     P        .              .
            .        R   C       P.             .              .
            .        R C       P    .           .              .
            .        RC       P     .           .              .
            .        RC       P     .           .              .
            .        C        P     .           .          . CR
            .        C     P        .           .          . CR
40.000 ------------C-----P--------------------------------- CR
            .        C     P        .              .       . CR
            .        C     P        .              .       . CR
            .        C     P        .              .       . CR
            .        C     P        .              .       . CR
            .        C     P        .              .       . CR
            .        C     P        .              .       . CR
            .        C     P        .              .       . CR
            .        C     P        .              .       . CR
            .        C     P        .              .       . CR
60.000 ------------C-----P--------------------------------- CR
```

Figure 7. Response to year-long 10% increase in slaughter rate

Figure 7 continued

EXERCISE 1

```
        PP=P,HP=H,SLAU=S

     0.8000        0.8500        0.9000        0.9500        1.0000  P
     .19000        .22000        .25000        .28000        .31000  H
     4000.0T       4100.0T       4200.0T       4300.0T       4400.0T S
0.0000 S-----------------------------------------------H---P
       S              .             .             .         H   P
       S              .             .             .         H   P
       S              .             .             .         H   P
       S              .             .             .         H   P
       S              .             .             .         H   P
       .              .             .             .         H   P PS
       .              .          H . P            .             S
       .              .       H     P .           .             S
       .              .    H     P   .            .             S
20.000 ---------------H---P---------------------------------S
       .              H    P         .            .             S
       S            . H      P       .            .             .
       S              .             .          H     P         .
       S              .             .            .    H    P   .
       S              .             .            .   H     P .
       S              .             .            .    H     P.
       S              .             .            .    H     P.
       S              .             .            .    H   P
       S              .             .            .    H   P
40.000 S-----------------------------------------------H---P
       S              .             .             .         H   P
       S              .             .             .         H   P
       S              .             .             .         H   P
       S              .             .             .         H   P
       S              .             .             .         H   P
       S              .             .             .         H   P
       S              .             .             .         H   P
       S              .             .             .         H   P
60.000 S-----------------------------------------------H---P
```

EXERCISE 1

```
        PORK=P,CON=C,PROD=R

       260.00M       270.00M       280.00M       290.00M       300.00M P
       475.00M       525.00M       575.00M       625.00M       675.00M CR
 0.0000 ------------------------------------------C-----------------P CR
        .             .             .            C        .        P CR
        .             .             .            C        .        P CR
        .             .             .            C        .        P CR
        .             .             .            C        .        P CR
        .             .             .            C        .        P CR
        .             .             .            R        .     C  P
        .         P   .             .            RC       .        .
        .       P     .             .            C        .        . CR
        .P            .             .            C        .        . CR
20.000 -P--------------------------------------C-------------------- CR
        P             .             .            C        .        . CR
        P             .        C    .            R        .        .
        .             .             .            CR       .  P     .
        .             .             .            C        .     P  . CR
        .             .             .            C        .     P  . CR
        .             .             .            C        .      P. CR
        .             .             .            C        .      P. CR
        .             .             .            C        .        P CR
        .             .             .            C        .        P CR
40.000 -----------------------------------------C-----------------P CR
        .             .             .            C        .        P CR
        .             .             .            C        .        P CR
        .             .             .            C        .        P CR
        .             .             .            C        .        P CR
        .             .             .            C        .        P CR
        .             .             .            C        .        P CR
        .             .             .            C        .        P CR
        .             .             .            C        .        P CR
60.000 -----------------------------------------C-----------------P CR
```

Figure 8. Response to year-long 10% increase in the pork consumption per person constant

Figure 8 continued

```
EXERCISE 1

        PP=P,HP=H,SLAU=S

     0.9500        1.0000        1.0500        1.1000        1.1500   P
     .27000        .30000        .33000        .36000        .39000   H
        0.0T       2000.0T       4000.0T       6000.0T       8000.0T  S
 0.0000  ------------P-----------S----------------------- PH
    .             P             S                .       . PH
    .             P             S                .       . PH
    .             P             S                .       . PH
    .             P             S                .       . PH
    .             P             S                .       . PH
    .             P             S                .       . PH
    .             .             S                .     P . PH
    .             .             S                .       P . PH
    .             .             S                .       P . PH
 20.000  -----------------------S----------------------P---- PH
    .             .             S                .       P . PH
    .             .             S                .       P . PH
    .             . P           S                .       . PH
    .             P             S                .       . PH
    .             P             S                .       . PH
    .             P             S                .       . PH
    .             P             S                .       . PH
    .             P             S                .       . PH
    .             P             S                .       . PH
 40.000  ------------P-----------S----------------------- PH
    .             P             S                .       . PH
    .             P             S                .       . PH
    .             P             S                .       . PH
    .             P             S                .       . PH
    .             P             S                .       . PH
    .             P             S                .       . PH
    .             P             S                .       . PH
    .             P             S                .       . PH
    .             P             S                .       . PH
 60.000  ------------P-----------S----------------------- PH
```

CHAPTER 22

Exercise 4

The accompanying figure shows no variation of release rate during the beginning of the simulation and only slight change by the end. The seasonal table for W1 has a decline from 1.0 to 0.7 steadily as TIME goes from 0 to 90 days. By the end of the run, i.e., TIME = 60, W1 has only declined to 0.83. Therefore RLSFLW is still primarily reflecting the constant LTINFL, and only slightly including an effect of ADIINF (which fully reflects the STEP by $\approx$ TIME = 36).

```
EX4 - VARIABLE WEIGHTINGS


INFLOW=I RLSFLW=F DIINFL=D ADIINF=4    RES=R

      0.5000         0.7000         0.9000          1.1000          1.3000  IFD4
      27.500         32.500         37.500          42.500          47.500  R
 0.0000 - - - R - - - - - - - - - - - - - I - - - - - - - - - - - - IFD4
       .       R       .             .      FI       .            . ID4
       .       R       .             .      FI       .            . ID4
       .         R     .             .     F D       .        I   . D4
       .          R    .             .     F 4   D       .     I   .
       .           R   .             .     F   4     . D      I   .
       .           R . .             .     F       4   . D I      .
       .             R.              .     F         4 .    I     . ID
       .              R              .     F          4.     I    . ID
       .              . R            .   F           .4      I    . ID
20.000 - - - - - - - - R - - - - - -F- - - - - 4 - - I - - - - ID
       .             .    R          . F          .  4    I      . ID
       .             .      R        . F          .  4   I       . ID
       .             .        R     .F            .  4  I        . ID
       .             .         R    .F            .  4 I         . ID
       .             .          R  F              .  4 I         . ID
       .             .             F.             .  4 I         . FR,ID
       .             .             F.R            .    4I        . ID
       .             .           F . R            .    4I        . ID
       .             .           F .    R         .    4I        . ID
40.000 - - - - - - - - - - - -F- - - -R- - - - - - -4I - - - ID
       .             .           F .       R      .    4I        . ID
       .             .          F .          R    .    I        . ID4
       .             .          F .            R  .    I        . ID4
       .             .         F  .              R.    I        . ID4
       .             .         F  .              . R   I        . ID4
       .             .        F   .              .  R  I        . ID4
       .             .        F   .              .   RI         . ID4
       .             .       F    .              .    IR        . ID4
       .             .       F    .              .     I  R     . ID4
60.000 - - - - - - - - - - - F - - - - - - - - - - - - I - - R - ID4
```

```
*        EX4 - VARIABLE WEIGHTINGS
NOTE
NOTE     ----------------------------------
NOTE
R        INFLOW.KL=NFLOW+STEP(0.2,5)
NOTE         UPSTREAM INFLO (MILLION GALLONS/DAY)
C        NFLOW=1
NOTE         NORMAL FLOW (MILLION GALLONS/DAY)
L        RES.K=RES.J+(DT)(INFLOW.JK-RLSFLW.JK)
N        RES=30
NOTE         WATER IN RESERVOIR (MILLION GALLONS)
R        RLSFLW.KL=(W1.K)(LTINFL)+(W2.K)(ADIINF.K)
NOTE         RELEASED FLOW (MILLION GALLONS/DAY)
C        LTINFL=1
NOTE         LONG-TERM INFLOW (MILLION GALLONS/DAY)
A        W1.K=TABLE(TW1,TIME.K,0,360,90)
NOTE         WEIGHTING1 (DIMENSIONLESS)
T        TW1=1.0/0.7/0.3/0.4/0.75
NOTE         TABLE FOR WEIGHTING1
N        W2=1-W1
NOTE         WEIGHTING2 (DIMENSIONLESS)
L        ADIINF.K=ADIINF.J+(DT/AIDEL)(DIINFL.JK-ADIINF.J)
N        ADIINF=DIINFL
NOTE         AVERAGE DELAYED INFORMATION (MILLION GALLONS/DAY)
C        AIDEL=10
NOTE         AVERAGE INFORMATION DELAY (DAYS)
R        DIINFL.KL=DELAY3(INFLOW.JK,MSDL)
NOTE         DELAYED INFORMATION ON INFLOW (MILLION GALLONS/DAY)
C        MSDL=5
NOTE         MEASUREMENT DELAY (DAYS)
NOTE
NOTE         CONTROL STATEMENTS
NOTE
PLOT     INFLOW=I,RLSFLW=F,DIINFL=D,ADIINF/RES=R
SPEC     DT=1/PLTPER=2/LENGTH=60
RUN
```

Model for Exercise 4.

Exercise 7

a. L   AVINFL.K = SMOOTH(INFLOW.JK,10)
   N   AVINFL = LTINFL
   A   DESRES.K = (30)(AVINFL.K)
   A   RESGAP.K = DESRES.K − RES.K

Notice on the run for this exercise how DESRES slowly rises toward its new equilibrium value of $(30)(1.2) = 36$ million gallons.

To achieve this target, the release rate RLSFLW needs to temporarily hold back water from the users (even though the dam is receiving more from INFLOW!), gradually building up again as RES moves toward DESRES.

Ms. Perez is achieving her objectives at the expense of the users' desires for water supply.

Rerun this model for shorter averaging time in AVINFL; for longer averaging; for quicker response in DAFLW; for slower response.

```
*         EX7 - INTRODUCING VARIABLE GOALS
NOTE
NOTE
NOTE      ---------------------------------
NOTE
R         INFLOW.KL=NFLOW+STEP(0.2,5)
NOTE         20 PERCENT STEP INCREASE ABOVE NORMAL INFLOW
NOTE         AT TIME = 5 DAYS
C         NFLOW=1
NOTE         UPSTREAM INFLOW (MILLION GALLONS/DAY)
L         RES.K=RES.J+(DT)(INFLOW.JK-RLSFLW.JK)
N         RES=30
NOTE         WATER IN RESERVOIR (MILLION GALLONS)
R         RLSFLW.KL=LTINFL-DAFLW.K
NOTE         RELEASED FLOW (MILLION GALLONS/DAY)
C         LTINFL=1
NOTE         LONG-TERM INFLOW (MILLION GALLONS/DAY)
A         DAFLW.K=(0.20)(RESGAP.K)
NOTE         DAILY ADJUSTMENT TO FLOW (MILLION GALLONS/DAY)
A         RESGAP.K=DESRES.K-RES.K
NOTE         RESERVOIR GAP (MILLION GALLONS)
A         DESRES.K=(AVINFL.K)(30)
NOTE         DESIRED WATER IN RESERVOIR (MILLION GALLONS)
L         AVINFL.K=AVINFL.J+(DT/10)(INFLOW.JK-AVINFL.J)
N         AVINFL=LTINFL
NOTE          AVERAGE INFLOW (MILLION GALLONS/DAY)
NOTE
NOTE          CONTROL STATEMENTS
NOTE
PLOT      INFLOW=I,RLSFLW=F/RES=R,DESRES=D
SPEC      DT=1/PLTPER=1/LENGTH=30
RUN
```

EX7 - INTRODUCING VARIABLE GOALS

```
INFLOW=I RLSFLW=F    RES=R DESRES=D

     0.4500           0.6500           0.8500           1.0500           1.2500  IF
    29.000           31.000           33.000           35.000           37.000  RD
 0.0000 - - - - R - - - - - - - - - - - - - - - - - - I - - - - - - - - - RD,IF
              R         .                .            I  .              . RD,IF
              R         .                .            I  .              . RD,IF
              R         .                .            I  .              . RD,IF
              R         .                .            I  .              . RD,IF
              R         .                .            F  .            I . RD
            R  D .                .            F       .            I .
             R   .D              .F                    .            I .
             R .    D        D F  .                    .            I .
              .R       D   F .                         .            I .
 10.000 - - - - - - - - - R - - - D F - - - - - - - - - - - - - - - I - -
                     .         R       F .                         I . FD
                     .          R     F.D                          I .
                     .            R  F D                           I .
                     .             R F  D                          I .
                     .              . F    D                       I . FR
                     .              .  F   D                       I . FR
                     .              .   F R D                      I .
                     .              .    F R D                     I .
                     .              .    F RD                      I .
 20.000 - - - - - - - - - - - - - - - - - - - -F- R - - - - - - - I - - RD
                     .              .          F R.               I . RD
                     .              .          F DR               I .
                     .              .          F.DR               I .
                     .              .           F D R             I .
                     .              .           .FD  R            I .
                     .              .           . FD  R           I .
                     .              .           .  F    R   I . FD
                     .              .           .   F   R   I . FD
                     .              .           .   DF  R  I .
 30.000 - - - - - - - - - - - - - - - - - - - - - - - - -F- -R- I - - FD
```

Run for Exercise 7.

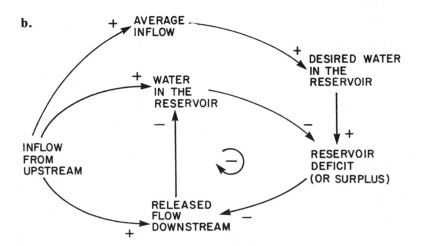

No new closed-loop has been added. But now variations in inflow will affect the release rate through the added leverage of change in the desired water in the reservoir. Furthermore, following the signs, an increase in inflow will lead to an amplified decrease in outflow, precisely the opposite of what the users would like.

Exercise 12

a.   When RES is less than DESRES, Ms. Perez wants to retain more water in the reservoir, i.e., release less. So the multiplier FMDR will be less than 1 in that case.

In preparing the illustrative multiplier shown, two points are known precisely. The multiplier is 1 when the ratio is 1. The multiplier is 0 (i.e., RLSFLW = 0) when the ratio is 0 (because then RES = 0).

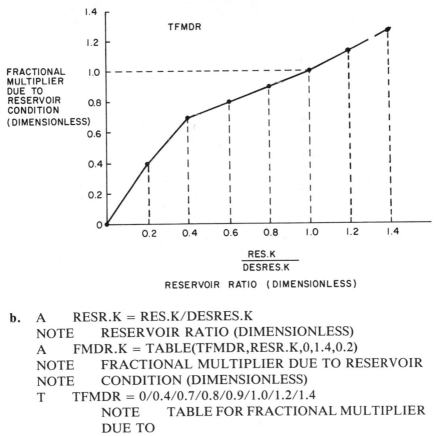

$$\frac{RES.K}{DESRES.K}$$

RESERVOIR RATIO (DIMENSIONLESS)

b.   A   RESR.K = RES.K/DESRES.K
      NOTE   RESERVOIR RATIO (DIMENSIONLESS)
      A   FMDR.K = TABLE(TFMDR,RESR.K,0,1.4,0.2)
      NOTE   FRACTIONAL MULTIPLIER DUE TO RESERVOIR
      NOTE   CONDITION (DIMENSIONLESS)
      T   TFMDR = 0/0.4/0.7/0.8/0.9/1.0/1.2/1.4
                NOTE   TABLE FOR FRACTIONAL MULTIPLIER
                DUE TO
      NOTE   RESERVOIR CONDITION
      R   RLSFLW.KL = (FMDR.K)(LTINFL)
                NOTE   RELEASED FLOW (MILLION
                GALLONS/DAY)

```
c.    *         EX12 - MORE MULTIPLIERS IN DECISION PROCESSES
      NOTE
      NOTE    --------------------------------
      NOTE
      R       INFLOW.KL=NFLOW+STEP(0.2,5)
      NOTE        20 PERCENT STEP INCREASE ABOVE NORMAL INFLOW
      NOTE        AT TIME = 5 DAYS
      C       NFLOW=1
      NOTE        UPSTREAM INFLOW (MILLION GALLONS/DAY)
      L       RES.K=RES.J+(DT)(INFLOW.JK-RLSFLW.JK)
      N       RES=30
      NOTE        WATER IN RESERVOIR (MILLION GALLONS)
      R       RLSFLW.KL=(LTINFL)(FMDR.K)
      NOTE        RELEASED FLOW (MILLION GALLONS/DAY)
      A       FMDR.K=TABLE(TFMDR,RESR.K,0,1.4,0.2)
      NOTE        FRACTIONAL MULTIPLIER DUE TO RESERVOIR CONDITIONS (DIMENSION
      T       TFMDR=0.0/0.4/0.6/0.75/0.90/1.00/1.10/1.20
      NOTE        TABLE FOR FRACTIONAL MULTIPLIER DUE TO RESERVOIR CONDITIONS
      A       RESR.K=RES.K/DESRES.K
      NOTE        RESERVOIR RATIO (DIMENSIONLESS)
      A       FMDD.K=TABLE(TFMDD,DR.K,0,1.4,0.2)
      NOTE        FRACTIONAL MULTIPLIER DUE TO DRAINAGE CONDITION
      NOTE        (DIMENSIONLESS)
      T       TFMDD=1.6/1.5/1.3/1.1/1.05/1.0/0.9/0.6
      NOTE        TABLE FOR FRACTIONAL MULTIPLIER DUE TO
      NOTE        DRAINAGE CONDITION
      A        DR.K=DRG.K/DESDRG.K
      NOTE        DRAINAGE RATIO (DIMENSIONLESS)
      C       LTINFL=1
      NOTE        LONG-TERM INFLOW (MILLION GALLONS/DAY)
      A       DESDRG.K=(50)(AVINFL.K)
      NOTE        DESIRED WATER IN DRAINAGE BASIN (MILLION GALLONS)
      L       AVINFL.K=AVINFL.J+(DT/10)(INFLOW.JK-AVINFL.J)
      N       AVINFL=INFLOW
      NOTE        AVERAGE INFLOW (MILLION GALLONS/DAY)
      L       DRG.K=DRG.J+(DT)(RLSFLW.JK-DRGRT.JK)
      N       DRG=50
      NOTE        WATER IN DRAINAGE BASIN (MILLION GALLONS)
      R       DRGRT.KL=(0.02)(DRG.K)
      NOTE        DRAINAGE RATE (MILLION GALLONS/DAY)
      A       DESRES.K=(30)(AVINFL.K)
      NOTE        DESIRED RESEVOIR VOLUMN (MILLION GALLONS)
      NOTE
      NOTE        CONTROL STATEMENTS
      NOTE
      PLOT    INFLOW=I,RLSFLW=F,DRGRT=T/RES=R,DRG=G,DESDRG=S
      SPEC    DT=1/PLTPER=2/LENGTH=60
      RUN
```

```
EX12 - MORE MULTIPLIERS IN DECISION PROCESSES


INFLOW=I RLSFLW=F  DRGRT=T    RES=R     DRG=G DESDRG=S

      0.8500          0.9500         1.0500        1.1500        1.2500  IFT
      20.000          30.000         40.000        50.000        60.000  RGS
  0.0000 - - - - - - - R - - - I - - - - - - - - - - - G - - - - - - - GS,IFT
      .                R      I          .                G           . GS,IFT
      .                R      I          .                G           . GS,IFT
      .                R      FT         .                G S     I     .
      .               .R  F   T          .                G     S  I     .
      .               . R F   T          .                G       SI     .
      .               . R F   T          .                G       I    . IS
      .               .  RF   T          .                G       I S    .
      .               .   F   T          .                G       I  S   . FR
      .               .   FR  T          .                G       I   S   .
 20.000 - - - - - - - -F-T- - - - - - - - - - G - - - I - -S- - FR
      .                .  FRT            .                G       I   S  .
      .                .  FT             .               G.       I     S . TR
      .                .  FT             .               G.       I     S . TR
      .                .   FR            .               G.       I      S. FT
      .                .  TFR            .               G.       I      S.
      .                .  TFR,           .               G.       I      S.
      .                .  T FR           .                G       I      S.
      .                .  T  FR          .                G       I      S.
      .                .  T   F          .                G       I      S FR
 40.000 - - - - - - - -T- -FR - - - - - - - G - - - I - - - S
      .                .    T   F        .                G       I       S FR
      .                .    T    F        .               G       I       S FR
      .                .    T    F        .               G       I       S FR
      .                .    T     F       .               G       I       S FR
      .                .    T    RF.                       G       I       S
      .                .     T    RF                       G       I       S
      .                .     T    RF                       G       I       S
      .                .     T     RF                     .G       I       S
      .                .      T    RF                     .G       I       S
 60.000 - - - - - - - - - - - T - - -RF - - - - - - -G- - - I - - - S
```

Run for Exercise 12c.

Under the comparable extreme described in the answer to Exercise 11, i.e., RESR = RES/DESRES = 30/36 = .83, FMDR would become .915. This is an 8.5-percent maximum decrease in RLSFLW, more noticeable to the downstream farmers than the 4-percent maximum increase generated in Exercise 11. But there is an important difference. As RESR is returned to 1.0, when RES = 36, RLSFLW returns toward and then above LTINFL, eventually to the full INFLOW value of 20 percent above LTINFL.

CHAPTER 23

Exercise 1

a.

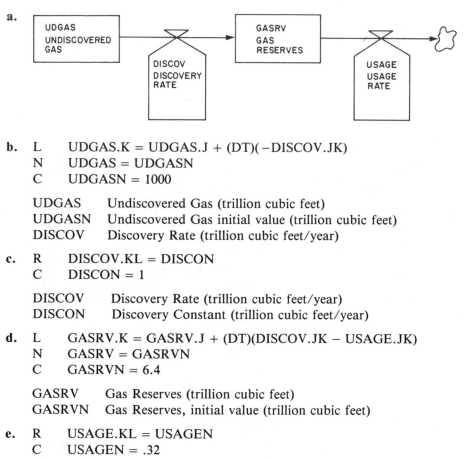

b. L   UDGAS.K = UDGAS.J + (DT)(−DISCOV.JK)
      N   UDGAS = UDGASN
      C   UDGASN = 1000

UDGAS   Undiscovered Gas (trillion cubic feet)
UDGASN   Undiscovered Gas initial value (trillion cubic feet)
DISCOV   Discovery Rate (trillion cubic feet/year)

c. R   DISCOV.KL = DISCON
      C   DISCON = 1

DISCOV   Discovery Rate (trillion cubic feet/year)
DISCON   Discovery Constant (trillion cubic feet/year)

d. L   GASRV.K = GASRV.J + (DT)(DISCOV.JK − USAGE.JK)
      N   GASRV = GASRVN
      C   GASRVN = 6.4

GASRV   Gas Reserves (trillion cubic feet)
GASRVN   Gas Reserves, initial value (trillion cubic feet)

e. R   USAGE.KL = USAGEN
      C   USAGEN = .32

USAGE   Usage Rate (trillion cubic feet/year)
USAGEN   Usage Rate constant (trillion cubic feet/year)

Exercise 2

a.   Model listing for Exercise 2(a).

```
*   NATURAL GAS MODEL
NOTE
NOTE        ************************************
NOTE
L   UDGAS. K=UDGAS. J+(DT)(-DISCOV. JK)
NOTE        UNDISCOVERED GAS (TRILLION CUBIC FEET)
N   UDGAS=UDGASN
C   UDGASN=1000
NOTE        UNDISCOVERED GAS, INITIAL VALUE (TRILLION CUBIC FEET)
R   DISCOV. KL=DISCON
C   DISCON=1
NOTE        DISCOVERY RATE (TRILLION CUBIC FEET/YEAR)
L   GASRV. K=GASRV. J+(DT)(DISCOV. JK-USAGE. JK)
NOTE        GAS RESERVES (TRILLION CUBIC FEET)
N   GASRV=GASRVN
C   GASRVN=6. 4
NOTE        GAS RESERVES, INITIAL VALUE (TRILLION CUBIC FEET)
R   USAGE. KL=USAGEN
C   USAGEN=0. 32
NOTE        USAGE RATE (TRILLION CUBIC FEET/YEAR)
NOTE
NOTE        *****       CONTROL STATEMENTS       *****
NOTE
N   TIME=1900
C   DT=0. 25
C   LENGTH=2050
C   PLTPER=3
PLOT   UDGAS=N(800, 1000)/GASRV=R
RUN
```

Exercise 5

a.

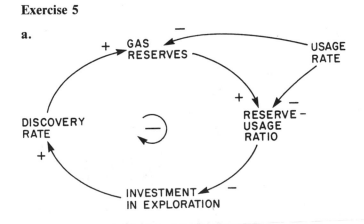

b.   If the reserve-usage ratio RUR = 10, RRUR = RUR/DRUR =
10/20 = 0.5. Since RRUR is less than 1, reserves are low.

c.

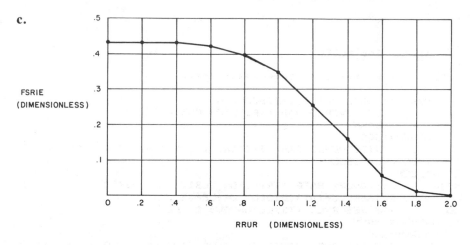

A   FSRIE.K = TABHL(FSRIET,RRUR.K,0,2,.2)
T   FSRIET = .43/.43/.43/.42/.39/.35/.26/.16/.05/.01/0

In the equation for FSRIE, TABHL stands for TABLE, High-Low.
It works like the TABLE function except at the extremes of the specified
horizontal axis. If the independent variable (in this case, RRUR) exceeds
the extremes indicated (here, below 0 or above 2.0), DYNAMO uses the
end-point dependent value (here, FSRIE = 0.43 or FSRIE = 0), without
generating an error message.

Exercise 8

a.

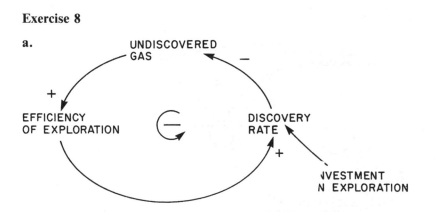

Exercise 10

a. When FUDGR = .25, EFFEXM = .125 (by interpolation from the TABLE function). Thus, EFFEXP = .125 × 67000 = 8375, and DCOST = 1/EFFEXP = $0.00012 per cubic foot.

b.
```
A    DCOST.K = 1/EFFEXP.K
A    PCOST.K = PCF*DCOST.K
C    PCF = 3
A    TCOST.K = DCOST.K + PCOST.K
```

DCOST	Discovery cost ($/trillion cubic feet)
EFFEXP	Efficiency of exploration (trillion cubic feet/$)
PCOST	Production cost ($/trillion cubic feet)
PCF	Production cost factor (dimensionless)
TCOST	Total cost ($/trillion cubic feet)

c. When FUDGR = 0.707, EFFEXM = 0.5, and EFFEXP = 33.5E-9. Thus, DCOST = $30 million and PCOST = $90 million. Therefore TCOST = DCOST + PCOST = $120 million.

d. Reserves would be exhausted in 1980, because RUR = GASRV/AUSAGE = 240/12 = 20 years.

e. If all "old" gas is used before any "new gas" is used, then gas discovered in 1960 would be used in 1980—since the "old" gas discovered before 1960 would be exhausted in 1980.

CHAPTER 24

Exercise 1

a. "The Detroit figures showed that crimes directly related to heroin climbed three percent . . . when the price of heroin rose by ten percent" (paragraph 6).

Paragraph 7 also describes a rise in the crime rate from a rise in heroin prices. The rise in crimes is a short-term response to higher heroin prices that result from shortages caused by police drug seizures.

Paragraph 11 claims that eliminating almost all heroin in the community would eliminate heroin-related crime—a long-term response to little or no heroin.

b., c., and **d.**

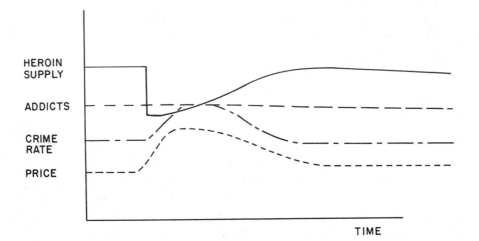

e. About 6 weeks until equilibrium.

f.

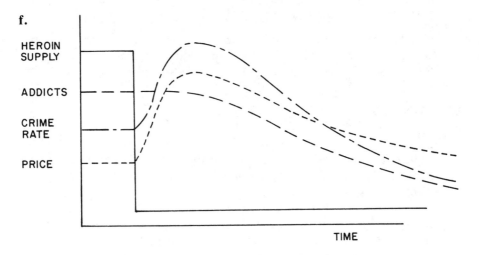

g. About 2 years.

Exercise 4

a.

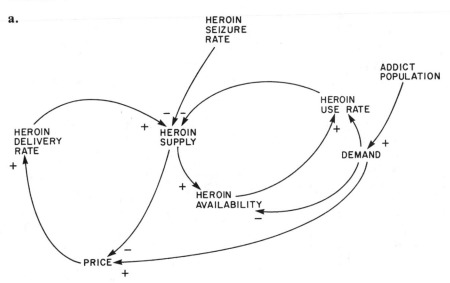

Price could be a function of Heroin Availability, since that is related to supply and demand.

b.

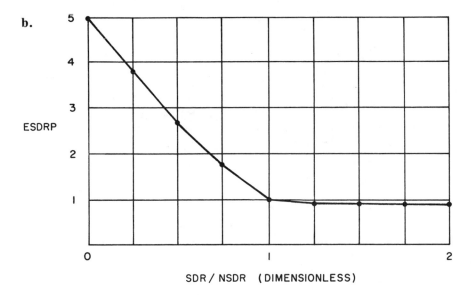

Slope and shape are important. Values are not as important. A reasonable value for PNORM is $10 a bag.

c.   Average Heroin Use Rate (bags/week)
      or
      Average Heroin Demand (bags/week)

d.   R   HDELIV.KL = AVUSE.K*EPHD.K
      A   AVUSE.K = SMOOTH(HUSE.JK,TAU)
      [or L AVUSE.K = AVUSE.J + (DT/TAU)(HUSE.JK − AVUSE.J)]
      C   TAU = some number of weeks

e.   L   PPRICE.K = PPRICE.J + (DT/TPP)(PRICE.J − PPRICE.J)
      [or A   PPRICE.K = SMOOTH(PRICE.K,TPP)]
      C   TPP = 4

f.

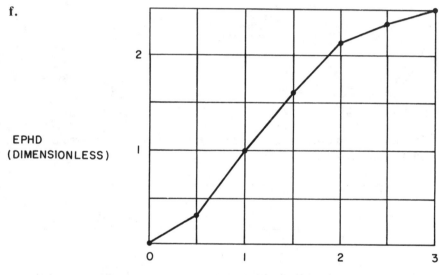

EPHD
(DIMENSIONLESS)

PRICE /PNORM (DIMENSIONLESS)

0 and 1 are critical points. Slope and shape are important.

      A   EPHD.K = TABLE (EPTAB, PRICE.K/PNORM,0,3,0.5)
      T   EPTAB = 0/.3/1/1.6/2.1/2.3/2.4

g.   A   MNEED.K = PRICE.K*HABIT
      A   FREQPA.K = (MNEED.K*FHSC)/RPC
      A   CRIMES.K = ADDCTS*FREQPA.K
      C   RPC = 60   ⎱
      C   FHSC = 0.8 ⎰ reasonable guesses

Units for FREQPA:

FREQPA = (MNEED*FHSC)/RPC = (dollars/ad-
dict/week)/(dollars per crime)

$$= \frac{1/(\text{addicts/week})}{1/\text{crimes}}$$

$$= \frac{\text{crimes}}{\text{addict/week}}$$

Exercise 7:

a. *One-Time Seizure:* Price and crime should quickly rise, as before. The addict population may drop a bit because of the lower heroin supply. But
then the overshoot in heroin observed in Model 2 will attract more addicts. The final result of a one-time seizure is actually *more* addicts and
more crime in Model 3.

Permanent increase: The behavior will be similar to Model 2, showing
some oscillations, but the lower heroin supply will support fewer addicts
in the long run, so the addict population will decline.

REFERENCES

Abelson, Philip A. "The Oil Price Spiral." *Science* 207 (February 29, 1980).

Alfeld, Louis E., and Graham, Alan K. *Introduction to Urban Dynamics*. Cambridge, Mass.: MIT Press, 1976.

Anderson, Jay Martin. "The Eutrophication of Lakes" in *Toward Global Equilibrium: Collected Papers*. Cambridge, Mass.: Wright-Allen Press, 1973, pp. 117–140.

Barbour, John. "Burros: Beasts of Burden Have Become Burden of Beasts in U.S. Deserts and Parks." *International Herald Tribune* (March 29–30, 1980).

Calhoun, B. F. "Population Density and Social Pathology." *Scientific American* 206.

Chalfant, Arnold R. "Ecology Primer." *National Wildlife* 10 (February 1972): 42–43.

Cole, H. S. D.; Freeman, Christopher; Jahoda, Marie; and Pavitt, K. L. R., eds. *Models of Doom*. New York: Universe Books, 1973.

Coyle, R. G. *Management System Dynamics*. London: John Wiley, 1977.

Deutsch, Karl. *The Nerves of Government*. New York: The Free Press, 1963.

Edmondson, W. T., and Lehman, John T. "The Effect of Changes in the Nutrient Income on the Condition of Lake Washington." *Limnology and Oceanography* (January 1981).

Forrester, Jay W. *Industrial Dynamics*. Cambridge, Mass.: MIT Press, 1961.

Forrester, Jay W. *Principles of Systems*. Cambridge, Mass.: MIT Press, 1968.

Forrester, Jay W. *Urban Dynamics*. Cambridge, Mass.: MIT Press, 1969.

Forrester, Jay W. *World Dynamics*. Cambridge, Mass.: MIT Press, 1973.

Forrester, Nathan B. *The Life Cycle of Economic Development*. Cambridge, Mass.: MIT Press, 1973.

Friedman, Saul. "When Heroin Supply Cut, Crime Rises, Says Report." *Boston Globe*, April 22, 1976.

Goodman, Michael R. *Study Notes in System Dynamics*. Cambridge, Mass.: MIT Press, 1974.

Grant, W. Vance, and Link, C. George. *Digest of Education Statistics*. Washington, D.C.: National Center for Education Statistics, U.S. Government Printing Office, 1979.

Hamilton, H. R.; Goldstone, S. E.; Milliman, J. W.; Pugh, A. L., III; Roberts, E. B.; and Zellner, A. *System Simulation for Regional Analysis: An Application to River Basin Planning.* Cambridge, Mass.: MIT Press, 1969.

Harsch, Joseph C. "Welfare Reform, the Stock Market, and Politics." *The Christian Science Monitor* (August 11, 1977).

Humphrey, Clifford C., and Evans, Robert G. *What's Ecology?* Northbrook, Ill.: Hibbard Press, 1971, pp. 8–10.

Jackson, James R. *History of Littleton.* Cambridge, Mass.: University Free Press, 1950.

Kormondy, Edward J. *Concepts of Ecology.* Englewood Cliffs, N.J.: Prentice-Hall, 1976.

Levin, Gilbert, and Roberts, Edward B. (with Hirsch, Gary B.; Kligler, Deborah S.; Roberts, Nancy; and Wilder, Jack). *The Dynamics of Human Service Delivery.* Cambridge, Mass.: Ballinger, 1976.

Levin, Gilbert; Roberts, Edward B.; and Hirsch, Gary B. *The Persistent Poppy: A Computer-Aided Search for Heroin Policy.* Cambridge, Mass.: Ballinger, 1975.

Lyneis, James M. *Corporate Planning and Policy Design: A System Dynamics Approach.* Cambridge, Mass.: MIT Press, 1980.

McBride, Stewart Dill. "Tidal Wave of Fishermen Endangers U.S. Lobster Beds." *The Christian Science Monitor* (August 15, 1975).

McHugh, J. L. *Fisheries and Fishery Resources of New York State.* Washington, D.C.: National Marine Fisheries Service, Department of Commerce, March 1977.

Mass, Nathaniel J. *Economic Cycles: An Analysis of Underlying Causes.* Cambridge, Mass.: MIT Press, 1975.

Mass, Nathaniel J., ed. *Readings in Urban Dynamics: Volume 1.* Cambridge, Mass.: MIT Press, 1974.

Meadows, Dennis L. *Dynamics of Commodity Production Cycles.* Cambridge, Mass.: MIT Press, 1970.

Meadows, Dennis L.; Behrens, William W., III; Meadows, Donella H.; Naill, Roger F.; Randers, Jorgen; and Zahn, Erich K. O. *Dynamics of Growth in a Finite World.* Cambridge, Mass.: MIT Press, 1974.

Meadows, Dennis L., and Meadows, Donella H., eds. *Toward Global Equilibrium: Collected Papers.* Cambridge, Mass.: MIT Press, 1973.

Meadows, Donella H.; Meadows, Dennis L.; Randers, Jorgen; and Behrens, William W., III. *The Limits to Growth.* New York: Universe Books, 1973.

Melillo, Jerry. *Ecology Primer.* West Haven, Conn.: Pendulum Press, Inc., 1972.

Myrdal, Gunnar. *Asian Drama: An Inquiry into the Poverty of Nations.* New York: The Twentieth Century Fund, 1968.

Naill, Roger. *Managing the Energy Transition.* Cambridge, Mass.: Ballinger, 1977.

O'Brien, Robert C. *Mrs. Frisby and the Rats of NIMH.* New York: Atheneum, 1971, pp. 169–170.

Popham, John N. "A Hard Look at 'Soft' Energy Path." *Birmingham News,* September 26, 1978.

Proceedings of Lexington Historical Society, vol. II. Lexington, Mass.: The Historical Society, 1900.

Pugh, Alexander L., III. *DYNAMO II User's Manual*. Cambridge, Mass.: MIT Press, 1976.

Pugh-Roberts Associates, Inc. *User Guide and Reference Manual for Micro-DYNAMO*. Reading, Mass.: Addison-Wesley, 1982.

Richardson, George P., and Pugh, Alexander L., III. *Introduction to System Dynamics Modeling with DYNAMO*. Cambridge, Mass.: MIT Press, 1981.

Roberts, Edward B. *The Dynamics of Research and Development*. New York: Harper and Row, 1964.

Roberts, Edward B., ed. *Managerial Applications of System Dynamics*. Cambridge, Mass.: MIT Press, 1978.

Samhaber, E. *Merchants Make History*. New York: The John Day Co., 1964.

Schmeck, Harold M., Jr. "Study Depicts Mayan Decline," *New York Times,* October 23, 1979.

Schroeder, Walter W., III; Sweeney, Robert E.; and Alfeld, Louis E., eds. *Readings in Urban Dynamics: Volume II*. Cambridge, Mass.: MIT Press, 1975.

Simon, Herbert A. "The New Science of Management Decision." In *The Shape of Automation*. New York: Harper and Row, 1965, pp. 53–111.

Sterling, Claire. "The Making of the Sub-Saharan Wasteland." *Atlantic Monthly* 233 (May 1974): 98–105.

Sullivan, Jim. "To Play or Not to Play." *Boston Globe,* April 3, 1980.

Von Bertalanffy, Ludwig. *General System Theory*. New York: George Braziller, 1968.

Weymar, F. Helmut. *The Dynamics of the World Cocoa Market*. Cambridge, Mass.: MIT Press, 1968.

Wiener, Norbert. *Cybernetics or Control and Communications in the Animal and the Machine*. New York: John Wiley, 1948.

INDEX